Gene Expression in Normal and Transformed Cells

NATO Advanced Science Institutes Series

A series of edited volumes comprising multifaceted studies of contemporary scientific issues by some of the best scientific minds in the world, assembled in cooperation with NATO Scientific Affairs Division.

This series is published by an international board of publishers in conjunction with NATO Scientific Affairs Division

A	**Life Sciences**	Plenum Publishing Corporation
B	**Physics**	New York and London
C	**Mathematical and Physical Sciences**	D. Reidel Publishing Company Dordrecht, Boston, and London
D	**Behavioral and Social Sciences**	Martinus Nijhoff Publishers The Hague, Boston, and London
E	**Applied Sciences**	
F	**Computer and Systems Sciences**	Springer Verlag Heidelberg, Berlin, and New York
G	**Ecological Sciences**	

Recent Volumes in Series A: Life Sciences

Volume 58—Arterial Pollution: An Integrated View on Atherosclerosis
edited by H. Peeters, G. A. Gresham, and R. Paoletti

Volume 59—The Applications of Laser Light Scattering to the Study of Biological Motion
edited by J. C. Earnshaw and M. W. Steer

Volume 60—The Use of Human Cells for the Evaluation of Risk from Physical and Chemical Agents
edited by Amleto Castellani

Volume 61—Genetic Engineering in Eukaryotes
edited by Paul F. Lurquin and Andris Kleinhofs

Volume 62—Heart Perfusion, Energetics, and Ischemia
edited by Leopold Dintenfass, Desmond G. Julian, and Geoffrey V. F. Seaman

Volume 63—Structure and Function of Plant Genomes
edited by Orio Ciferri and Leon Dure III

Volume 64—Gene Expression in Normal and Transformed Cells
edited by J. E. Celis and R. Bravo

Gene Expression in Normal and Transformed Cells

Edited by

J. E. Celis

Aarhus University
Aarhus, Denmark

and

R. Bravo

EMBL Laboratory
Heidelberg, Federal Republic of Germany

Plenum Press
New York and London
Published in cooperation with NATO Scientific Affairs Division

Proceedings of a NATO/Gulbenkian
Foundation-sponsored Summer School,
held May 25—June 4, 1982,
in Sintra-Estoril, Portugal

Library of Congress Cataloging in Publication Data

Main entry under title:

Gene expression in normal and transformed cells.

(NATO advanced science institutes series, Series A, Life sciences; v. 64)
"Published in cooperation with NATO Scientific Affairs Division."
"Proceedings of a NATO/Gulbenkian Foundation-sponsored summer school, held May
25—June 4, 1982, in Sintra-Estoril, Portugal"—Verso t.p.
Includes bibliographical references and index.
1. Gene expression—Congresses. 2. Cytogenetics—Congresses. 3. Cancer cells—
Congresses. I. Celis, J. E. II. Bravo, R. (Rodrigo) III. North Atlantic Treaty Organization.
Scientific Affairs Division. IV. North Atlantic Treaty Organization. V. Fundacão Calouste
Gulbenkian. VI. Series. [DNLM: 1. Cell transformation, Neoplastic—Congresses. 2.
Cytogenetics—Congresses. QZ 202 G326 1982]
QH450.G462 1983   574.87'3223   83-4145

ISBN-13: 978-1-4684-4543-5   e-ISBN-13: 978-1-4684-4541-1
DOI: 10.1007/978-1-4684-4541-1

© 1983 Plenum Press, New York
Softcover reprint of the hardcover 1st edition 1983

A Division of Plenum Publishing Corporation
233 Spring Street, New York, N.Y. 10013

PREFACE

This volume is based on the proceedings of a NATO-Gulbenkian Foundation sponsored Summer School held in May-June 1982 in Sintra-Estoril, Portugal.

Given the accelerated growth of knowledge in the field of eukaryotic gene expression, it seemed timely to hold a NATO Advanced Study Institute to discuss current developments in this area of biology and to evaluate the potential of emerging technologies such as gene transfer, recombinant DNA cloning and quantitative high resolution two-dimensional gel electrophoresis. The initial articles in this volume describe various differentiation models and address questions such as the relationships between differentiation and cell proliferation, biochemical changes accompanying differentiation, expression of differentiated gene products and their regulation as well as gene organization of cytoskeletal proteins. The second section describes properties of neoplastic cells, surveys current assays for transformation and offers some new insights into the mechanisms involved in carcinogenesis. The third part is dedicated to viral oncogenesis and to the role of *oncogenes* in cell transformation. Particular emphasis is given to the role of tyrosine kinases in cell transformation. The concluding section deals with various aspects of gene expression in normal and transformed cells with special emphasis given to studies using two dimensional gel electrophoresis, cell hybridization, gene transfer and immunological techniques.

We wish to express our appreciation to Dr. Maria C. Lechner who provided valuable advice and help concerning the organization of the meeting. We are also indebted to Ms Jonna Christensen and Ms Lisbeth Heilesen for their outstanding organization and administration of the meeting. Finally, we must express our deepest appreciation to Ms Jonna Christensen who patiently typed all the manuscripts.

November 1982   J.E. Celis
                                                       R. Bravo

CONTENTS

DIFFERENTIATION MODELS AND GENE EXPRESSION IN DIFFERENTIATED CELLS

GENE EXPRESSION IN NORMAL AND TRANSFORMED CELLS

MOLECULAR APPROACH TO THE STUDY OF NEURAL FUNCTION AND DIFFERENTIATION

M.M. Portier, B. Croizat, F. Berthelot, B. Edde,
D. Paulin# and F. Gros

*Laboratoire de Biochimie Cellulaire, Collège de France
Paris, France and #Département de Biologie Moléculaire
Institut Pasteur, Paris, France*

1. INTRODUCTION

Neurobiology constitutes one of the most challenging aspects of cellular and development biology due to the complexity of the central nervous system and to the diversity of the behavioral patterns among evolved eukaryotic organisms.

A fair understanding of the central nervous system at the molecular level with regard to its integrated funtions and its ontogenetic programme will require many more decades, given for example the enormous amount of synaptic connection existing in the cerebrum tissue. Yet, important achievements have recently been made in the study of neural cells due to a multidisciplinary approach derived from molecular genetics and immunology.

In the first part of this article, we shall attempt to show, based upon few examples, how recent progress of molecular biology, recombinant DNA studies and immunochemistry have provided new and important insights into some of the key questions concerning the problems of cellular interaction and synaptic transmission in central and peripheral tissues.

1

The second part will be more specifically devoted to nerve
cell differentiation (a problem under study in our laboratory) as
approached by the use of neuroblastoma cell lines or of cultured
primary neurons.

2. MOLECULAR GENETICS, IMMUNOLOGY AND THE STUDY OF NEURAL CELLS

(i) *Characterization of cells belonging to the nervous system*

Without going into details, the neural system, and more
particularly the brain tissue, underlies a marked cellular heterogeneity which was first underlined by neuroanatomical studies
which one has attempted to interpret on the basis of the classical
theories of histogenesis, some of which were put forward as early
as 1889, by people such as His (1) or Hardesty (2).

Neurons and the various types of glial cells are supposed to
originate from the primitive spongioblasts according to classical
theories. Whatever the validity of these theories may be, the
fact is that the nerve tissue can include cells as diverse as
bipolar or multipolar neurons, oligo dendrocytes (the myelinating
cell of the peripheral system), Schwann cells, various types of
astrocytes (some of the basic, cellular elements forming part of
the glial tissue in the central or peripheral systems), as well
as ependymal cells, fibroblasts, macrophages, microglia, lepto-
meningeal cells etc. In most instances, it is reasonably easy
to identify these cell types to morphological features, as was
largely done by embryologists and histologists in the past.

More recently, immunochemistry and particularly indirect
immunofluorescence technique has added on a new dimension to the
problem of cell typing. Type-specific antibodies were raised
against either a surface component or an internal protein component

that happens to be proper to the cell under concern. Antibodies can either be of the classical type (i.e. polyclonal), in which case the serum is saturated by heterologous antigens before use, or they can be monoclonal. In both cases, they are coupled to a fluorescent dye, or to peroxydase, or a double precipitation with an immuno-globulin directed against the first antibody is utilized (3, 4).

Table 1 lists a series of antigenic markers which have proved to be type-specific for nervous system cells in culture. These antigens can be proteinaceous, polysaccharidic or lipidic in nature (eg. Ran-I LETS, GFA, etc.) or they can be materialized as toxin-binding receptors (tetanus toxin receptors). Hence, with a battery of cell type-specific antibodies or of specific ligands one can proceed for instance to immunocytochemical characterization of the cells. A word of caution is to be placed, however, because the type of antigen expressed *in vitro* does not always correspond to the antigens accumulated *in vivo* (3).

Aside from the possibility of doing general cell "typing" by use of indirect immunofluorescence, the use of monoclonal antibodies has proved of great interest for it permits to distinguish among subclasses of neurons, according to the nature of the cognate antigenic determinant, to its concentration or surface distribution within a given cell. One of the best examples comes from the work of Barnstable (5) on the recognition of different cell types in the rat retina.

Retina in mammals can be regarded as composed of stratified arrays of neuronal and glial cells endowed with various functions. Most popular are the "cones" and "rods", the two major types of photosensitive cells containing the light-sensitive molecule, rhodopsin. Since sections can be made in the retinal tissue, it is

easy to do physiological assignment of cells according to their
relative position in the particular array under concern.

By immunizing female BALB/c mice with crude membrane pellets
from dissociated retina derived from CD rats, in the presence of a
Freund adjuvant, hybridoma clones could be obtained following
fusion of spleen cells with appropriate myelomas. Hybrid and clone
culture supernatants were tested for antibody activity using an
indirect assay and seven antibodies could be selected, based on
their lack of reactivity with rat thymocytes or fibroblasts.

A salient result from this work is illustrated in Figure 1
taken from Barnstable's work. Retina-specific monoclonal antibody
called RET-P$_1$ labels the whole of the photoreceptor layer cell
bodies, outer and inner segments (independent work indicates that
it reacts with an antigenic determinant present only on "rods" but
not on "cones", and distinct from rhodopsin itself), RET-P$_2$ labels
only outer segments, and RET-P$_3$ only the cell bodies. Three other
antibodies were found to react only with Müller (glial) cells. One
antibody (RET-N) not only reacted with retinal neurones, but also
with neuronal cells from other tissues. The chemical nature of
these antigens is still unknown, but it is clear that some photoreceptor cell-specific molecules appear to be widely distributed
at the surface of the cell, while others are restricted to discrete
areas.

Advantage can be taken of the unique specificity of monoclonal antibodies to approach some important aspects of the higher
level organization in neuronal tissues or systems.

a) For instance, the discovery by Nirenberg and his associates
(7) of an antigen playing an important role in the positional
information that is involved in the assembling of retinal cells

Table 1. *Major markers for nervous system cells in culture*

Cell type	Tetanus toxin	Ran-1	Thy-1	LETS	GFAP	GC	Phago-cytosis	FC-re-ceptors	Cilia
Neurones	+	-	+/-[a]	-	-	-	-	-	-
Schwann cells	-	+	-	-	-	-	-	-	-
Astrocytes	-	-	+/-[b]	-	+	-	-	-	-
Oligodendrocytes	-	-	-	-	-	+	-	-	-
Fibroblasts	-	-	+	+	-	-	-	-	-
Leptomeningeal cells	-	-	-	+	-	-	-	-	-
Ependymal cells	-	-	-	-	-	-	-	-	+
Macrophages and microglia	-	-	-	-	-	-	+	+	-

[a]Present in some but not all neurones in culture

[b]Present in some astrocytes after one week in culture

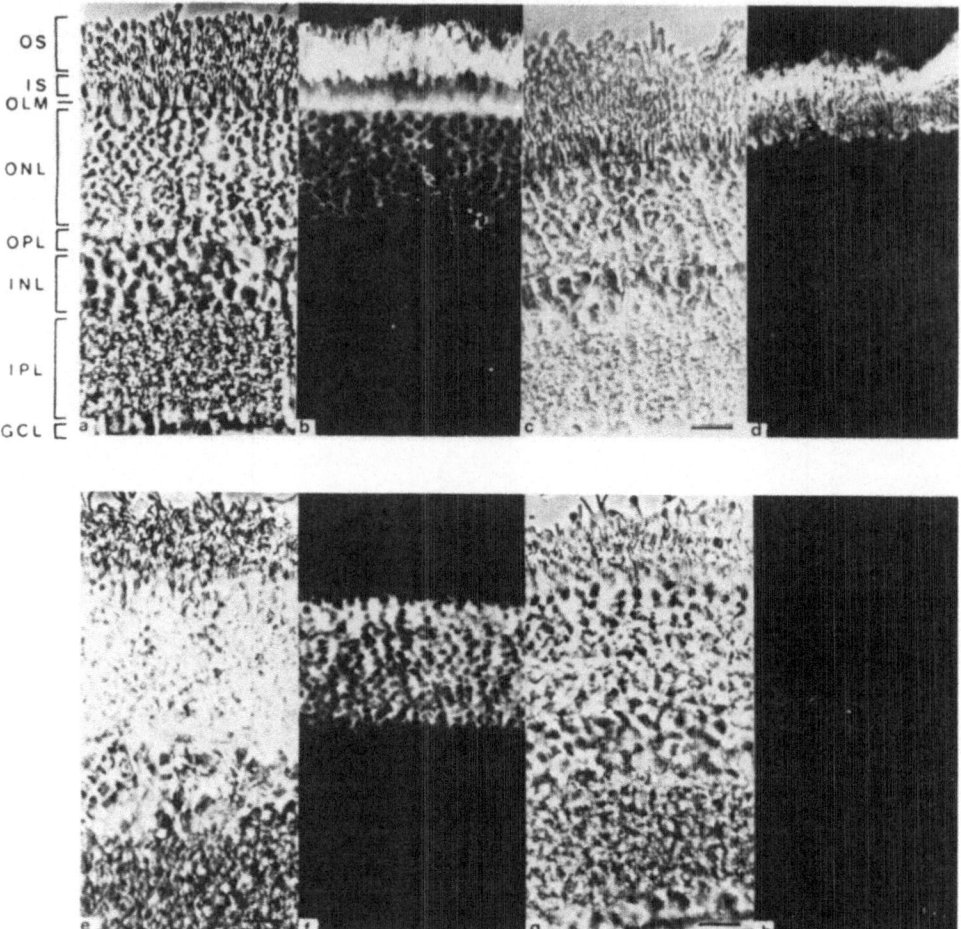

Fig. 1. *Indirect immunofluorescent labelling of sections of adult rat retina by antibodies RET-P₁ (a,b), RET-P₂ (c,d) and RET-P₃ (e,f) employing rhodamine-conjugated goat anti-mouse IgG, that had ben affinity-purified on mouse IgG-Sepharose 4B.* Sections (15 μm) were prepared on a freezing microtome using retinas that had been fixed with 1% paraformaldehyde, 0.1% glutaraldehyde for 1 hr, and treated with 30% (w/v) sucrose overnight. Nonspecific fluorescence (g,h) was determined using a monoclonal antibody against a human cell-surface glycoprotein as the primary antibody. (a, c, e, g) Phase-contrast and (b, d, f, h) epifluorescence micrographs are shown for each field. (OS) outer segment; (IS) Inner segment; (OLM) outer limiting membrane; (ONL) outer nuclear layer; (OPL) outer plexiform layer; (INL) inner nuclear layer; (IPL) inner plexiform layer; (GCL) ganglion cell layer. Scale bar, 20 μm.

during early embryogenesis illustrates how antibodies can be
utilized as probes for the study of pattern recognition elements.
Thus, by isolating batteries of monoclonal antibodies reacting
with well defined sections of chicken retina, sampled at different
phases of development, the authors were able to identify one
antibody whose distribution within the retinal tissue obeys a
strictly specified gradient since its concentration increases in
proportion to the square of the circumferential distance from the
ventroanterior pole of the gradient towards the dorsoposterior
pole (Figure 2).

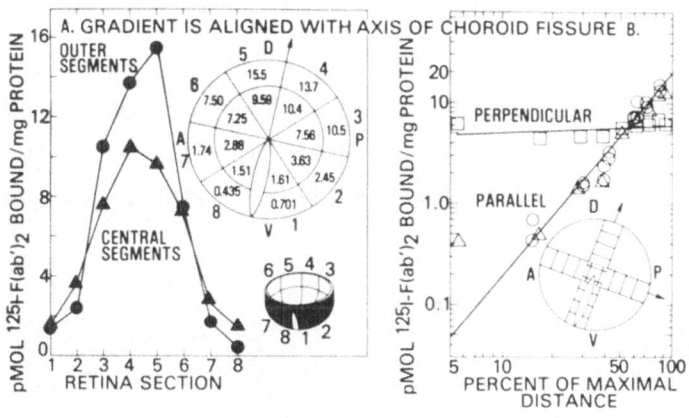

Fig. 2. *Orientation of the TOP gradient.* Specific binding of
$[^{125}I]F(ab')_2$ (pmoles per mg of protein) is shown within the
appropriate segment of retina in A and on the ordinate in A and B.
(A) Each left retina was cut into eight 45° sectors (7.25 mm in
length), which were divided into central (4.9 mm) and outer (2.35
mm) segments. (B) Demonstration that TOP concentration detected
is a function of the square of distance from the ventroanterior
margin of the retina. (Δ) Strips of retina 2.5 mm wide running
from the ventroanterior margin (0% distance) to the dorsoposterior
margin of the retina (14.5 mm = 100% of maximal distance), parallel
to the choroid fissure, were removed from eight retinas (left eyes)
and each was cut into nine 1.5 mm segments as shown. ($\Box$) Strips
of retina from anterior (0%) to posterior (100%) margins of retina
were prepared and assayed as above. (O) The data from A. The length
of the arc from the ventral pole of the gradient to the centre of
each segment was calculated assuming the retina to be a hemisphere
and using equations relating angles and sides of spherical triangles.

b) The other example also deals with the problem of how a
given neuron, or a pair of identical neurons, can occupy a finite
position with a particular ganglion and what kind of interaction
it establishes with functionally related neurons during development.
Work by Zipser (4, 8) indicates that one can identify a unique
neuron among thousands of accompanying neural cells taking advantage
of the fact that each neuron presumably is in contact and communi-
cates with other neurons by means of surface markers that presumably
are unique or highly specific to this neuron. This might have some
important bearing on integrated behaviours, for behaviours presumably
depend upon very specific interneuronal connections.

This work was carried out with the neural apparatus of the
leech, a system whose electrophysiological properties and development has given rise to extensive studies (9). This apparatus
comprises chains of ganglia connected in many ways by lateral or

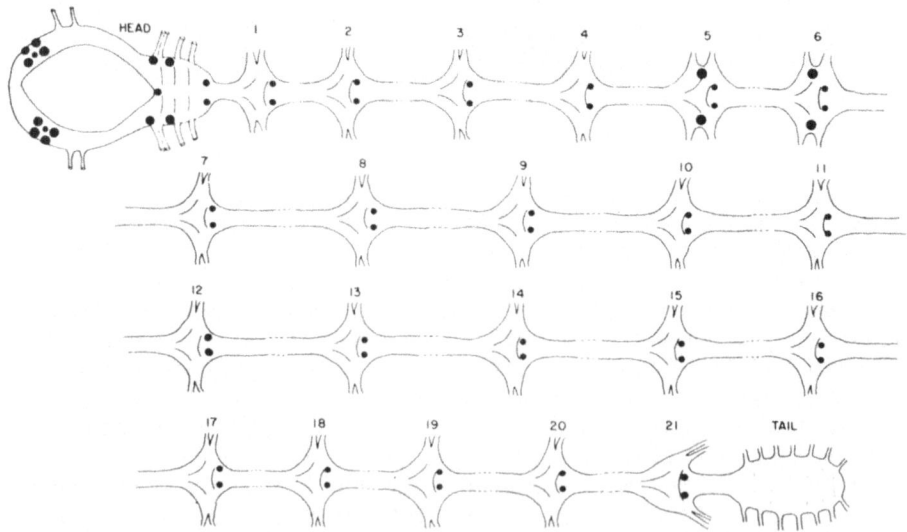

Fig. 3. *Map of Lan3-1-positive cell bodies in the entire nerve
cord of the leech.* This diagram illustrates the symmetry and re-
petitive organization of this simple nervous system. The head map
is still partial, and the tail ganglion also contains cell bodies,
but their exact subganglionic location has not yet been determined.

longitudinal bundles of nerve fibres (Figure 3), each ganglion, including no less than 400 neurons. The function of some of these neurons has been well specified, some being involved in the mating process, others belonging to the mechano-sensory system or being sensitive to heat,pain, etc.

Among 400 hybridomas obtained in immunizing mice with the total neural system of the leech, 64% were found to manufacture antibodies that specifically bind to neurons. 40 of them did react with only one pair of identical neurons, located bilaterally within all the ganglia or certain of them (see Figures 4 and 5).

In conclusion it is clear that immunological techniques have proven of great help in approaching the problem of neural organization and development, not speaking of the possibility, largely utilized, to monitor for specific neurotransmitters or to localize synaptics vesicles (10, 11).

(ii) *Function of the neural system – the chemistry of transmitters*

The molecular basis for synaptic transmission, at least in higher vertebrates, is related to the sending by the nerve endings of chemical messages that act, post-synaptically either on other neurons and glands, or on muscles. These messages are usually called "transmitters" and they can be regarded as a particular class of short-lived hormones acting at a very short distance from the place where they are synthesized (12).

Schematically, either the receptor is located in close vicinity to a ionophore (in which case an allosteric change in the ionophore conformation will cause an increase in the inward ionic flux) or it is closely associated with an adenylcyclase complex in such a way that, after formation of a transmitter-receptor complex, this will cause increased synthesis of cAMP. By the intermediary

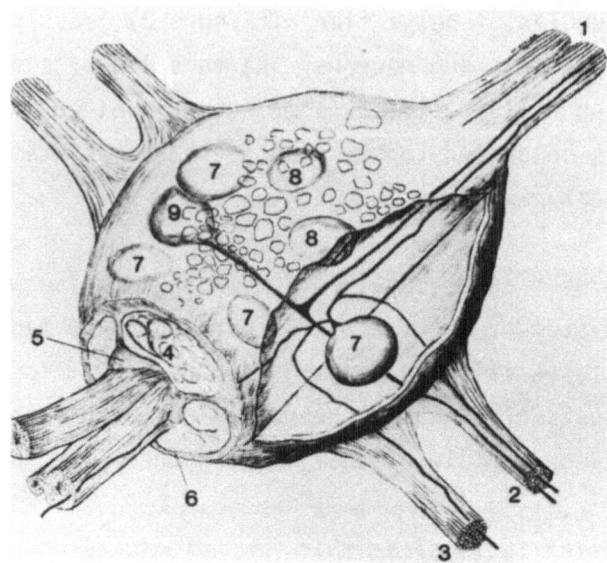

Fig. 4. *Diagram of a leech midbody ganglion.* (1) Connective; (2) anterior root; (3) posterior root; (4) neuronal cell bodies in glial packages; (5) the beginning of the neuropil where synapses occur; (6) capsule; (7) two pairs of bilaterally symmetrical pressure cells; (8) the pair of large Retzius cells are shown in the background of several hundred neuronal cell bodies; (9) one of the two lateral penile evertor (PE) cells (the other has been dissected away). Each cell body has a characteristic location and number of axons. Note that the PE motor neuron projects into con-contralateral roots and that the sensory pressure cell has ipsi-lateral projections. Monoclonal antibodies were raised by immunizing mice with the entire leech nerve cord. Both the P3-X63-8Ag and SP2 cell lines were used as the myeloma parent. Hybrid cell lines were tested by direct immunohistochemical screening on the leech nerve cord. Interesting lines were cloned in soft agar. The leech nerve cords used in screening were fixed in 4% paraformaldehyde and 0.1 M phosphate buffer (pH 7.4) for 30 min at room temperature or in Bouin's fixative for 4 hr at room temperature. After washing in 0.05 M phosphate buffer (pH 7.4) and 0.9% sodium chloride, the connective-tissue capsules were cut; incubations with antibody and washes were in phosphate-buffered saline and 0.2% saponin.

involvement of a cAMP-dependent protein kinase, the membrane properties of the postsynaptic neuron will be modified, so as to cause stimulation or inhibition of its activity.

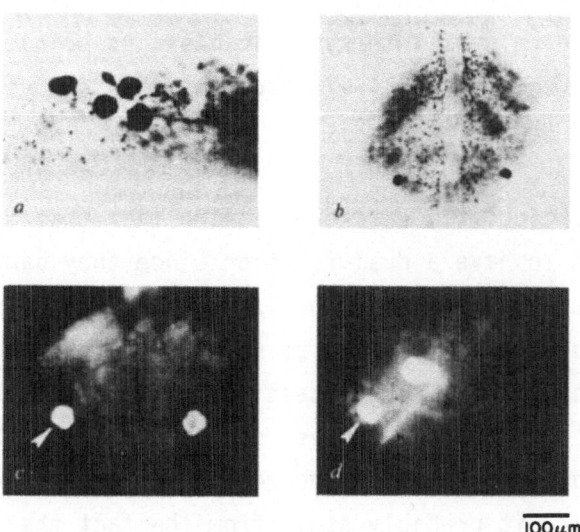

100μm

Fig. 5. *Specific staining with Lan3-1.* The monoclonal antibody
was visualized immunocytochemically using peroxidase-conjugated
(a, b) or rhodamine-conjugated (c) second antibody. (a) shows
cell bodies in the right supraoesophageal ganglion stained with
Lan3-1. (b) shows a bilaterally symmetrical pair of reactive cell
bodies in a typical midbody ganglion. In addition to the two
deeply stained cell bodies the neuropile contains a large number
of stained beaded processes (varicosities) which extend into one
or more axons in the connective. (c) shows two larger cell bodies
which reproducibly occurs in the fifth and sixth ganglia stained
with a rhodamine-conjugated second antibody. The two smaller cell
bodies are also present but lie in a different focal plane. (d)
shows the same ganglion as (c) but viewed under FITC optics
which reveals the presence of microinjected fluorescent dye lucifer
yellow. The left lateral PE cell (marked by an arrow) was identified
by a unique synaptic relationship to the morphologically and
physiologically identifiable rostral penile evertor cell. The other
cell labelled by lucifer yellow as a control is the Retzius cell.

For long, only a limited number of substances were known that
were able to mediate synaptic transmission. Examples were acetylcholine, monoaminergic substances (norepinephrin, dopamin, serotonin,
histamin), plus certain aminoacids (glutamic acid, glycin, aspartate,
as well as γ-aminobutyric acid, taurine, etc. But during the 5-6
years or so, the list of neurotransmitters has considerably

enlarged, and a great variety of peptides (neuropeptides) were
found to fulfill the role of intersynaptic messages being, in many
instances, released concomitantly with some of the "canonical"
transmitters listed above (Table 2) (13).

What looked particularly striking was the fact that most of
these peptides seem to have a dual function since they can act as
typical neurotransmitters being synthesized by neural cells of the
brain in minute amounts, or they can be secreted in much larger
amounts by elements of the gastro-intestinal tract where they
function as typical hormones (examples: secretin, cholecystokinin,
gastrin, etc.) (Table 3). Although the reason for this situation
is not totally clear, it probably stems from the fact that, in
primitive cells, very likely, a same substance was able to act as
a short or long-distance signal, examples of this sort being known,
particularly in protozoons.

Whatever it may be, the field of neuropeptide chemistry is
expanding rapidly (13, 14) and everyone is familiar with the
discovery of classes of brain peptides which can act on the same
receptors as morphin or its derivatives, the most typical of which
belonging to the enkephalin or endorphin families.

.A point of particular concern in the present article is that
this field has received great impetus not only from classical
peptide chemistry but also, and more unexpectedly, from recombinant
DNA technology.

Not only genes coding for many of these peptides have been
cloned but, more interestingly, use of cDNA probes have permitted
to specify, in most cases, the genome sequence corresponding to
the long size precursors of these peptides, a task otherwise
difficult to achieve in view of the short half-life of these
precursor molecules.

Table 2. *Canonical transmitter substances and their key biosynthetic enzymes*

Transmitter	Enzymes
Acetylcholine	Choline acetyltransferase (specific)
Biogenic amines	
Dopamine	Tyrosine hydrolase (specific)
Norepinephrine	Tyrosine hydrolase and dopamine β-hydrolase (specific)
Serotonin	Tryptophan hydrolase (specific)
Histamine	Histidine decarboxylase (specific uncertain)
Amino-acids	
γ-Aminobutyric acid (GABA)	Glutamic acid decarboxylase (probably specific)
Glycine	General metabolism (specific pathway undetermined)
Glutamine	General metabolism (specific pathway undetermined)

Table 3. *Neuroactive peptides*[a]

Gut-brain peptides
   Vasoactive intestinal polypeptide (VIP)
   Cholecystokinin octapeptide (CCK-8)
   Substance P
   Neurotensin
   Methionine enkephalin
   Leucine enkephalin
   Insulin
   Glucagon

Hypothalamic-releasing hormones
   Thyrotropin-releasing (TRH)
   Luteinizing hormone-releasing hormone (LHRH)
   Somatostatin (growth hormone release-inhibiting factor, SRIF)

Pituitary peptides
   Adrenocorticotropin (ACTH)
   β-Endorphin
   α-Melanocyte-stimulating hormone (α-MSH)

Others
   Angiotensin II
   Bradykinin
   Vasopressin
   Oxytocin
   Carnosine
   Bombesin

---

[a]Taken from ref. 13.

The most striking example is provided by the cloning of the
large precursors to peptides with morphin-like activity. One of
these precursors called Proopiomelanocortin (POMPC) is normally
processed to produce ACTH, LPH and endorphin. Not only study of the
cDNA sequence (which corresponds to the total coding sequence of
the precursor) can confirm data previously obtained from peptide
chemistry, but it has given rise to important observations regarding
the region preceding the ACTH sequence. In particular, one could
identify a sequence: His - Phe - Arg - Trp that is characteristic
of melanophoric (MSH) peptides and already found in the α and β

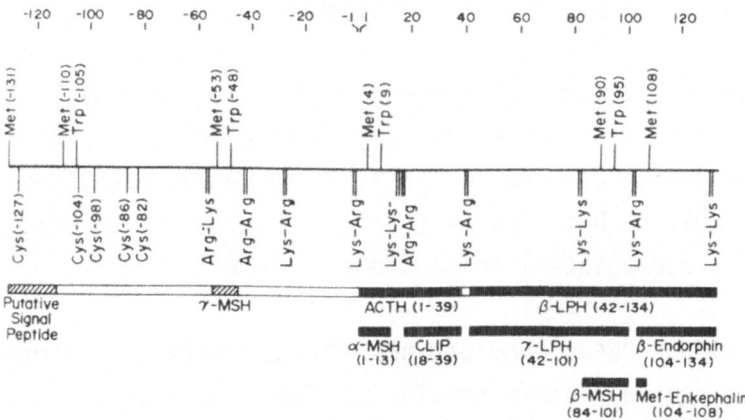

Fig. 6. *Schematic representation of the structure of bovine
ACTH-β-LPH precursor.* Characteristic amino acid residues are shown,
and the positions of the methionine, tryptophan and cysteine residues are given in parentheses. The location of the translational
initiation site at the methionine residue at position -131 is
assumed (see the text). The closed bars represent the regions for
which the amino acid sequence was known, and the open and the
shaded bars represent the regions for which the amino acid sequence
has been predicted from the nucleotide sequence of the ACTH-β-LPH
precursor mRNA. The locations of known components peptides are shown
by closed bars; the amino acid numbers are given in parentheses.
The locations of γ-MSH and the putative signal peptide are indicated
by shaded bars; the termini of these peptides are not definitive.

MSH sequences. A melanophoric-like sequence was also identified
between residues 111 and 105 (15). All these data have contributed
to derive the basic organization of the POMPC precursor, a sequence
including four repeats of the MSH type, presumably arising by gene
duplication from an ancestral sequence.

Figure 6 illustrates the peptide structure of the large
endorphin precursor as derived from recombinant DNA data, a long
polypeptide which, by sequential processing, can generate α, β
and γ-MSH, ACTH, β-LPH and β-endorphins.

Of considerable interest is the fact that, although the β-
endorphin moiety includes in its continuity the enkephalin sequence, enkephalin peptides are not generated by processing of
POMPC. Rather they are formed in brain and in the adrenal medulla
by cleavage of a very large precursor whose sequence was recently
elucidated thanks to the genetic engineering approach. This precursor includes several copies of Met- and Leu-enkephalin molecules
and is probably so built up as to release large stoichiometric
amounts of these opiate-like peptides (Figure 7) (16).

Recombinant DNA technology has also permitted to determine
the genomic sequence corresponding to POMPC. Worthy of notice is
the fact that the β-endorphin sequence contains no "intron", a
situation which made it possible to express "active" mammalian
endorphin in *E. coli* cells transformed with a λ lac-endorphin
recombinant phage. A β-galactosidase - β-endorphin hybrid protein
was expressed. After isolation and chemical cleavage, it generated
active β-endorphin (17).

Other goals have thus far been achieved due to the utilization
of the recombinant DNA technology in the field of neuroscience: for
instance, not only other neuropeptide sequences have been cloned,

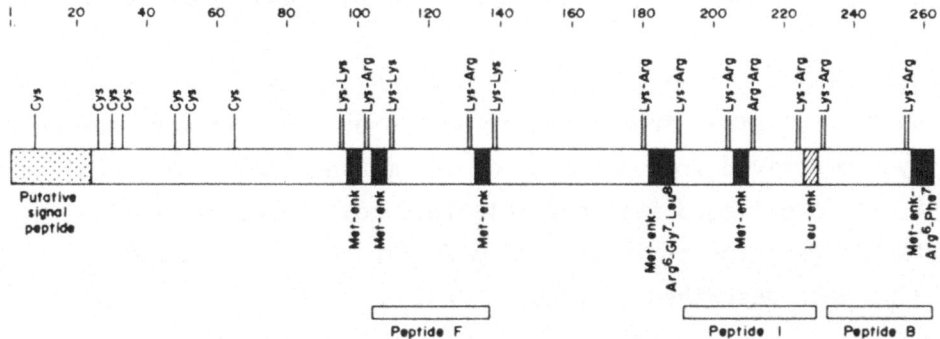

Fig. 7. *Schematic representation of the structure of bovine pre-proenkephalin.* The sequences of Met-enkephalin, Met-enkephalin-Arg[6]-Phe[7] and Met-enkephalin-Arg[6]-Gly[7]-Leu[8] are indicated by closed boxes, the sequence of Leu-enkephalin by a shaded box and the putative signal peptide by a stippled box. All the paired basic amino acid residues and cysteine residues are shown. Amino acid numbers are given above. The known peptide structures, peptide F (residues 104-137), peptide I (residues 192-230) and peptide B (residues 233-263) are displayed underneath by open bars; the known peptides representing partial sequences of peptide I - peptide E (residues 206-230), BAM-22P (residues 206-227) BAM-20P (residues 206-225) and BAM-12P (residues 206-217) - are not shown.

but more recently cDNA complementary sequences corresponding to the major neurotoxin-binding ACh receptor subunit (18) or to tyrosine hydroxylase (19) have also been obtained. This opens the way to important investigation at a molecular level, particularly as far as neurogenesis is concerned, a topic which we will examine next.

2. THE DIFFERENTIATION OF NEURAL CELLS

As it is always the case with most somatic tissues or cells, neuronal differentiation (neurogenesis) has been tackled both *in vivo* and *in vitro*. In particular, the use of neural established lines which, when placed in appropriate conditions, acquire the phenotypic properties of mature neurons has proved of interest to molecular biologists.

The systems most investigated at this time and illustrated are:

a) *Neuroblastoma* from human or murine origin, the most commonly used system derived from a tumour of the mouse neural crest, called C-1300. It is believed that neuroblastoma most frequently arise in sympathetic ganglion or in adrenal medullar cells, but other origins have also been described (20, 20a, 20b).

b) *Clonal cell lines* of the central nervous system (CNS) originating from nitrosomethyl urea-induced tumours (21).

c) *Neuroblastoma X somatic cell hybrids:* eg. Nb x glioma (22), or Nb x L cells (23) hybrids have been largely utilized, the first one being of considerable interest for it displays surface receptors to opiates.

d) *Pheochromocytoma.* They come from tumours of sympathetic ganglions. Cells of the PC-12 line respond well to the addition of NGF, contrary to most neuroblasoma (24).

e) *Transformed neurosecretory cell lines:* eg. hypothalamic cell lines releasing neurophysins or vasopressins (25).

f) *Peptidergic cell lines:* eg. AtT-20 is a line cloned from an ACTH secreting tumour of the pituitary, which was of great use to study the processing of the ACTH precursor (26).

(i) *Neuroblastoma differentiation generalities*

Neuroblastoma can display two main developmental states. Either they appear like round, immature neuroblasts: this is so when cultivated in suspension conditions in a serum-containing medium (Figure 8) or as neurite-bearing cells in which case they

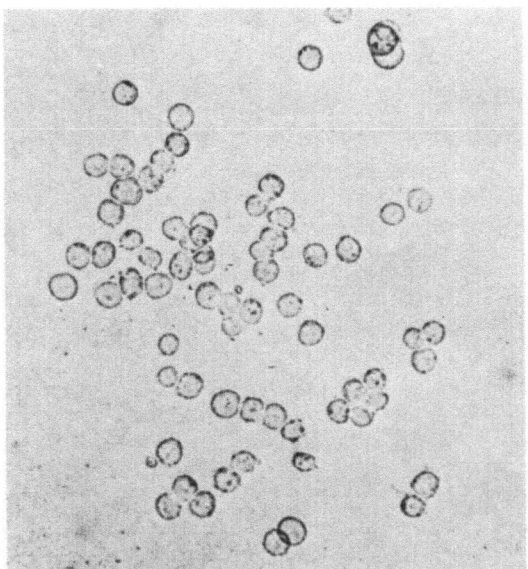

Fig. 8. *Morphology of neuroblastoma.* Round cell population 24 hr
after transfer from a Petri dish to a tissue culture dish.

acquire both the electrophysiological and biochemical properties
typical of mature neurons (Figure 9). This developmental transition
is achieved when post-mitotic suspension-grown neuroblastoma are
transferred onto a solid support in monolayer conditions, within
a serum-free medium. Many substances (referred to as "inducers")
can also cause morphological differentiation, some acting even in
the presence of serum. A variety of substances or physical effects
can cause induction (X-ray inactivation, simple serum withdrawal,
addition of prostaglandins or cyclic AMP analogs, DMSO, hexamethylene bis-acetamide, butyrate, etc.).

     In spite of their chemical diversity, these inducers all cause
cessation of DNA synthesis: i.e. cells enter post-mitosis before
differentiating. Yet, two main categories of effects can be observed:
in some instances, neuroblastoma cells aquire both the morphological

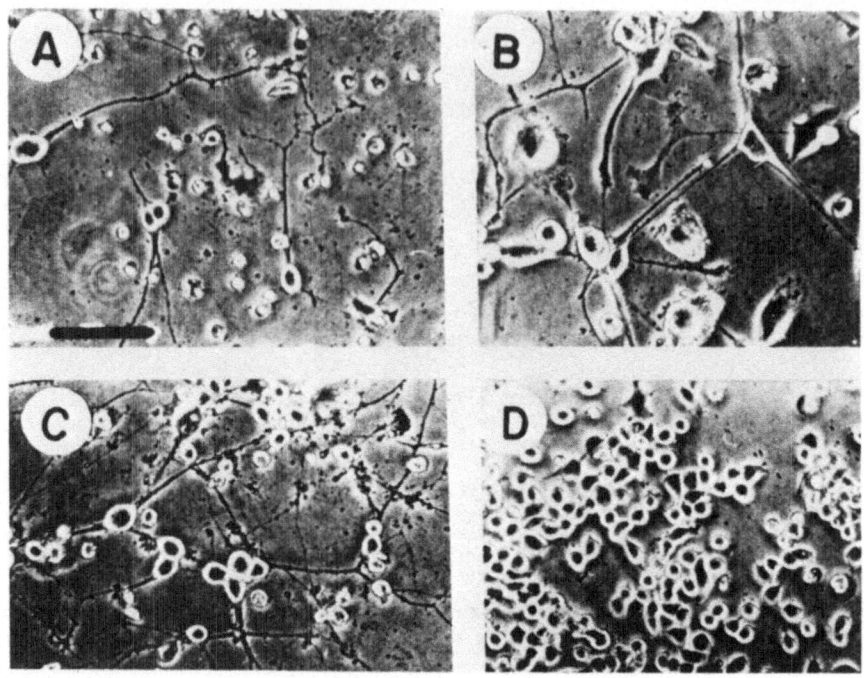

Fig. 9. *Axon-dendrite formation by (A) cholinergic clone NS-20; (B) adrenergic clone N_1E; (C) inactive clone N-18; and (D) inactive clone N_1A-103, which does not form axons or dendrites.* Cells were incubated in growth medium without serum for five days to stimulate neurite formation. The scale shown in A applies to all panels, and corresponds to 10 μm.

and biochemical properties of mature neurons; in others, inducing agents cause biochemical differentiation but neurite formation does not take place (this is particularly so with Na butyrate). Non-inducible Nb variants have also been isolated (27). For instance, strain NA 103, whatever the inducer, expresses most of the enzymatic and biochemical characteristics of wild-type parent strains but fails to extend neurites and lacks excitability properties characteristic of morphologically differentiated cells.

Biochemical differentiation can be assessed according to many criteria. Differentiated neuroblastoma harbor new surface antigens or various specific enzymes involved in the synthesis of the appropriate transmitters. Of particular interest is the γ enolase subunit, also referred to as protein 14.3.2. It is an isozymic form that is uniquely expressed in neural cells. During conversion of dividing neuroblasts to mature neurons, the α-enolase subunit ceases to be formed and an α to γ transition is observed in adult neurons (28, 29, 29a, 29b). Induced neuroblastoma sometimes only express part of the neurogenic programme and lack certain of the properties of mature neurons. Such is the case for the ability to establish synaptic connections.

(ii) *Work from the laboratory*

Using neuroblastoma cell lines as well as cultivated primary neurons or developing brain as models, our laboratory has been essentially engaged in the study of neurogenesis. Global approaches included questions such as the changes in poly-A$^+$ mRNA complexity (30, 31) or in the distribution of cytoskeletal proteins (32), but more specific challenge of neurogenesis could be sought by looking at changes in the level of defined protein markers (14.3.2, cholinesterase, tyrosine hydrolase, scorpion venom receptors, etc.) (33-36).

As an illustration, we will presently report on data recently obtained with the use of a novel neuroblastoma inducer, for its study seems to have direct bearing on the early events accompanying neurogenesis, in relation to changes in the cytoskeleton apparatus. We shall next turn to recent findings concerning later events.

(iia) *Early events in neurogenesis. Comparative effects of CCA and other inducers*

Pharmacological studies aimed at analyzing *in vitro* effects of drugs endowed with antianorexic and anticonvulsive properties

have led to the discovery of a new series of compounds capable of inducing
neuroblastoma differentiation (37). By analyzing their mode of
action, new insights could be obtained on early neurogenesis *in
vitro*. One of the substances that proved most active in our hands
was a simple cyclohexane derivative called cyclohexane-carboxylic
acid (CCA). Cells of the N1E-115 strain, when grown in suspension
in the presence of 7.5% calf serum, appear like round-shaped
immature neuroblasts (Figure 8). By contrast, when transferred onto
the surface of a plastic dish in the absence of serum, they cease
to divide and begin to extend fine ramified neurites usually
distributed in a bipolar fashion (Figure 9). If transfer occurs
in the presence of CCA (0.1%) plus serum, monolayer cells also
produce significant extensions. It is to be noted that these
extensions are multipolar and that the cell bodies are considerably
flattened suggesting better adhesion to the substratum (Figure 10).
That these "extensions" are not simply retractile pseudopodia but
typical neurites can be assessed by use of appropriate antineuro-

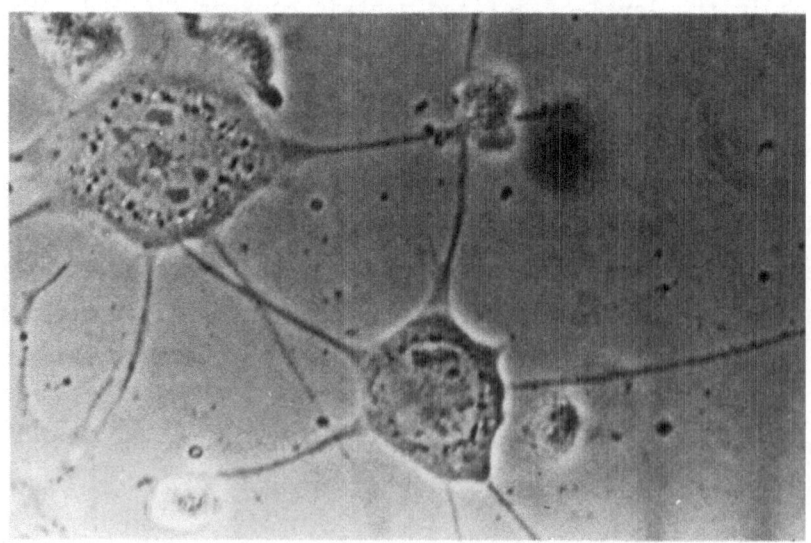

Fig. 10. *Morphology of N1E cells grown in the presence of CCA.*

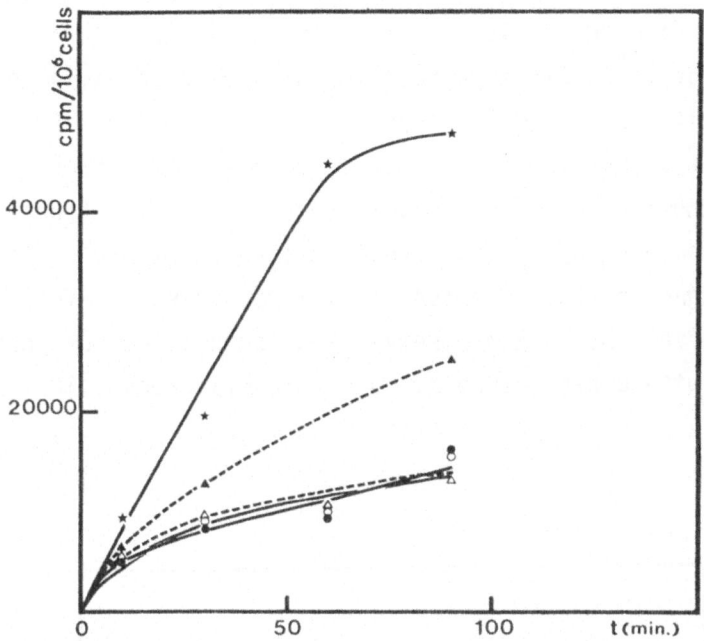

Fig. 11. *Cellular incorporation of [^{14}C] 2-deoxyglucose.* Total radioactivity was measured in the 12 000 g supernatant and normalized to 10^6 cells. Cells were counted with a haemocytometer. Each value is the average of four successive countings. * — * CCA treated cells; o — o serum free cultures; ● — ● Me$_2$SO treated cells; Δ --- Δ growing cells in logarithmic phase; ▲ --- ▲ confluent cells in stationary phase.

filament protein antibodies in an immunofluorescent assay. CCA-induced cells harbor some of the biochemical characteristics typical of normally induced neuroblastoma. Tyrosine hydroxylase synthesis is stimulated as is the accumulation of voltage-dependent Na$^+$ ionophores (data not shown).

From the metabolic stand point, an interesting observation was made: because CCA was known to exert some protective effects in the brain of animals under anoxia, it was believed that the drug could influence oxygen uptake in neural cells. To challenge for this hypothesis uptake of 2-deoxyglucose, a glucose analogue that is phosphorylated and can complex with glucose isomerase, without conversion to fructose, was analyzed. Figure 11. illustrates the kinetics of [^{14}C] 2-deoxyglucose total uptake. Measurement of radioactivity in a 20,000 g supernatant derived from NP-40 lysate shows that CCA exhibits a marked stimulatory effect, as compared to undifferentiated neuroblastoma or to cells treated with other inducers. By ethanol precipitation, only the radio-activity engaged in the isomerase complex was measured. The CCA stimulatory effect was again observed and found to be more pro-nounced than with other inducers. These effects precede in time the appearance of neurite outgrowth (Figure 12).

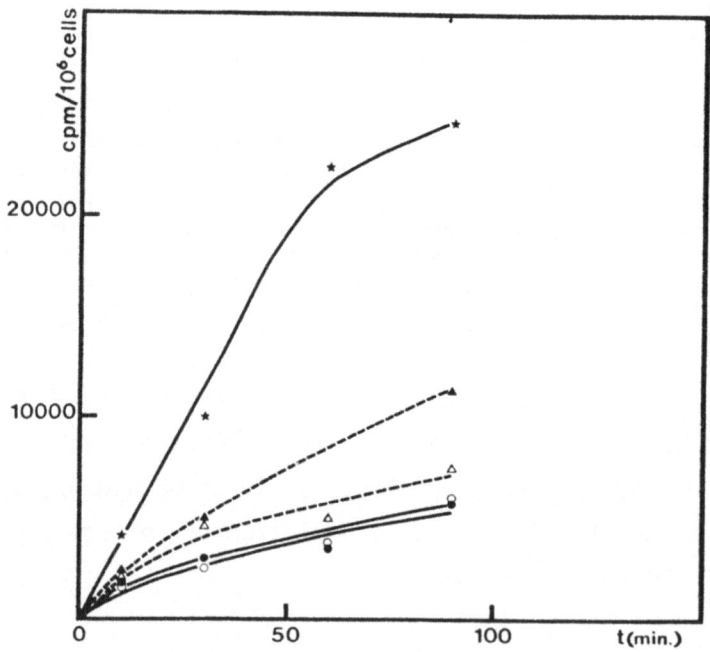

Fig. 12. *Measurement of ethanol precipitated radioactivity.* Ethanol precipitated radioactivity from the 12 000 g supernatant was measured and normalised to 10^6 cells. * ── * CCA treated cells; o ── o serum free cultures; ● ── ● Me$_2$SO treated cells; Δ --- Δ growing cells in logarithmic phase; ▲ --- ▲ confluent cells in stationary phase.

(iib) *Study on cytoskeletal and membrane bound proteins*

Work from several laboratories (Littauer *et al.* (38); Chan and Baxter (39), Dahl and Wiebel (40), Shelanski (41) has shown that brain maturation or *in vitro* terminal differentiation of neuroblastoma is paralleled by modulations in the rates of synthesis of, isotubulins, actins, neurofilament proteins, as well as by the expression of specific isoforms. We have thus explored the CCA effects at this level, comparison being made with other inducers.

Usually neuroblastoma cells induced or not were labelled with [^{35}S]-methionine and then lysed with Nonidet P40. After centrifugation of the lysate at 12,000 g, proteins were analyzed by two dimensional gel electrophoresis both in the supernatant and in the pellet fractions.

Figure 13 illustrates a general pattern of [^{35}S]-methionine labelled proteins from a total (CCA-induced) lysate: we have been particularly concerned with proteins such as actin, α and β tubulin, vimentin, α-actinin, vinculin (130 Kd) as well as with two pellet-associated (and presumably membrane-bound) proteins, designated as "Y" and "Z" respectively.

Table 4 summarizes a large amount of data obtained under a variety of inducing conditions. The data refer to the relative contents and presumably relative rates of synthesis of a certain number of relevant protein markers. Figure 14 permits to visualize the data from the preceding table. The following conclusions can be drawn:

When NIE 115 cells are maintained in monolayer conditions and induced by CCA in the presence of serum, rates of (total) tubulin and actin synthesis are reduced. By contrast, a large increase is observed in rates of synthesis and in the accumulation of Z, Y and vimentin.

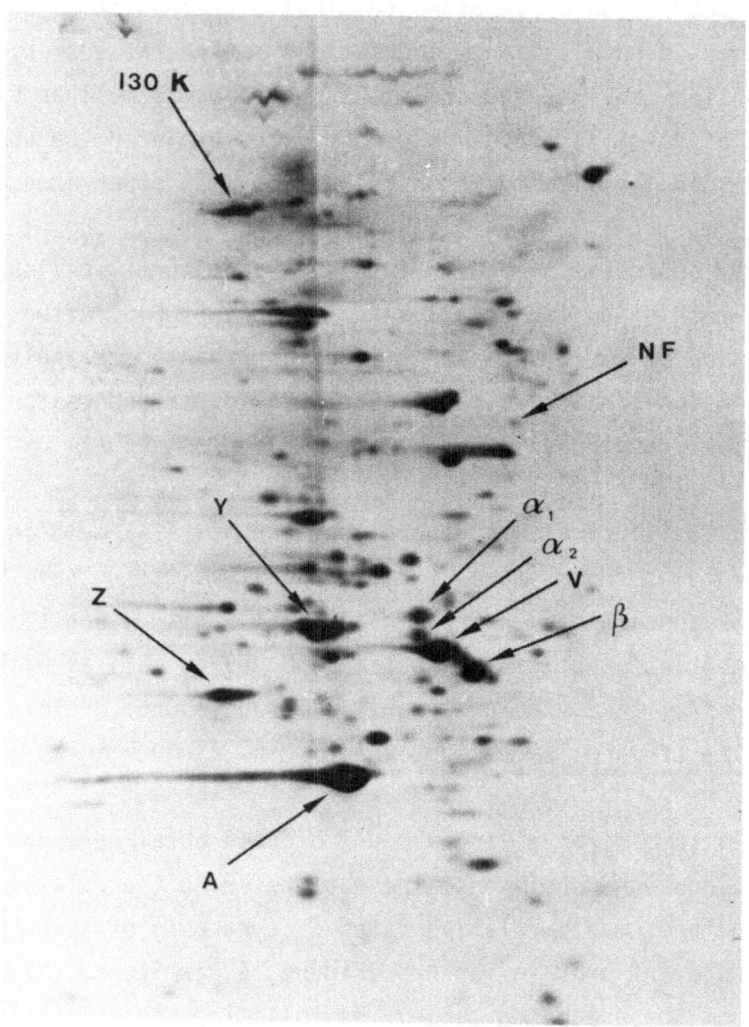

Fig. 13. *Two dimensional gel electrophoresis of CCA induced neuroblastoma cells.*

Table 4. *Relative contents of relevant protein markers*

Inducing agent and/or conditions of cultures	Isotubulins					Actin	Z	Y	Vi	130 K	NF70 K
	α_1	α_2	β	6	Total						
Monolayer cells											
CCA	0.65	0.84	0.52	0.75	0.61	0.60	2.24	5.21	2.83	4.8	2.02
DMSO	0.65	0.64	0.26	0.64	0.46	0.70	1.05	0.83	0.62	1.12	1.02
Serum deprivation	0.70	0.96	0.38	0.89	0.61	1.02	1.67	1.96	0.79	1.9	0.96
Suspension cells											
CCA	0.43	0.52	0.22	0.43	0.34	0.44	1.05	1.58	0.79	1.01	0.89
Serum containing medium	1.09	0.96	0.76	1.04	0.94	0.84	1.19	1.25	0.77	ND	ND
Serum deprivation	0.74	0.80	0.87	1.11	0.92	0.45	1.19	1.25	0.40	ND	ND

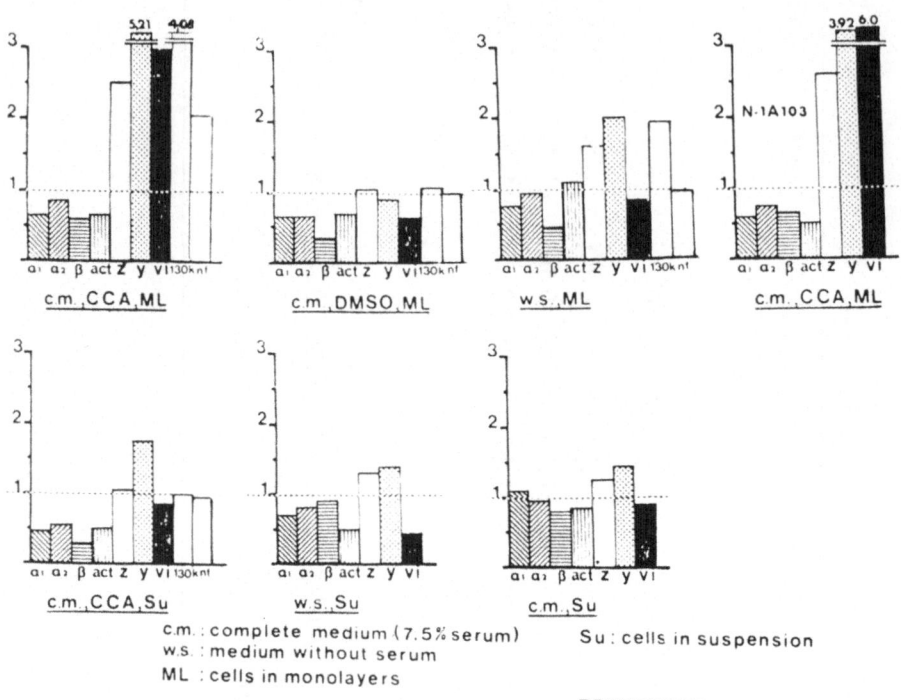

Fig. 14. *Summary of relative proportions of relevant protein markers.*

- The same situation applies to the variant strain N1A-103 which is not capable of extending neurites, but can express all the biochemical markers characteristic of induced neuroblastoma.

- The presence of serum is required to observe the stimulatory effect of CCA on vimentin and Z protein. This is not so, however, with regard to the reduced level of total tubulin and actin synthesis.

- Increased rate of synthesis of vimentin, Z and Y in cells

induced by CCA plus serum, cannot be observed when cells remain in suspension. Thus, adhesion to a solid substratum is apparent requisite or has to take place concomitantly.

- Other inducers or inducing conditions (serum withdrawal or DMSO addition) produce similar qualitative effects with regard to the negative modulation of actins and tubulins, or (as far as serum withdrawal) with regard to the increase in Y and Z. However, there is no, or very little, positive modulation at the level of Y and Z with DMSO as an inducer.

- Most importantly (and this has been verified a number of times), CCA effects are irreversible: if CCA is added to cells maintained in suspension (under conditions where no differentiation occurs) and if those cells are transferred to a solid substratum plus serum, in the absence of CCA, the biochemical modulations are the same as if CCA had remained present after transfer. We have also checked that capacity to extend neurites is acquired by cells which have remained in contact with CCA.

Most of these data can be summarized and interpreted as shown in Figure 15.

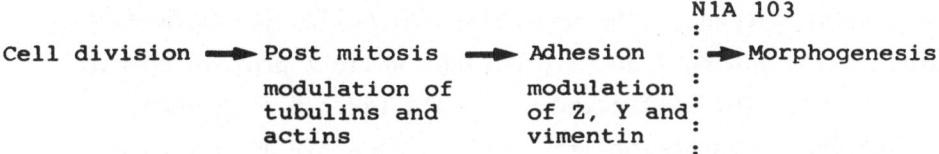

Fig. 15. *Modulation of protein markers.*

What we designate here as negative modulation in the level of tubulin and cytoplasmic actin could be associated with the change in the mitotic cycle of preinduced neuroblastoma, since it is observed with post-mitotic suspension-grown cells or in the N1A-103

variant. Positive modulation at the level of vimentin, Y, Z and
130 Kd protein could be related to the process of cell adhesion
since it was clearly shown that it involves transfer onto a solid
substratum. But since they are observed even in the variant NIA-103,
these effects would precede the step of neurite "sprouting", and
could thus be regarded as early events, perhaps preparatory to
neurite elongation.

The rather large accumulation of proteins of this latter group
(as opposed to their modest, and probably reversible increase with
other inducers) could be in relation with the irreversibility of
CCA induction (we recall that with other inducers neurite outgrowth
is reversible).

The case of vimentin deserves particular comments: (1) First,
the fact that neuroblastoma cells, induced or not, contain vimentin
was confirmed by *in situ* immuno-cytochemistry (data not shown).
(2) It has long been reported that vimentin is absent from neuronal
cells. This is probably correct as far as mature, post-mitotic and
immobile neurons are concerned, but we believe this is not so at
all stages of neurogenesis. In recent work, we were able to show
that primary rat neurons, when put in culture, contain vimentin
(42). Thus vimentin expression is not (necessarily) related to
the neoplastic nature of the neuroblastoma cells. (3) We are in
the process of cloning Y and Z proteins. While Z protein appears
to increase also in CCA-treated myoblasts (42), Y is probably
a membrane-bound neuron-specific marker. Recently it was found
to accumulate in brain from CCA-treated neonate rats (42). Its
identification by immunochemical technique is under way.

(ii) *Late events in neurogenesis – Positive modulation of certain*
      tubulins
      We recall that the effects which we have thus far analyzed

are, developmentally speaking, anterior to the phenomenon of neurite outgrowth and seen to accompany to a large extent events such as cell adhesion, topological distribution of the nucleus, formation of microtubule organization centres (from which neurotubules elongate).

It was interesting to search for more ultimate markers associated with terminal neurogenesis and directly connected with the phenomenon of neurite extension. Results from work carried out in our laboratory by Eddé, Jeantet and their colleagues (43) seem to provide a reasonable approach.

It is well known that, in the brain, tubulin is, by far, the most abundant protein, amounting to 30-50% of total cytoplasmic proteins (Shelanski *et al.* (44)). Tubulin has been involved in a variety of neural functions including axoplasmic flow and depolarization-associated exocytosis (45).

During the past 5 years, several authors have reported on the very high degree of tubulin polymorphism in brain contrasting to what is observed in other organs. Recent work from our laboratory (46, 47) as well as from others (48) indicates that as many as 20 distinct forms would be observable in the brain, based on iso-electrofocusing data, the greatest heterogeneity being observed during brain maturation in developing mammals. Recombinant DNA studies would suggest the existence of at least 8-10 tubulin-related sequences (49, 49a, 49b).

Recently, we have followed the fate of isotubulin micro-heterogeneity in developing mouse neuroblastoma by comparing wild-type clone N1E-115 with the variant N1A-103.

Figure 16 indicates results from a two dimensional gel elec-

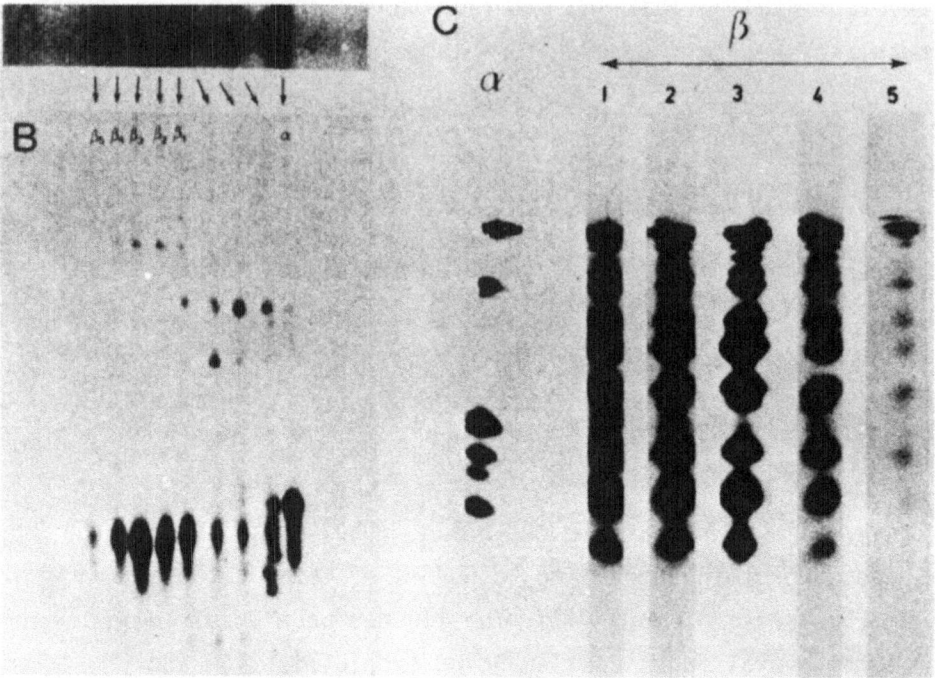

Fig. 16. *Analysis of N1E-115 tubulin: (A) isoelectric focusing (only tubulin region is presented); (B) SDS-electrophoresis of each band of the tubulin region separated after staining of the focused gel; (C) peptide maps of the proteins which present pI and MW values corresponding to tubulin.* These maps were compared with those obtained for α and β purified brain tubulin (data not shown). Digestion was performed with *S. aureus* protease V8 (6 ng/slot).

trophoretic analysis of phosphocellulose-purified tubulin from differentiated N1E-115 cells. The technique involves isoelectro-focusing in one dimension, cutting each band and running it separately in a second direction, on a SDS-acrylamide gel, so as to minimize overlapping and smearing of the bands. Six forms could be detected whose apparent molecular weights and protease digestion patterns show that they correspond to 5 β isotubulins (named by us β_1 to β_5) plus an α-type isotubuin.

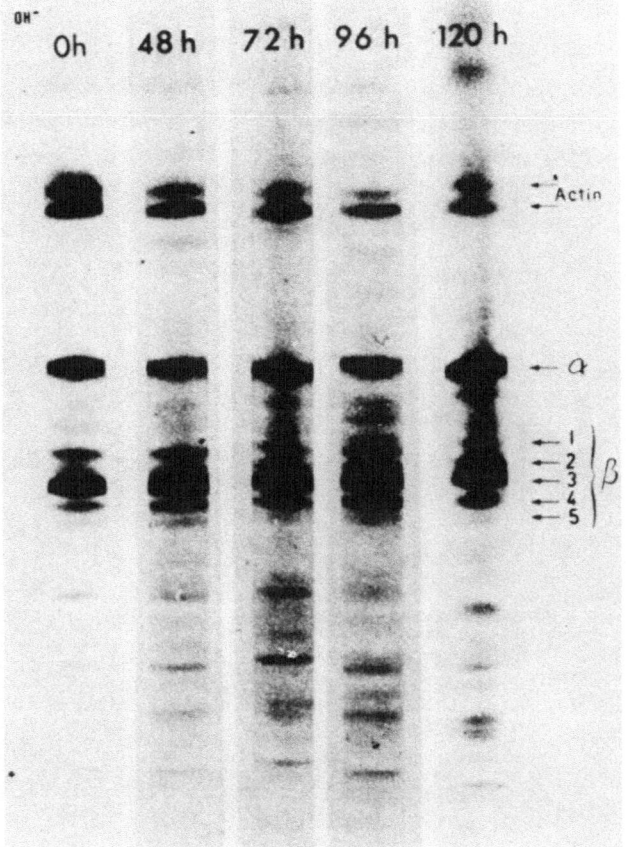

Fig. 17. *Isoelectric focusing of partially purified N1E-115 tubulin at different times of the culture in serum-free medium.* Oh, refers to exponentially growing cells.

When a similar analysis was performed on undifferentiated neuroblastoma cells, only 5 isotubulins were observed; the band corresponding to β_2 was lacking: β_2 appeared only the 3rd day and its level in the cell seemed to go in parallel with the onset of morphological differentiation (Figure 17).

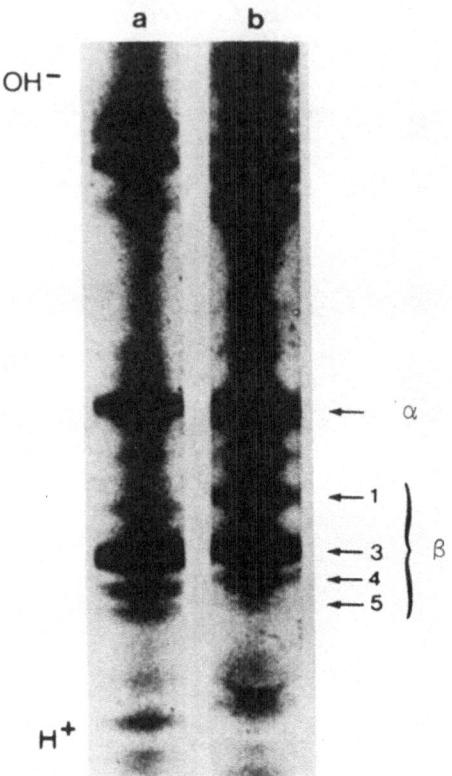

Fig. 18. *Isoelectric focusing of partially purified tubulin from N1A-103 cells.* Exponentially growing (a) and serum starved (b) cells.

Quite remarkably, when the same experiment was done with an induced N1A-103 strain, the β_2 form could not be detected (Figure 18). Since induction of this variant strain causes appearance of all the biochemical markers characteristic of the wild-type strain but is not paralleled by neurite extension, this strongly suggests that appearance of the β_2 isoform would be correlated with morphological differentiation.

The mechanism responsible for β_2 appearance was further investigated. It is known that isotubulin microheterogeneity not only reflects the existence of many distinct gene products but also some post-translational modifications (glycosylation, tyrosylation, etc.) (50, 51, 52).

In order to explore to what extent the existence of multiple tubulin bands in differentiating cells was due to the expression of new genes or to a modification of preexisting gene products, some pulse-chase experiments were carried out. If, following transfer of N1E-115 cells in monolayer conditions, in the absence

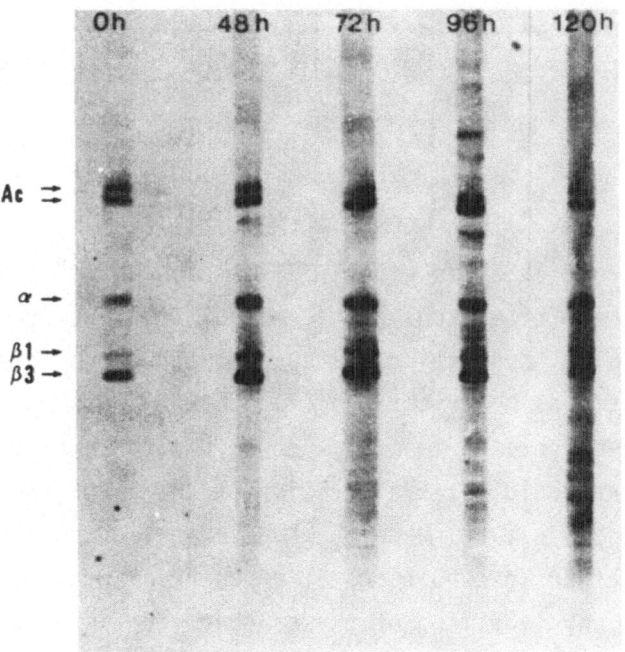

Fig. 19. *Autoradiographs of the gels presented in Figure 17; Ac: actin.*

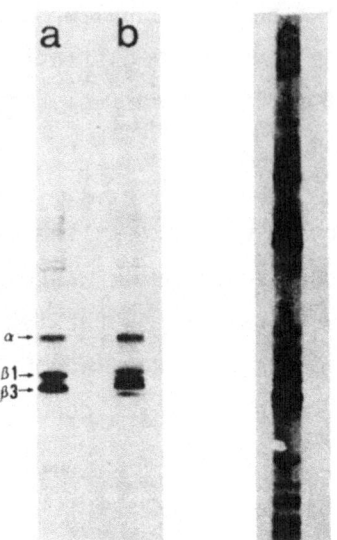

Fig. 20. *Pulse-chase experiments: isoelectric focusing of partially purified tubulin from N1E-115 cells: in the differentiated stage; (a) 3 hr pulse-labelling; (b) 3 hr pulse-labelling followed by a chase for 24 hr; in exponentially growth conditions; (c) 3 hr pulse-labelling followed by a chase for 24 hr.* Autoradiographs.

of serum, cells were labelled with [^{35}S]-methionine for periods of 3 hr at different times, only α, β_1 and β_3 tubulins incorporated the tracer initially, suggesting that β_2, β_4 and β_5 would arise from posttranslational modifications (Figure 19). Supporting this view, if a pulse-chase experiment was done in differentiating neuroblastoma, one could observe efficient transfer of the label from β_1, β_3 into β_2, β_4, β_5. When the same experiment was repeated with cells maintained in conditions of exponential growth, no accumulation nor any labelling of β_2 was observed (Figure 20).

The scheme of Figure 21 attempts to summarize our present view about the developmental control of early and late markers

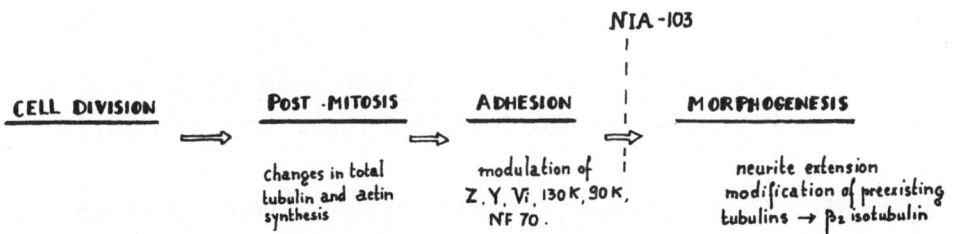

Fig. 21. *Developmental control of early and late markers during neuroblastoma differentiation.*

during neuroblastoma differentiation. Work is currently done to characterize the nature of the modification which generates β_2 isotubulin.

In conclusion, we can see that certain aspects of neurogenesis can probably be easily approached by means of neural cell lines, but it is clear, however, that no firm conclusion can be drawn until the salient data will be confirmed by analysis of primary neurons in culture or by studies on developing brain. This problem is currently being investigated in our laboratory.

3. ACKNOWLEDGEMENTS

This work was supported by grants from the Centre National de la Recherche Scientifique, the Délégation Générale a la Recherche Scientifique et Technique, the Institut National de la Santé et de la Recherche Médicale, the Commissariat à l'Energie Atomique, the Fondation pour la Recherche Médicale Française, the Ligue Nationale Française contre le Cancer, the Muscular Dystrophy Association and the SANOFI.

4. REFERENCES

1) HIS, W. (1889). Arch. Anat. Entroicklungsgeschichte, p. 49
   cited in "Histogenes of the Central Nervous System" (1967),
   Jan Langman.

2) HARDESTY, J. (1904). Ann. J. Anat. 3, 229. Cited in "Histogenesis of the Central Nervous System" (1957), Jan Langman

3) MIRSKY, R. (1980). Cell-type-specific markers in nervous
   system cultures. TINS, 3, No. 8, 190.

4) ZIPSER, B. and McKAY, R. (1981). Monoclonal antibodies specific
   for identifiable leech neurons. In: Monoclonal Antibodies to
   Neural Antigens, (eds. R. McKay, M.C. Raff, L. Reichardt)
   Vol. 2, p. 91. Cold Spring Harbor Reports in Neurosciences,
   Cold Spring Harbor, N.Y.

5) BARNSTABLE, C.J. (1980). Monoclonal antibodies which recognize
   different cell types in the rat retina. Nature, 286, 23.

6) BARNSTABLE, C.J. (1981). Developmental studies of rat retina
   cells using cell-type-specific monoclonal antibodies. In:
   Monoclonal Antibodies to Neural Antigens, op. cit. p. 219.

7) TRISLER, G.D., SCHNEIDER, M.D., and NIRENBERG, M. (1981). A topo-
   graphic gradient of molecules in retina can be used to identify neuron position. Proc. Natl. Acad. Sci. USA, 78, 2145.

8) ZIPSER, B. and McKAY, R. (1981). Monoclonal antibodies
   distinguish identifiable neurons in the leech. Nature, 289,
   549.

9) STENT, G.S. and WEISBLAT, D.A. (1982). The development of a
   simple nervous system. Sci. Amer. 241, No. 1, 135.

10) MATTHEW, W.D., REICHARDT, L.F. and TSAVALER, L. (1981).
    Monoclonal antibodies to synaptic membranes and vesicles.
    In: Monoclonal Antibodies to Neural Antigens, op. cit. p. 163.

11) DE BLAS, A.L., BUSIS, N.A. and NIRENBERG, M. (1981). Monoclo-
    nal antibodies to synaptosomal membrane molecules. In: Mono-
    clonal Antibodies to Neural Antigens, op. cit. p. 181.

12) SCHWARTZ, J.H. (1981). Chemical basis of synaptic transmission In: Principles of Neural Sciences, p. 107.

13) SNYDER, S.H. (1980) Brain peptides as neurotransmitters. Science, 209, 976.

14) ACHER, R. (1981). Evolution of neuropeptides. TINS 4, No. 9, 225.

15) NAKANISHI, A., INOVE, A., KITA, T., NAKAMURA, M., CHANG, A., COHEN, S., and NUMA, S. (1979). Nucleotide sequence of cloned cDNA for bovine corticotropin - β lipotropin precursor. Nature, 278, 423.

16) COMB, M., SEEBURG, P.H., ADELMAN, J., EIDEN, L. and HERBERT, E. (1982). Primary structure of the human Met and Leu enkephalin precursor and its mRNA. Nature, 295, 663.

17) SHINE, J., FETTES, I., LAN, N.C.Y., ROBERT, J.L. and BAXTER, J.D. (1980). Expression of cloned β-endorphin gene sequences by *Escherichia coli*. Nature, 285, 456.

18) GIRAUDAT, J., DEVILLERS-THIERY, A., ROUGEON, F., AUFFRAY, C. and CHANGEUX, J.P. (1982). Identification of a cDNA clone coding for the acetylcholine binding subunit of *torpedo marmorata* acetylcholine receptor. EMBO Journal, in press

19) MALLET, J., personal communications

20) AUGUSTI-TOCCO, G. and SATO, G (1969). Establishment of functional clonal lines of neuron from mouse neuroblastoma. Proc. Natl. Acad. Sci. USA, 64, 311.

20a) SCHUBERT, D., HUMPHREYS, S., BARONI, C. and COHN, M. (1969). *In vitro* differentiation of a mouse neuroblastoma. Proc. Natl. Acad. Sci. USA., 64, 316.

20b) For a general review cf. DE LAAT, S.W. and VAN DER SAAG, P.T. (1982). The plasma membrane as a regulatory site in growth and differentiation of neuroblastoma cells. In: International Review of Cytology, 74, 1.

21) STALLCUP, N.B. and COHN, M. (1979). Cell-specific antisera as reagents for studying the nervous system. TINS,

22) NELSON, P., CLIFFORD, C. and NIRENBERG, M. (1976). Synapse
    formation between clonal neuroblastoma x glioma hybrid cells
    and striated muscle cells. Proc. Natl. Acad. Sci. USA., 76,
    123.

23) MINNA, J., GLAZER, D. and NIRENBERG, M. (1972). Genetic
    dissection of neural properties using somatic cell hybrids.
    Nature New Biol. 235, 225.

24) GREENE, L.A. and SHOOTER, E.M. (1980). Ann. Review. Neurosci.
    3, 353.

25) DE VITRY, F. (1977). Growth and differentiation of a primitive
    nervous cell line after *in vivo* transplantation into syngeneic
    mice. Nature, 267, 48.

26) ROBERTS, J.L., PHILIPPS, M.A., ROSA, P.A. and HERBERT, E.
    (1978). Steps involved in the processing of common precursor
    forms of adrenocorticotropin and endorphin in cultures of
    mouse pituitary cells. Biochemistry, 17, 3619.

27) AMANO, T., RICHELSON, E. and NIRENBERG, M. (1972). Neurotrans-
    mitter synthesis by neuroblastoma clones. Proc. Natl. Acad.
    Sci. USA, 60, 258.

28) LEGAULT-DEMARE, L., ZEITOUN, Y., LANDO, D., LAMANDE, N.,
    GRASSO, A. and GROS, F. (1980). Expression of a specific
    neuronal protein 14-3-2 during *in vitro* differentiation
    of neuroblastoma cells. Exp. Cell Res., 125, 233.

29) PICKEL, V.M., REIS, D.J., MARANGOS, P.J., ZOUZELY NEURATH,
    C. (1976). Immunocytochemical localization of nervous
    system specific proteins (NSP-R) in rat brain. Brain Res.,
    105, 184.

29a) ZOUZELY NEURATH, C. and KELLER, A. (1977). Mechanisms of
    regulation and special function of protein synthesis in the
    brain. (eds. S. Roberts, A. Latjha and W.H. Gispen), Elsevier/
    North Holland Biomedical Press, Amsterdam, p. 279.

29b) MOORE, B.W. (1972). Chemistry and biology of two proteins S100
    and 14-3-2 specific to the nervous system. Int. Rev. Neurobiol.
    15, 215.

30) FELSANI, A., BERTHELOT, F., GROS, F. and CROIZAT, B. (1978).
    Complexity of polysomal poly(A) RNA in undifferentiated and
    differentiated neuroblastoma cells. Eur. J. Biochem,92, 569.

31) BERTHELOT, F., GROS, F. and CROIZAT, B. (1980). Complexity
    of polysomal poly(A) RNA in different developmental stages
    of non-differentiating neuroblastoma clone. FEBS Lett. 122,
    109.

32) GROS, F., CROIZAT, B., PORTIER, M-M., BERTHELOT, F. and
    FELSANI, A. (1982). The regulation of gene expression during
    terminal neurogenesis. In: Molecular Genetic Neurosciences,
    (eds. F.O. Schmidt, S.J. Bird and F.E. Bloom), p. 335, Raven
    Press, New York.

33) LEGAULT-DEMARE, L., LAMANDE, N., ZEITOUN, Y., GROS, F., SCARNA,
    H., KELLER, A., LANDO, D. and COUSIN, M.A. (1981). Transition
    between isozymic forms of enolase during *in vitro* differentia-
    tion of neuroblastoma cells. Neurochemistry Int., 3, No 5,
    303.

34) LAZAR, M. and VIGNY, M. (1980). Modulation of the distribution
    of acetylcholinesterase molecular forms in a murine neuro-
    blastoma x sympathetic ganglion cell hybrid cell lines. J.
    Neurochem., 35, 1067.

35) THIBAULT, J., VIDAL, D. and GROS, F. (1981). *In vitro* transla-
    tion of mRNA from rat pheochromocytoma tumours, characteri-
    zation of tyrosine hydrolase. Biochem. Biophys. Res. Comm.
    99, 960.

36) BERWALD-NETTER, Y., MOUTOT, N.M., KOULAKOFF, A. and COURAUD,
    F. (1981). Na^+ channel associated scorpion toxin receptor
    sites as probes for neuronal evolution *in vivo* and *in vitro*.
    Proc. Natl. Acad. Sci. USA, 78, 1245.

37) CROIZAT, B., BERTHELOT, F., FERRANDES, B., EYMARD, P. and
    SAHUQUILLO, C. (1979). Differenciation morphologique du
    neuroblastome par l'acide 1-methyl cyclohexane carbosilique
    (CCA) et certains dérivés en C_1. C.R. Acad. Sci. Paris, 289,
    1283.

38) GOZES, I., SAYA, D. and LITTAUER, U.Z. (1979). Tubulin micro-
    heterogeneity in neuroblastoma and glioma cell lines differs
    from that of the brain. Brain Res., 171, 171.

39) CHAN, V. and BAXTER, C. (1979). Compartments of tubulins and
    tubulin-like proteins in differentiating neuroblastoma cells.
    Brain Res., 174, 135.

40) DAHL, J.L. and WEIBEL, V.J. (1979). Changes in tubulin hetero-
    geneity during postnatal development of rat brain. Biochem.
    Biophys. Res. Comm. 86, 822.

41) SHELANSKI, M.L. and LIEM, R.K.H. (1979). Neurofilaments. J.
    Neurochem., 33, 5.

42) PORTIER, M.M. and CROIZAT, B. Personal Communications

43) EDDE, B., JEANTET, C.and GROS, F. (1981). One β-tubulin sub-
    unit accumulates during neurite outgrowth in mouse neuro-
    blastoma cells. Biochem. Biophys. Res. Comm. 103, 1035.

44) SHELANSKI, M.L. and FEIT, H. (1972). In: The Structure and
    Functions of the Nervous Tissue, (ed. G.A. Bourne), 6, p.47,
    Academic Press, New York.

45) THAO, N.B., WOOTEN, G.H., AXELROD, J. and KOPIN, I.J. (1972).
    Inhibition of release of dopamine-β-hydrolase and norepine-
    phrine from synpathetic nerves by colchicine, vinblastine,
    or cytochalasin-B. Proc. Natl. Acad. Sci. USA, 69, 520.

46) DENOULET, P., EDDE, B., JEANTET, C. and GROS, F. (1982)
    Evolution of tubulin heterogeneity during mouse brain de-
    velopment. Biochimie, 64, 165.

47) EDDE, B., PORTIER, M-M., SAHUQUILLO, C., JEANTET, C. and
    GROS, F. (1982). Changes in some cytoskeletal proteins
    during neuroblastoma cell differentiation. Biochimie,
    64, 141

48) GINZBURG, I., BECHAR, L., GIVAL, D. and LITTAUER, U.Z. (1981)
    The nucleotide sequence of rat α-tubulin: 3'-end characteris-
    tics, and evolutionaly conservation. Nucl. Acids Res., 9, 2691

49) CLEVELAND, D.W., LOPATA, M.A., McDONALD, R.J., COWAN, W.J., RUTTER, W.J. and KIRSCHNER, M.W. (1980). Number and evolutionary conservation of α and β tubulin and cytoplasmic β and γ actin genes using specific cDNA probes. Cell, 20, 95.

49a) SANCHEZ, F., NATZIC, J., CLEVELAND, D.W., KIRSCHNER, M.W. and McCARTHY, B. (1980) A dispersed multigene family encoding tubulin in *Drosophila melanogaster*. Cell, 22, 845.

49b) SILFLOW, C.D. and ROENBAUM, J.L. (1981). Multiple α and β tubulin genes in *Chlamydomonas* and regulation of tubulin in RNA levels after deflagellation. Cell 24, 81.

50) PIPERNO, G. and LUCK, D.J. (1976). Phosphorylation of axonemal proteins in *Chlamydomonas reinhardtii*. J. Biol. Chem., 251, 2161

51) RAYBIN, D. and FLAVIN, M. (1977). Modification of tubulin by tyrosylation in cells and extracts and its effect on assembly *in vitro*. J. Cell Biol., 73, 492.

52) FEIT, H. and SHELANSKI, M.L. (1975) Is tubulin a glycoprotein? Biochem. Biophys. Res. Comm. 66, 920.

CELLULAR SYSTEMS AND ASPECTS OF PROTEIN SYNTHESIS IN THE STUDY OF MUSCLE CELL DIFFERENTIATION

R.G. Whalen

Département de Biologie Moléculaire
Institut Pasteur
25, Rue du Dr. Roux
75724 Paris, France

1. INTRODUCTION

Among those systems that molecular biologists have chosen to study terminal differentiation, the genesis of skeletal muscle cells provides many of the advantages required to facilitate experimentation. Mononucleate cells can be grown in cell culture, and the dynamic process of cell fusion followed microscopically. The fusion of these mononucleate cells into multinucleate structures known as myotubes provides a dramatic visual indication that terminal differentiation is taking place (1). The fact that this process takes place in cell culture, allows the experimenter to intervene to modify culture conditions and thus attempt to modify the processes of myogenesis. This is crucial to the dissection of the relations between cell proliferation and cell differentiation. These two phenomena are mutually exclusive in the case of skeletal muscle cells and their myogenic precursors, the myoblasts.

Many of the proteins specific to muscle tissue begin to be

synthesized during the time that cells fuse (1). These proteins
include enzymes, the contractile proteins that provide the mechanical
force for muscle contraction (2), and membrane proteins such as
specific receptors for the neurotransmitter acetycholine (3). Muscle
cells also secrete a basement membrane which has several
characteristics that distinguish it from the basement membrane of
other cell types (4). Thus several categories of proteins are
available as subjects in the study of protein biosynthesis. Some
of these proteins, notably the contractile proteins, are found in
different isoforms not only in different types of muscle tissue
but also in non-muscle cells (5). These homologous isoforms make
up protein "families" and raise interesting questions concerning
the organization of the multiple genes coding for these proteins
and the manner in which they are expressed differentially (6).

From a very early point after the formation of myotubes, the
differentiated muscle cell is in intimate contact with another
cell type, the motor neuron. The neuron not only controls the
contractile activity of the muscle fiber, but also exerts a
"trophic" influence on the morphological maturation of the fiber.
In some cases, the nerve is implicated in determining the types
of contractile protein isoforms accumulated by the muscle (7, 8).
The occurrence of this heterotypic cell-cell contact and the
problems of intercellular communication are further major features
of the biology of the muscle cell.

In this article, I first discuss the myogenic cell systems
currently used and the way in which myogenesis can be manipulated.
These studies have given some insight as to the controlling factors
of myogenesis. Second, I discuss some of the contractile proteins
commonly used as markers in the myogenic system, and summarize
recent results on the qualitative and quantitative features of
contractile protein synthesis during myogenesis.

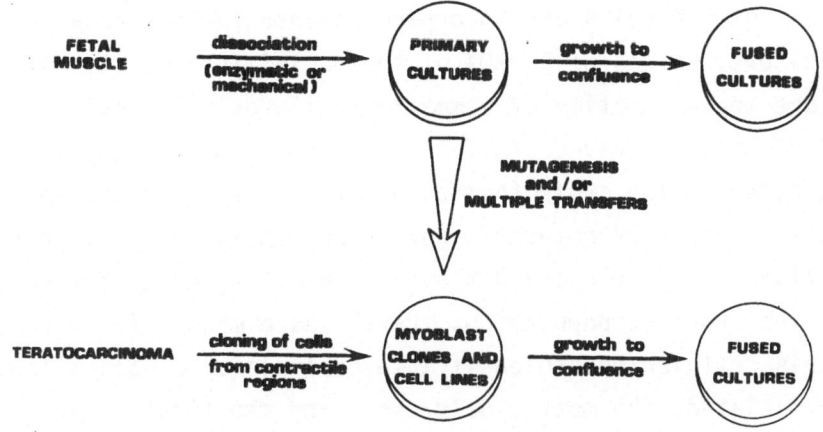

Fig. 1. *Schematic representation of myogenic cell systems.*

2. MYOGENIC CELL SYSTEMS

Figure 1 shows some aspects of the myogenic cell culture
systems. Mononucleate cells can be obtained from fetal or newborn
mammals - rats and mice are the most commonly used - or from avian
embryos - chickens and quails are the most popular. These mononucleate
cells are released from tissue by enzymatic treatment (trypsin,
collagenase) or by mechanical means (vigorous pipetting), as
commonly done to prepare "primary" cultures from other tissue
types. Cells grow in a standard nutrient medium usually supplemented
with fetal calf serum as a source of growth factors. These cells
will normally undergo several cycles of cell division and then
begin to fuse. The onset of fusion seems to be dependent on
reaching confluency for mammalian cells whereas avian cells will
fuse even at lower cell density, as though the number of cell
divisions preceeding fusion in culture was pre-determined. These
dividing mononucleate cells are called "myoblasts" although since
their myogenic potential is apparent only retrospectively (as a
result of observing fusion and/or biochemical differentiation) they
are also referred to as "presumptive myoblasts" (9).

Not all of the cells are incorporated into fusions in a typical primary culture. Values of 50-80% of the nuclei incorporated into myotubes are representative of mammalian cultures while avian cultures are better behaved and fusion of 95% or more can often be achieved. Although the cells that do not fuse are often called fibroblasts, their true nature is not known. It is possible that the conditions of culture are not appropriate for fusion for some subset of the myoblast populations. Partly as a means of circumventing this cellular heterogeneity, several clonal myoblast lines have been isolated. The most widely used, and the first to be isolated, is the rat cell line L6 obtained by Yaffé (10). L6 was produced by treating primary cultured of rat myoblasts with a chemical mutagen, methylcholanthrene. Of the cells that survived the treatment, some were myogenic and could be cloned to provide cells that are theoretically homogeneous in phenotype. However, L6 cells do not fuse to 100% in confluent cultures, although the myogenic nature of 100% of the cells in a population can be demonstrated by seeding them at clonal density and scoring those colonies that eventually develop fusions. Subsequent to the isolation of L6, some rat cell lines (e.g. L8) were obtained simply by repeated passaging of primary culture cells (11). This latter approach has also been successful for obtaining mouse cell lines, especially by the group of Hauschka (12).

Finally, a myogenic cell line has been obtained from a mouse teratocarcinoma. These germ line tumours form solid tumours of differentiated cell types when grown in mice. By culturing the cells of such a tumour in primary culture and cloning of cells from a myogenic region, it proved possible to obtain a myogenic clone (13).

Other approaches can be used to obtain pure cultures of fused myotubes. If cytosine arabinoside is added to cultures at the time when myoblasts are beginning to fuse, then those cells that are

still dividing will be killed by the drug. In this way, cultures
in which greater than 99% of the nuclei are in fusions are obtained
(14). This method works best with avian myoblast cultures. It is
also possible to remove cells from the culture dish by trypsin
treatment just at the time when small myotubes have formed. The
suspension is then allowed to sediment through a gradient of serum
of culture media. The small myotubes are bigger than the mononucleate myoblasts and fall to the bottom. When these myotubes are
returned to culture, they attach to the dish. Over 99.9% of the
nuclei were found in myotubes following this purification (19).

3. FUSION DOES NOT TRIGGER BIOCHEMICAL DIFFERENTIATION IN MYOBLASTS

About the time that myoblasts begin to fuse in culture the
expression of many muscle-specific functions can be measured biochemically. More discussion of these markers of myogenic differentiation is presented below. The temporal correlation of the appearance
of these markers with fusion naturally raises the question as to
whether the fusion process provokes the initiation of expression of
the muscle phenotype. The answer to this question seems to be clearly
no. The experiments which answer the question are a first example of
the usefulness of manipulating the conditions of myogenic cell cultures.

Fusion can be blocked in two ways that are relevant to this
discussion. First, the level of calcium can be lowered from the
usual ca. 2 mM to 0.2-0.3 mM either by adding EGTA or by preparing
special media. Second, the addition of *cytochalasin B* to myoblasts
at the time they would normally begin to fuse inhibits the fusion
process. For the avian primary culture system, many groups have
now reported experiments in which different biochemical parameters
of myogenic differentiation were measured in fused-blocked cultures
and found to increase with kinetics similar to those observed
in cultures allowed to fuse normally (16). This was true whether

fusion was inhibited by lowering calcium levels or by using *cytochalasin B*. In rat myoblast cells, *cytochalasin B* results in blocked fusion and the mononucleate cells accumulate muscle specific enzymes as well as large amounts of contractile protein. However, both fusion and biochemical differentiation are blocked if the calcium levels are lowered (16). This latter result does not detract from the general conclusion that fusion does not trigger the expression of other muscle-specific parameters. It probably indicates that, in rat cells at least, calcium inhibits more than just the process of fusion, which is hardly surprising.

4.   OTHER MEANS OF MANIPULATING MYOGENESIS

Several agents have been found that maintain myoblast cells in the proliferative phase. Dimethylsulfoxide (DMSO) for example inhibits fusion in cultures of L8 cells at concentration of 1-2% (Figure 2). Unlike low calcium levels or *cytochalasin B*, this chemical does not dissociate fusion and differentiation but rather produces a population of cells which continue to proliferate (17). The synthesis of muscle-specific contractile proteins is not detected in these cells. Upon removal of DMSO, however, the capacity to fuse and to differentiate biochemically is restored to these cells.

A similar situation has been described in which a purified mitogen, "fibroblast growth factor" (FGF), will also cause mouse myoblasts to continue to proliferate provided that the levels of FGF are not allowed to diminish as a result of their metabolism by the cells (18). If myoblasts are transferred to a medium which lacks FGF and which had previously been used to grow cells (referred to as "conditioned medium"), the myoblasts will then fuse. The mitogens normally found in fresh medium supplemented with serum have presumably been removed by the "conditioning" of the medium by the cells. If FGF is added to this conditioned medium then

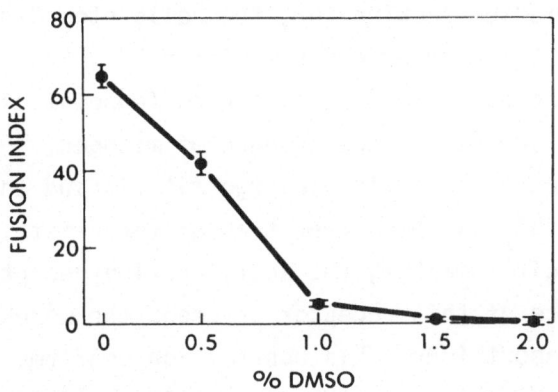

Fig. 2. *Concentration dependence of DMSO inhibition of cell fusion on L8 myoblast cells.* From Blau and Epstein (17).

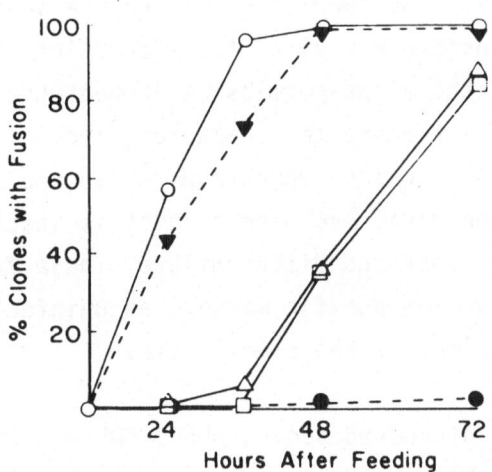

Fig. 3. *Delay of fusion in mouse myoblasts by FGF.* Cells were fed fresh medium (●), conditioned medium (CM)(o), CM plus 1 ng/ml FGF (▼), CM plus 10 ng/ml FGF (Δ) or CM plus 100 ng/ml FGF (□). From Linkhart *et al.* (18).

fusion is delayed by a length of time proportional to the amount of
FGF added (Figure 3). If FGF is abruptly removed by changing to
more conditioned medium lacking FGF, the cells stop dividing and
go on to fuse. At the time of removal of the FGF, the cell populatio
is growing asynchronously, and cells can be found at different
points in the cell cycle. In the absence of mitogen, the cells
complete the current cell cycle, undergo mitosis and enter into
the G1 phase. In this G1 phase they then differentiate. Measurements
of one muscle-specific marker, the acetylcholine receptor, show
that the appearance of this receptor preceeds the morphological
act of fusion by about 4 hr. This observation confirms the idea
that fusion does not trigger biochemical differentiation. The
appearance of the contractile protein myosin has also been shown
to preceed fusion in quail myoblast cultures (19).

Another means of manipulating myogenesis involves infection
of avian myoblasts with a mutant of Rous sarcoma virus. RSVts68
(reviewed in 20). This virus carries a temperature-sensitive mutatior
in the *onc* gene responsible for viral transformation. Upon growing
infected myoblasts at the virus-permissive temperature (34-35°C),
the myoblasts exhibit a transformed character, and in particular
they proliferate without fusing. Upon transfer to the non-permissive
temperature (41°C), the viral *onc* gene product is inactivated and
the myoblasts will now fuse and differentiate. These fused cells
accumulate the same muscle-specific markers as uninfected primary
cultures and to approximately the same levels.

The three agents discussed above, DMSO, FGF and RSVts68 -
presumably acting in three rather different ways - all cause
myoblast cells to continue proliferating and all inhibit differentiation. These results therefore reinforce the general idea that
proliferation and differentiation in the myoblast system are
mutually exclusive.

5. POSSIBLE SIGNALS CAUSING CELLS TO WITHDRAW FROM THE CELL CYCLE

What events or signals either provoke the cessation of proliferation in myoblasts or induce myogenic differentiation? A major clue to the answer of this question comes from experiments reported by Zalin and colleagues. A transient rise in the intra-cellular level of cyclic AMP was found in chicken myoblast cultures which preceeds the onset of fusion by 5-6 hr (21). The duration of this peak was approximately 1 hr and the amount of cAMP rose to 10-15 times the basal levels (Figure 4). The prostaglandin PGE_1 can provoke a similar short-lived increase in cAMP, and PGE_1 treatment can advance the onset of fusion in a subpopulation of myoblasts.

Careful analysis of the cell cycle parameters in PGE_1-treated cultures led to the conclusion that the subpopulation of myoblasts

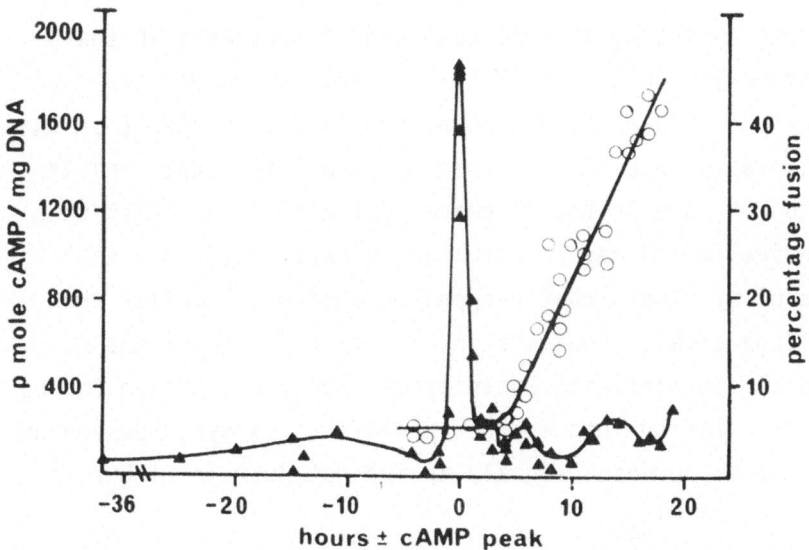

Fig. 4. *Time course of intracellular cAMP levels (▲) and fusion (o) in chick myoblast cultures.* Values obtained from 5 cultures. From Zalin and Montague (21).

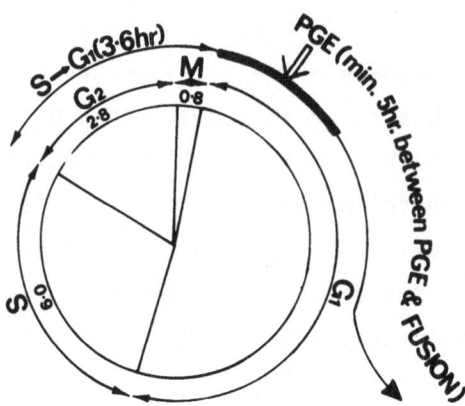

Fig. 5. *Model of the myoblast cell cycle illustrating duration of the various phases and the times involved in its response to* PGE_1. From Zalin (22).

which respond by fusing is that composed of myoblasts in the G1 phase of the cell cycle (22). These results are summarized schematically in Figure 5. The model proposed as a result of these experiments takes into account that the myoblasts that fuse in response to PGE_1 are in the G1 phase, but also illustrates that the responsive period may be situated in early G1, since the percentage of myoblasts that respond by fusing is smaller than the percentage of all myoblasts in G1. The division of the G1 phase into two compartments would predict that cells have the option of fusing or entering into another proliferative cycle depending on whether environmental signals arrive not just in G1 but at a defined period in G1.

Prolonging the total time spent in G1 may as a consequence prolong the responsive period, and therefore the probability that a myoblast will fuse would increase as the length of G1 increases. This idea has been advanced by Koningsberg and colleagues based on their data on avian myoblasts (23). The length of the G1 phase does in fact increase for cells growing in tissue culture or even in the embryo. In the framework of this model, it can be considered that fusion occurs in a stochastic fashion and that as a result of fusion the myoblasts are restrained from further proliferative cycles. The mean length of the "terminal" G1 phase (i.e. the G1 of cells that subsequently fuse) is 13-14 hr for quail myoblasts.

Recent observations in our laboratory (24) suggest that substantially lengthening the G1 phase of rat L6 myoblasts is not sufficient to provoke cell fusion. This conclusion is based on experiments in which L6 cells were grown in Ham's F12 medium supplemented with 10% fetal calf serum. The cells grow to confluency and become quiescent (i.e. they no longer synthesize DNA). Most of the cells in these cultures are in the G1 phase of the cell cycle. Over 70% of the cells have a G1 phase of at least 96 hr. Nonetheless, the cells do not fuse (even if the calcium concentration of the medium is increased) nor do they differentiate biochemically. Part of the explanation for these results may be that Ham's F12 medium is depleted of some component required for growth and fusion.

6. CONTRACTILE PROTEINS AS MARKERS FOR MYOGENIC DIFFERENTIATION

The molecular basis of contraction and its control by calcium is found in the properties of the contractile proteins present in skeletal muscle tissues. Essentially, myosin and actin can form filaments which interact and move relative to each other. The energy for this process is furnished by the myosin-mediated hydrolysis of ATP which is stimulated by actin during the interaction of these two protein filament systems. Calcium sensitivity is conferred upon this system by the proteins known as tropomyosin and troponin, which bind to the actin filaments. Although the

precise details are still being worked out (25), it would seem
that tropomyosin acts by sterically blocking the interaction of
myosin and actin. This inhibition can be relieved by troponin,
which is in fact composed of three subunits each playing a unique
role in the system. When the motor nerve initiates a concentration
in the muscle, the calcium concentration in the muscle cytoplasm
increases from 10^{-7}M to 10^{-5}M. The excess calcium binds to tro-
ponin-C (TN-C) and the fixation of calcium by TN-C allows it to
bind strongly to the inhibitory troponin subunit, TN-1. The
formation of the complex TN-C:TN-I releases the third troponin
subunit TN-T, which at low calcium concentrations is bound to TN-I.
The TN-T released from TN-I can bind to tropomyosin and in doing
so relieves the steric inhibition of actin-myosin interaction. The
arrangement of all these proteins into the structural unit of
contraction, the sarcomere, allows contraction to occur by the
movement of the two filaments systems relative to each other in
what has come to be known as the "sliding filament" model of
contraction.

As mentioned in the Introduction, these contractile proteins
begin to be synthesized in quantity in developing muscle cell at
about the time that fusion takes place. This quantitative increase,
to be discussed in more detail below (section 7), is only part
of the story. All eukaryotic cells contain contractile proteins
either as polypeptides with strict structural and functional
analogy to the muscle proteins (e.g. actin, tropomyosin and myosin;
see ref. 5) or as polypeptides with similar functions and structures
but performing a larger range of tasks in the cell (e.g. calmodulin,
which is a calcium binding protein found in non-muscle cells and
which has amino acid sequence homology to TN-C; ref. 26).

In the particular case of actin, the homology of the protein
found in non-muscle cells is so great compared to muscle actin,

Skeletal muscle: B1-ASP-GLU-ASP-GLU-THR-THR-ALA-LEU-VAL-CYS-ASP-
ASN-GLY-SER-GLY-LEU-VAL-LYS   (α-actin)

Cardiac muscle : B1-ASP-ASP-GLU-GLU-THR-THR-ALA-LEU-VAL-CYS-ASP-
ASN-GLY-SER-GLY-LEU-VAL-LYS   (α-actin)

Vascular smooth muscle: B1-GLU-GLU-GLU-ASP-SER-THR-ALA-LEU-VAL-CYS-ASP-
ASN-GLY-SER-GLY-LEU-CYS-LYS   (α-actin)

Visceral smooth muscle: B1-GLU-GLU-GLU-THR-THR-ALA-LEU-VAL-CYS-ASP-ASN-
GLY-SER-GLY-LEU-CYS-LYS   (γ-actin)

Cytoplasmic (non-muscle) : B1-ASP-ASP-ASP-ILE-ALA-ALA-LEU-VAL-VAL-ASP-ASN-
GLY-SER-GLY-MET-CYS-LYS   (β-actin)

B1-GLU-GLU-GLU-ILE-ALA-ALA-LEU-VAL-ILE-ASP-ASN-
GLY-SER-GLY-MET-CYS-LYS   (γ-actin)

Fig. 6. *Comparisons of the amino acid sequences of the N-terminal tryptic peptides of different mammalian actins.* The designations α , β and γ refer to the results of isoelectric focusing (see text and ref. 29). "B1" refers to the N-terminal blocking group, probably acetyl. From Vandekerckhove and Weber (27, 28).

that for many years it seemed likely that the two were identical. However, a combination of experimental approaches, especially the powerful strategy of combining protein chemistry and amino acid sequencing elaborated by Vandekerckhove and Weber (27, 28), has shown not only that non-muscle actin is different from the muscle protein, but has defined the extent of variation in the family of actin proteins. Figure 6 gives a summary of the sequence evidence of Vandekerckhove and Weber which defines six actin polypeptides in warm-blooded vertebrates. This Figure shows only the N-terminal sequences which carry the most characteristic differences distinguishing each actin type. This region is the only one where amino acid substitutions take place among charged amino acids although in all cases the N-terminal peptide is very acidic. The remaining 357 amino acids of the C-terminal portion of the actin

	Skeletal	Cardiac	Visceral	Vascular
Skeletal	-	4	6	8
Cardiac	4	-	4	6
Visceral smooth	6	4	-	3
Vascular smooth	8	6	3	-
β-*cytoplasmic*	25	23	23	23
γ-*cytoplasmic*	24	22	20	22

Fig. 7. *The number of amino acid exchanges between different mammalian actins.* From Vandekerckhove and Weber (27, 28).

polypeptides contain amino acid differences but all of the substitutions are very conservative (e.g. Ile/Val; Leu/Met; Ser/Thr).
Figure 7 gives the number of amino acid differences between the
six actin forms.

Although the studies of Vandekerckhove and Weber defined the
precise sequence differences among the different actins, the
existence of several isoforms was first demonstrated in 1976 (29,
30). Using the two dimensional gel electrophoresis technique of
O'Farrell (31), the three major types of actin (α, β and γ)
differing slightly in isoelectric point were defined. The β and
γ forms are found in all non-muscle cells while skeletal and
cardiac muscle had the α form. In myoblast cells, β- and γ-actin
are also found and α-actin appears in fused cultures (29, 30, 32
and Figure 8). Thus the myoblast, although committed to becoming
a myotube, does not synthesize or accumulate the muscle actin isoform but rather contains actins typical of other non-muscle cells.
This situation is a general result in the myoblast-myotype system.
For those contractile proteins or other structural proteins such as
intermediate filaments found in myoblast cells, it is the non-muscle
isoform that is present; the muscle isoforms appear when cells fuse.
Thus none of these proteins provides a marker which distinguishes
myoblasts from other non-muscle cells.

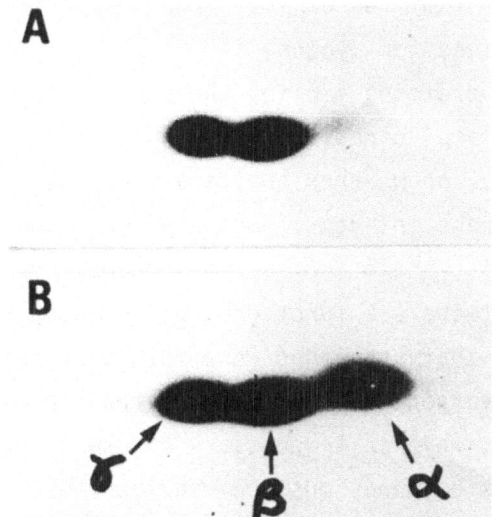

Fig. 8. *Separation by two dimensional gel electrophoresis of the actin isoforms in bovine myoblasts (A) or myotype (B) cultures.* From Whalen *et al.* (32).

The level of detailed information that one has for the actin protein family has not been attained for any of the other contractile proteins. In general, however, it is clear that myosin and tropomyosin exist in multiple forms and that the non-muscle form(s) are different from the muscle forms. In addition, it can be demonstrated in skeletal muscle that the physiologically different types of muscle fibers contain different myosin (and other contractile protein) forms. Cardiac ventricular and atrial myosin are also different from the skeletal muscle molecules. Developing skeletal muscle also have apparently unique forms present at the fetal and neonatal stages (33). Although comparative amino acid sequence studies of myosin heavy chains are rare because of the large size of the molecule, the presence of the myosin protein in muscle tissue has allowed the survey of myosins from different muscles to be undertaken by other protein chemical methods (34).

The existence of families or "isoforms" of contractile proteins
has several implications for myogenesis. Firstly, the existence of
distinct but related proteins implies the existence of at least as
large a number of related genes. Whether these genes are clustered
on a given chromosome, or whether they are expressed as groups
are questions of current interest. Secondly, when two proteins such
as skeletal and cardiac actin are so highly homologous, does their
existence mean that there are functional differences between the
two, or does it mean that the genes were formed by duplication and
have subsequently diverged? If there are no functional differences
at the protein level, why is it necessary to use two genes? Does
this mean that blocks of genes must be regulated separately?
Thirdly, what are the "architectural" problems associated with the
de novo formation of the sarcomeric structures? Are they formed
completely independent of the cytoskeleton that characterizes the
myoblast, or does the sarcomere form using the cytoskeleton as a
scaffold? When sarcomeres are already formed in the developing
muscle and the myosin type is subsequently changed (e.g. when
adult myosin replaces neonatal myosin), how is the "old" myosin
removed and the "new" myosin inserted. There are as yet no good
answers to any of these questions.

7. QUANTITATIVE ASPECTS OF THE SYNTHESIS OF CONTRACTILE PROTEINS
    DURING MYOGENESIS

The availability of the myogenic cell system in which differentiation takes place in cell culture naturally led to many
quantitative measurements of the proteins accumulated or synthesized
as myoblasts fuse and form myotubes. Earlier studies (reviewed in
ref. 1) often measured enzyme activities (e.g. aldolase, phosphorylase, creatine phosphokinase, acetylcholinesterase) that increased
during myogenesis. In the case of the acetylcholine receptor, its
accumulation and halflife could be assayed using [^{125}I]-bungarotoxin,
however, when synthesis was measured it was necessary to carry out

a complete purification for each time point (35).

Among the contractile proteins, the myosin heavy chain
(M_r 200 Kd) was the most easily analyzed for rates of synthesis
since its large size and unique solubility properties made it
possible to measure incorporation of radioactivity into the heavy
chain band on SDS-polyacrylamide gels. Very careful studies of
myosin heavy chain synthesis were carried out by Paterson and
Strohman (36) studying chicken myoblasts, and Emerson and Beckner
(19) who studied quail myoblasts. In both cases, very large increases
in the amount of heavy chain synthesized were demonstrated. For the
quail system, the rate of heavy chain synthesis in fused cultures
was calculated to be 25,000 to 30,000 molecules/myotubes nucleus
per minute. The rate in cultures of dividing myoblasts was 300
molecules/nucleus. In such cultures, a very small percentage of
nuclei are found in fusions (ca. 1%); if these myotubes synthesize
heavy chain at the same rate as myotubes synthesize heavy chain
at the same rate as myotubes in fused cultures then the value of
300 molecules/nucleus could be entirely due to the presence of
the premature fusions in the myoblast culture. The pattern of
heavy chain synthesis is different in different myoblast cell
systems. In the avian systems discussed above, the rate of heavy
chain synthesis remains at a high level for the entire post-
fusion period studied (generally a period of 4-5 days). In rat
L6 myoblasts, the synthesis of heavy chain is transiently induced
from less than 1% up to 25% of total protein synthesis. However,
the rate of synthesis then decreases to almost the initial 1% level
as the heavy chain protein becomes accumulated to the level of (25%
of total protein) of mature myotubes (37). Whether these system
differences are real, or whether they are simply related to the
length of time that the cultures were followed, or whether they
have any *in vivo* parallels, are questions that remain to be
answered.

With the development of two dimensional gel electrophoresis
(31, 38) as a powerful tool for the analysis of protein synthesis
in total cellular extracts, a comprehensive approach to the study
of quantitative differences between myoblasts and myotubes could
be undertaken. The study of Garrels (15) illustrates some interesting
features of myoblast differentiation. It is important to recognize
that Garrels worked with pure populations of L6 myotubes, prepared
by gravity sedimentation (see Section 2). This was probably a
significant factor in accentuating the potential differences between
myoblasts and myotubes. Because 'two dimensional electrophoresis
does not require prior purification of proteins in order to measure
their synthesis, it proved possible to quantitate synthesis for
several proteins not previously studied. Because this technique
separates many of the known muscle and non-muscle contractile
protein isoforms, there is no ambiguity as to which form the
synthesis values apply.

As expected from previous studies, Garrels found that a number
of quantitative differences can be found when comparing myoblasts
and myotubes. Because of the resolution of the analytical technique
used, the increases in the rates of synthesis were considerably
larger than those derived using other techniques. For example, the
rates of synthesis of the muscle α- and β-tropomyosins and one of
the myosin light chains were found to increase by at least 1000
fold. Other proteins, including α-actin and the embryonic form of
myosin light chain (39), increased 300 and 600 fold, respectively
(see Figure 9). Only in the case of α-actin was the final percentage of synthesis greater than 1%, thus illustrating the power of
this technique in measuring changes in rates of synthesis for
proteins having such low levels of synthesis.

Another very important point to emerge from the work of
Garrels was that several proteins detected in myoblasts are not

Protein	% Incorporation	Ratio (myotubes/myoblasts)
α–*Actin*		
Myoblasts	<0.0041	>300
Myotubes	1.24	
α–*Tropomyosin*		
Myoblasts	<0.00014	>1100
Myotubes	0.15	
β–*Tropomyosin*		
Myoblasts	<0.00014	>3400
Myotubes	0.47	
$LC1_{emb}$		
Myoblasts	0.00061	610
Myotubes	0.37	
LC2		
Myoblasts	<0.00021	>1800
Myotubes	0.38	

Fig. 9. *Major increases in protein synthesis during L6 differentiation*. The major changes occurring during L6 differentiation among known contractile proteins were quantitated by two dimensional gel electrophoresis. The % incorporation is the cpm in each spot divided by the total cpm applied to each gel. The abbreviation $LC1_{emb}$ refers to the embryonic myosin light chain and LC2 refers to the other muscle type myosin light chain synthesized by L6 cells (see ref. 39). From Garrels (15).

synthesized at a detectable rate in purified myotubes (Figure 10). This result is certainly expected, but there had been very few previously documented samples of this phenomenon for the myoblast-myotube transition. This repression of synthesis can be quantitatively as dramatic as the induction of synthesis in myotubes (Figure 10). Foe example, one species of collagen detected on the gels decreased its rate of synthesis over 1000 fold during myogenesis. In general, about 4% of all proteins analyzed by Garrels increased their rate of synthesis in myotubes, and 4% decreased their rate. It should be noted that only about 200 protein species were analyzed.

Protein	% Incorporation	Ratio (myotubes/myoblasts)
sm—Tropomyosin		
Myoblasts	0.066	>1/180
Myotubes	<0.00037	
nm—Tromomyosin		
Myoblasts	0.041	1/110
Myotubes	0.00037	
nm'—tropomyosin		
Myoblasts	0.061	>1/160
Myotubes	<0.00037	
C1—Collagen		
Myoblasts	0.41	>1/1100
Myotubes	<0.00037	

Fig. 10. *Major decreases in protein synthesis during L6 differentiation*. Values were obtained as described in Figure 9. The abbreviations refer to smooth muscle (sm) and non-muscle (nm, nm') tropomyosin types. C1-collagen refers to a major spot in a series of related spots seen on the two dimensional gels. From Garrels (15).

This is about 10 fold below the number of proteins that can be analyzed using the two dimensional gel technique. Furthermore, the computerized analysis system was, relatively speaking, a very primitive one, requiring operator-directed matching of spots. Current technology is much more advanced and many interesting results can be expected by following up this approach.

8. CONCLUSIONS

The cellular systems currently available to study muscle development allow intervention by the experimenter to control the events of terminal differentiation. Such manipulation is essential if one is to understand the mechanisms underlying the processes of differentiation. Myoblasts in cell culture are an excellent illustration of the exclusively relation between proliferation

and the differentiated state. Further advances in techniques of
cell culture might permit a determination of which nutrient components are involved in making the decision to proliferate or to
fuse and differentiate.

The muscle cell system is rich in specific protein markers,
an essential feature to probe a developing system. Although the
contractile proteins, by virtue of being well-characterized in
muscle tissue and relatively abundant in the cell culture systems,
have been extensively studied, although many other markers remain
to be explored. Current progress in the cloning of contractile
protein genes ensures that the muscle cell will yield many answers
to questions concerning gene organization and expression. The clear
advantage of the contractile protein complex in this regard is that
families of genes are being explored and that the genes must be
expressed in some coordinate way to produce the proteins in the
stoichiometric fashion required in the sarcomere.

9. REFERENCES

1) MERLIE, J.P., BUCKINGHAM, M.E. and WHALEN, R.G. (1977). Mole-
   cular aspects of myogenesis. Current Topics in Developmental
   Biology (eds. A. Monroy, and A.A. Moscona), Vol. 11, p. 61,
   Academic Press.

2) ADELSTEIN, R.S. and EISENBERG, E. (1980). Regulation and kinetics of the actin-myosin-ATP interaction. Ann. Rev. Biochem.
   49, 921.

3) JANSEN, J.K.S. and LOMO, T. (1981). Development of neuro-
   muscular connections. Trends Neurosci., 4, 178.

4) SANES, J.R. and HALL, Z.W. (1979). Antibodies that bind
   specifically to synaptic sites on muscle fiber basal lamina.
   J. Cell Biol., 83, 357.

5) POLLARD, T.D. (1981). Cytoplasmic contractile proteins. J. Cell
   Biol., 91, 156s.

6) FIRTEL, R.A. (1981). Multigene families encoding actin and tubulin. Cell, 24, 6.

7) SALMONS, S. and HENRIKSSON, J. (1981). The adaptive response of skeletal muscle to increased use. Muscle & Nerve, 4, 94.

8) JOLESZ, F. and SRETER, F.A. (1981). Development, innervation, and activity-pattern induced changes in skeletal muscle. Ann. Rev. Physiol. 43, 531.

9) ABBOTT, J., SCHILTZ, J., DIENSTMAN, S. and HOLTZER, H. (1974). The phenotypic complexity of myogenic clones. Proc. Natl. Acad. Sci. USA, 71, 1506.

10) YAFFÉ, D. (1968). Retention of differentiation potentialities during prolonged cultivation of myogenic cells. Proc. Natl. Acad. Sci. USA, 61, 477.

11) YAFFÉ, D. and SAXEL, O. (1977). A myogenic cell line with altered serum requirements for differentiation. Differentiation 7, 159.

12) HAUSCHKA, S.D., CLEGG, C.H., LINKART, T.A. and LIM, R.W. (1977) Mouse myogenesis: Karyotypic, morphological proliferative, and biochemical analysis of permanent clonal cell lines and their subclonal variants. J. Cell Biol., 75, 383a.

13) JAKOB, H., BUCKINGHAM, M.E., COHEN, A., DUPONT, L., FISZMAN, M. and JACOB, F. (1978). A skeletal muscle cell line isolated from a mouse teratocarcinoma undergoes apparently normal differentiation *in vitro*. Exp. Cell Res., 114, 403.

14) CHI, J.C.H., RUBINSTEIN, N., STRAHS, K. and HOLTZER, H. (1975). Synthesis of myosin heavy and light chains in muscle cultures. J. Cell Biol., 67, 523.

15) GARRELS, J.I. (1979). Changes in protein synthesis during myogenesis in a clonal cell lines. Dev. Biol., 73, 134.

16) KONIECZNY, S.F., McKAY, J. and COLEMAN, J.R. (1982). Isolation and characterization of terminally differentiated chicken and rat skeletal muscle myoblasts. Dev. Biol., 91, 11.

17) BLAU, H.M. and EPSTEIN, C.J. (1979). Manipulation of myogenesis
    in vitro: Reversible inhibition by DMSO. Cell, 17, 95.

18) LINKHART, T.A., CLEGG, C.H. and HAUSCHKA, S.D. (1980). Control
    of mouse myoblast commitment to terminal differentiation by
    mitogens. J. Supramolec. Struc. 14, 483.

19) EMERSON, C.P., Jr. and BECKNER, S.K. (1975). Activation of
    myosin synthesis in fusing and mononucleated myoblasts. J.
    Mol. Biol., 93, 431.

20) FISZMAN, M.Y. (1982). Viral transformation and differentiation
    of muscle cells in culture. Biochemistry of Cellular Regulation,
    Vol. III (ed. M.E. Buckingham), p. 197, CRC Press.

21) ZALIN, R.J. and MONTAGUE, W. (1974). Changes in adenylate
    cyclase, cyclic AMP and protein kinase levels in chick
    myoblasts, and their relationship to differentiation. Cell,
    2, 103.

22) ZALIN, R.J. (1979). The cell cycle, myoblast differentiation
    and prostaglandin as a development signal. Dev. Biol., 71, 274.

23) KONIGSBERG, I.R., SOLLMANN, P.A. and MIXTER, L.O. (1978). The
    duration of the terminal G1 in fusing myoblasts. Develop. Biol.
    63, 11.

24) PINSET, C., MÉTÉZEAU, P. and WHALEN, R.G. (1982). Induction
    of differentiation in the myogenic cell line L6. Fifth
    International Congress on Neuromuscular Diseases, Sept. 12-17,
    1982, Marseilles, France.

25) TAYLOR, K.A. and AMOS, L.A. (1981). A new model for the geo-
    metry of the binding of myosin crossbridges to muscle thin
    filaments. J. Mol. Biol., 147, 297.

26) KLEE, C.B., CROUCH, T.H. and RICHMAN, P.G. (1980). Calmodulin,
    Ann. Rev. Biochem. 49, 489

27) VANDEKERCKHOVE, J. and WEBER, K. (1978). At least six different
    actins are expressed in a higher mammal: An analysis based on
    the amino acid sequence of the amino-terminal tryptic peptide.
    J. Mol. Biol. 126, 783.

28) VANDEKERCKHOVE, J. and WEBER, K. (1979). The complete amino acid sequences from bovine aorta, bovine heart, bovine fast skeletal muscle and rabbit slow skeletal muscle. Differentiation, 14, 123.

29) WHALEN, R.G., BUTLER-BROWNE, G.S. and GROS, F. (1976). Protein synthesis and actin heterogeneity in calf muscle cells in culture. Proc. Natl. Acad. Sci. USA, 73, 2018.

30) GARRELS, J.I. and GIBSON, W. (1976). Identification and characterization of multiple forms of actin. Cell, 9, 793.

31) O'FARRELL, P.H. (1975). High resolution two dimensional electrophoresis of proteins. J. Biol. Chem. 250, 4007.

32) WHALEN, R.G., BUTLER-BROWNE, G.S., SELL, S. and GROS, F. (1979). Transitions in contractile protein isozymes during muscle development. Biochimie, 61, 625.

33) WHALEN, R.G., SELL, S.M., BUTLER-BROWNE, G.S., SCHWARTZ, K., BOUVERET, P. and PINSET-HÄRSTRÖM, I. (1981). Three myosin heavy chain isozymes appear sequentially in rat muscle development. Nature, 292, 805.

34) WHALEN, R.G., BUGAISKY, L.B., BUTLER-BROWNE, G.S., PINSET-HÄRSTRÖM, I., SCHWARTZ, K. and SELL, S.M. (1982). Characterization of myosin isoenzymes appearing during rat muscle development. Cold Spring Harbor Monograph on Molecular and Cellular Control of Muscle Development, in press.

35) MERLIE, J.P., CHANGEUX, J.-P. and GROS, F. (1978). Skeletal muscle acetylcholine receptor. Purification, characterization and turnover in muscle cell cultures. J. Biol. Chem. 253, 2882.

36) PATERSON, B. and STROHMAN, R. (1972). Myosin synthesis in cultures of differentiating chick embryo skeletal muscle. Dev. Biol. 29, 113.

37) BENOFF, S. and NADAL-GINARD, B. (1978). Transient induction of poly(A)-short myosin heavy chain messenger RNA during terminal differentiation of L6E9 myoblasts. J. Mol. Biol. 140, 283.

38) GARRELS, J.I. (1979). Two dimensional gel electrophoresis
     and computer analysis of proteins by clonal cell lines.
     J. Biol. Chem. 254, 7961.

39) WHALEN, R.G., BUTLER-BROWNE, G.S. and GROS, F. (1978).
     Identification of a novel form of myosin light chain present
     in embryonic muscle tissue and cultured muscle cells. J. Mol.
     Biol., 126, 415.

ORGANIZATION OF MUSCLE-SPECIFIC GENES IN THE RODENT GENOME

H. Czosnek, Y. Carmon, M. Shani, U. Nudel,
P.E. Barker*, F.H. Ruddle* and D. Yaffe

*Department of Cell Biology, The Weizmann Institute of
Science 76100 Rehovot, Israel and *Department of
Biology, Yale University, New Haven, Conn. 06511, USA*

1. INTRODUCTION

In vitro myogenesis offers and experimental model for the
study of the molecular mechanisms involved in gene expression
during cell differentiation. Terminal differentiation of muscle
cells is characterized by the fusion of mononucleated myoblasts
into multinucleated fibers. The morphological changes are
accompanied by biochemical modifications, including the onset or
a great increase in the synthesis of the major muscle contractile
proteins, their regulatory polypeptides and enzymes needed to
produce the energy for muscle contraction (reviewed in ref. 1).
Several of these proteins (actin, myosin heavy chain, tropomyosin,
creatine kinase) have been shown to be members of families of
closely related isoforms, some of which are muscle-specific and
are synthesized during terminal differentiation, others are present
in many cell types. The capacity of myogenic cells (as well as
other precursor cells) to proliferate during extended periods
without expressing the genes involved in terminal differentiation
(2, 3), indicate the existence of mechanisms which retain the
latent program of gene expression, and mechanisms of gene activa-

71

tion which recognize these genes. Studying the structure and organization of such genes sets in chromatin may provide information on the nature of the control mechanisms. As a part of such studies we investigated the following questions:

a) Whether in proliferating precursor cells, genes which are programmed to be expressed later in development are in a specific conformational state which renders them preferentially sensitive to digestion with nucleolytic enzymes.

b) Whether muscle specific genes are clustered in a single chromatin domain.

2.   DNAase I SENSITIVITY OF MUSCLE-SPECIFIC GENES DURING MYOGENESIS

Investigations using a variety of cell types have shown that active genes have altered chromatin structures which render them preferentially sensitive to digestion by nucleases (4, reviewed in 5). Several studies suggested that this preferential sensitivity may reflect the potentiality of genes to be expressed in a cell rather than their actual transcriptional activity (6, 7).

Cultures of the rat myogenic cell line L8 (3) were used to examine the question of whether in proliferating precursor cells, genes which are programmed to be expressed later in development differ in DNAase I sensitivity from genes which are never expressed in these cells.

Nuclei were isolated from cultures of cloned populations of proliferating myoblasts of the myogenic cell line L8. Parallel cultures from the same cell line were induced to differentiate by changing the nutritional medium, and the nuclei were isolated 100 hr later. At this stage, the great majority of the cells have fused into multinucleated fibers. Nuclei were also isolated from rat brain to provide information on the DNAase I sensitivity of muscle-specific genes in a non-muscle tissue. The nuclei were incubated with increasing concentrations of DNAase I; the isolated DNA was digested with *Eco* R1, the fragments separated by electrophoresis and transferred onto nitrocellulose filters. The blotted DNA was hybridized with cloned probes for the rat skeletal muscle α-actin (plasmid p749) and for the cytoplasmic β-actin (plasmid p72) (8-10). A probe for a rat immunoglobulin constant region gene (plasmid B1, gift from I. Schechter), was used as an internal control to measure the DNAase I sensitivity of a gene not expressed in the muscle tissue.

As can be seen from Figure 1.1, the chromatin region containing the myosin light chain 2 gene is much more sensitive to nuclease in the differentiated muscle cells than in the proliferating myoblasts. In the brain this gene seems to be mildly sensitive, although its expression at the RNA level was undetectable (8). The DNAase I sensitivity of the skeletal muscle and β-actin genes was examined in a similar manner (Figure 1.2). As can be seen, the DNA fragment containing the cytoplasmic β-actin gene is very sensitive to DNAase I in all three cell types. However, the DNA fragment containing the skeletal muscle actin is DNAase I sensitive only in nuclei isolated from differentiated muscle cells (11).

The changes in DNAase I sensitivity of these genes associated with terminal differentiation, indicate a qualitative change in their transcriptional activity, which occurs at this stage of differentiation.

The fact that in the proliferating muscle precursor cells, genes which are programmed to be expressed during terminal differentiation are indistinguishable in their DNAase I sensitivity from genes which are not expressed in this cell lineage, shows that,

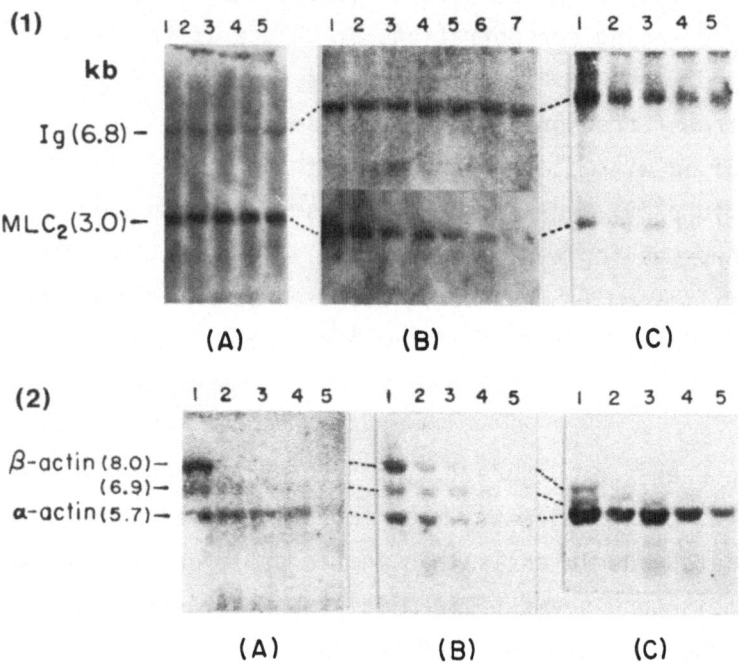

Fig. 1. *Preferential DNAase I sensitivity of the myosin light chain 2 and skeletal muscle actin genes in differentiated cultures of the rat myogenic cell line L8.* Nuclei from proliferating myoblasts (A), differentiated cultures (B), and rat brain (C), were incubated with increasing concentrations of DNAase I (samples 1-5 or 1-7), as described in ref. 11. The DNA was extracted, digested with *Eco* R1, electrophoresed in 1% agarose gels and blotted onto nitrocellulose paper. After hybridization with specific probes as indicated below, the blots were washed in 0.1 x SSC at 70° and fluorographed. (1). Blot hybridization with a probe (plasmid p103) specific for sequences of the rat myosin light chain 2 (MLC2) and with a probe (plasmid B1) specific for a rat immunoglobulin C_k gene (Ig). The blots in (A) and (C) were hybridized with a mixture of the two probes; blot (B) was hybridized with the Ig probe, washed in alkali and then hybridized with the MLC2 probe. (2) Blots of DNA samples identical to those analyzed above were hybridized with a mixture of probes specific for sequences of the rat skeletal muscle actin (α-actin) and for the cytoplasmic β-actin (β-actin). (From Carmon *et al.*, ref. 11).

the altered state of these genes reflected in preferential DNAase I sensitivity is not associated with the maintenance of this latent program of gene expression through many cell generations.

In a number of genes, it was found that in addition to the preferential sensitivity to DNAase I of the entire region of the structural gene a defined small region hypersensitive to DNAase I, usually 5' to the gene. It has been suggested that these hypersensitive sites are involved in the control of transcription and that this particular chromatin structure might be established before initiation of transcription, perhaps related to cell commitment (12). It was therefore of interest to test the DNAase I sensitivity of the DNA region upstream to a gene expressed during myogenesis. We have isolated from a rat genomic DNA library a recombinant phage containing the skeletal muscle actin gene (10). The insert contains the entire coding region of the gene located at the 3' end and 11 Kb flanking the gene on its 5'side. Fragments of this insert were used to probe the DNAase I sensitivity of the 5' region flanking the skeletal muscle gene. The results show that the DNAase I sensitive region ends between 40 and 700 bp upstream from the structural gene (including the CAP site and the TATA box). No DNAase I hypersensitive sites were detected 5' to the skeletal muscle actin gene in proliferating as well as in differentiated muscle cells (11).

3. THE GENES CODING FOR THE MUSCLE CONTRACTILE PROTEINS, MYOSIN HEAVY CHAIN, MYOSIN LIGHT CHAIN 2 AND SKELETAL MUSCLE ACTIN ARE LOCATED ON THREE DIFFERENT CHROMOSOMES

The apparent coordinate expression of the genes coding for the major proteins of the muscle contractile apparatus during terminal differentiation of muscle cells, may reflect a regulatory mechanism acting on a group of genes clustered in a single chromosomal region. The assignment of the muscle-specific genes to

chromosomes may provide some information on the type of controls
involved in their expression in muscle cells.

Blots of restricted DNA from a battery of cloned mouse-
Chinese hamster somatic cell hybrids, each containing different
subsets of mouse chromosomes, were hybridized with cloned DNA
probes specific for the myosin heavy chain, myosin light chain 2
and skeletal muscle actin. Hybridizable mouse DNA fragments
appeared on a background of hybridizable hamster DNA fragments,
comparing the chromosomal composition of the hybrid cell lines
with the presence or absence of a given hybridizable mouse fragment
enables the identification of the mouse chromosome containing the
locus of the probed gene (13).

(i) *All detectable members of the myosin heavy chain gene family
are on mouse chromosome 11*

Investigations at the protein level indicate the existence
of different isoforms of skeletal muscle myosins, which appear
sequentially during development (embryonic, neonatal and adult
types) (14). In addition, there are smooth muscle, cardiac and
nonmuscle myosins (reviewed in ref. 15). To detect myosin heavy
chain (MHC) DNA sequences, we used plasmid p82 which contains
a cDNA insert which hybridizes specifically with rat MHC mRNA
(9, 16). This probe hybridizes to 5 rat genomic *Eco* R1 fragments.
At least 3 of these fragments have been shown to belong to three
different non allelic MHC genes (9). One of these genes has been
identified tentatively as the adult muscle MHC gene (unpublished).
Hybridization of this probe to *Eco* R1 digested mouse DNA resulted
in 5 major radioactive bands, differing in size from those found
in Chinese hamster (17). These bands could be detected only in
the hybrid cell line mFE11 (ref. 17 and Table 1). Since this is
the only cell line which contains chromosome 11, the results indicate that all detectable MHC genes are located on chromosome 11

(17). Similar results were obtained using as probes, fragments
isolated from a recombinant bacteriophage containing a segment
of a MHC gene which was identified tentatively as the rat adult
skeletal muscle MHC (Figure 2A, Table 1).

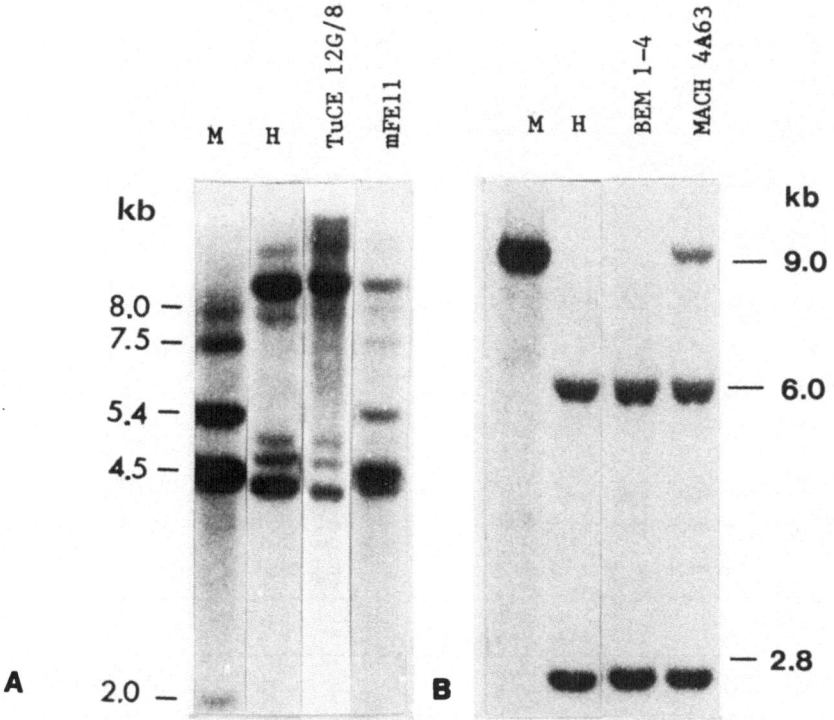

Fig 2. *Assignment of the myosin heavy chain and the myosin light
chain 2 genes to mouse chromosomes.* Blots of *Eco* R1-digested genomic
DNA from parental and hybrid cell lines were hybridized with DNA
probes specific for myosin heavy chain gene sequences* (A), and
for the myosin light chain 2 gene** (B). The patterns of hybridizable
DNA fragments of mouse (M), Chinese hamster (H) and of representa-
tives of negative and positive hybrid cell lines are shown. The
complete analysis is summarized in Table 1. For more details see
ref. 17.

* A 4.3 Kb *Eco* R1 DNA fragment from the recombinant phage MHC 15
containing sequences of the rat adult skeletal muscle; same results
were obtained using plasmic p82.

** Insert of plasmid p103.

Table 1.　*Mouse – hamster somatic cell hybrids*

Hybrid	Mouse chromosome number																			
	1	2	3	4	5	6	7	8	9	10	11	12	13	14	15	16	17	18	19	x
BEM 1-6	+	+	+	+	-	+	-	+	+	+	-	+	+	+	+	+	+	+	+	+
BEM 1-4	+	+	+	-	+	+	-	+	-	+	-	+	+	+	+	+	+	+	+	+
MACH 7A13-3B3	-	+	-	-	+	-	+	-	+	-	-	+	+	+	+	+	+	+	+	-
MACH 4A63	-	+	-	-	-	-	+	-	-	-	-	+	+	-	+	+	+	+	+	-
MACH 4A64-A1	+	+	-	-	-	-	+	-	-	-	-	+	-	-	+	-	+	-	+	-
MACH 4B31Az3	-	+	-	-	-	-	+	+	-	-	-	-	-	-	-	+	+	+	-	-
MACH 2A2	+	+	+	-	-	+	+	+	+	+	-	+	+	+	+	+	+	+	+	-
ECm4e	-	-	-	-	-	-	-	-	-	-	-	-	-	+	+	-	-	-	-	-
MAE32	-	-	-	-	-	-	-	-	-	-	-	-	-	-	+	-	-	-	-	+
MACH 2A2-B1	-	+	+	+	-	+	+	+	+	+	-	+	-	+	+	+	+	-	-	+
MACH 2A2-A1	+	+	+	-	-	+	+	+	+	+	-	+	+	+	+	+	+	+	+	+
MACH 2A2-C2	+	+	+	-	-	-	+	+	+	+	-	+	+	-	+	+	+	-	+	+
TuCE 12G/8	-	-	-	-	-	-	-	-	-	-	+	+	+	-	+	+	+	+	+	+
TuCE 12G/5	-	+	-	-	-	-	-	-	-	+	-	+	+	+	+	+	+	+	-	+
mFE11	+	+	+	+	+	+	+	+	+	+	+	+	+	+	+	+	+	+	+	+

Hybrid	Hybridization				Hybrid	MHC	MLC2	α-ACT
	MHC	MLC2	α-ACT					
BEM 1-6	(-)	(-)	(+)		MAE32	(-)	nd	(-)
BEM 1-4	(-)	(-)	(+)		MACH 2A2-B1	(-)	nd	(+)
MACH 7A13-3B3	(-)	(+)	(-)		MACH 2A2-A1	(-)	nd	(+)
MACH 4A63	(-)	(+)	(-)		MACH 2A2-C2	(-)	nd	(+)
MACH 4A64-A1	(-)	(+)	(-)		TuCE 12G/8	(-)	nd	(-)
MACH 4B31Az3	(-)	(+)	(-)		TuCE 12G/5	(-)	nd	(-)
MACH 2A2	(-)	(+)	(+)		mFE11	(+)	nd	nd
ECm4e	(-)	(-)	(-)					

(ii) *The myosin chain 2 is on mouse chromosome 7*

A probe containing an insert of DNA complementary to rat skeletal muscle myosin light chain (MLC2) mRNA was hybridized with a blot containing *Eco* Rl restricted DNA from hybrid cell lines and from mouse and hamster parental lines. The hybridization pattern indicates that the MLC2 gene is located on mouse chromosome 7 (Figure 2B, Table 1).

(iii) *The skeletal muscle actin and several other actin DNA sequences are located on mouse chromosome 3*

The actins constitute a family of highly conserved proteins. At least six different vertebrate actins have been identified by amino acid sequence analysis: The skeletal muscle (α-actin), the heart, stomach and aorta smooth muscles and the non-muscle cytoplasmic actins (β- and γ-actins) (18). These proteins are the products of closely related non allelic genes. During muscle cell differentiation, there is a switch from the non-muscle β-, and γ-actin, to the muscle-specific α-actin (16, 19).

The organization of actin genes in the mouse genomes was investigated using a probe (p749) which hybridizes at low stringency with DNA sequences of muscle and non-muscle actins. Under very stringent conditions, it hybridizes specifically to the skeletal muscle actin (10). Plasmic p749 was hybridized to a blot of *Eco* Rl-

Table 1. Karyotypes of mouse-Chinese hamster hybrid cell lines tested for the presence of mouse genes coding for muscle contractile proteins. The somatic cell hybrids used in this study and their karyotypic analysis have been described in detail elsewhere (27, 28). Chromosomes with a frequency of $\geq$ 20% were scored + and with a frequency of < 20% were scored -. The number of copies of mouse chromosomes in each of the hybrid cell lines is given in ref. 17. The presence (+) or absence (-) of mouse DNA fragments containing sequences which hybridized with probes for the genes encoding the myosin heavy chain (MHC), myosin light chain 2 (MLC2) and skeletal muscle actin (α-ACT) is indicated at the bottom of the table (nd = not determined).

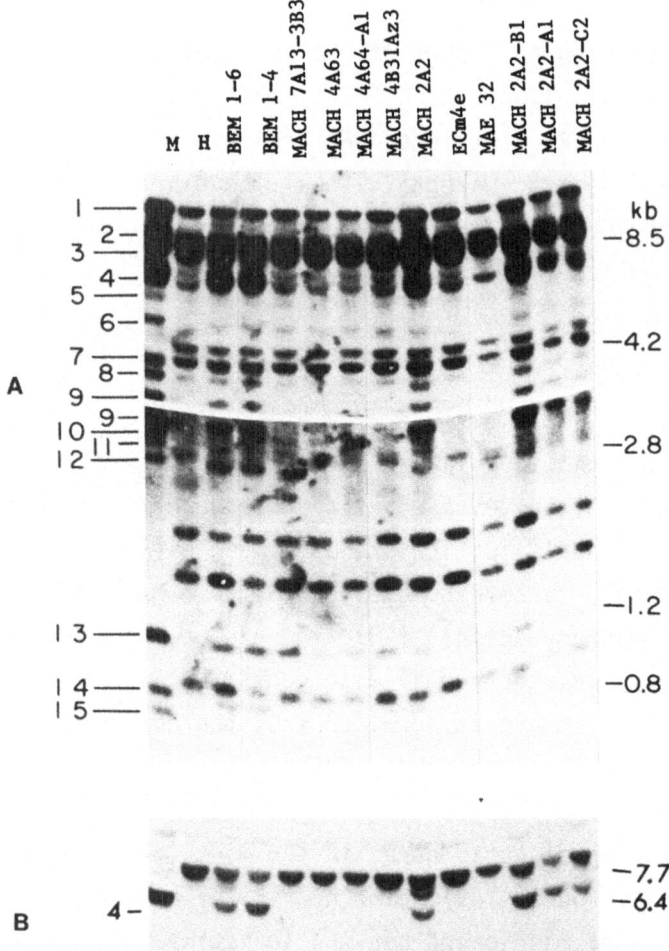

Fig. 3. *Assignments of the skeletal muscle actin and of other actin DNA sequences to mouse chromosomes.* (A) - A blot of *Eco* Rl-digested genomic DNA from parental and hybrid cell lines was hybridized with the insert of plasmid p749. After hybridization, the blot was washed at low stringency; the upper part of the blot was fluorographed for 1 day, the lower part for 4 days. (B) - The blot shown in A was washed at high stringency and fluorographed in order to identify the DNA fragment containing the skeletal muscle actin gene. Numbers on the left refers to hybridizable DNA fragments of mouse origin (from Czosnek *et al.*, ref. 17).

digested DNA from mouse-hamster hybrid and parental cell lines.
At low stringency washing conditions, at least 15 radioactive bands
of mouse DNA were obtained (Figure 2A). Among the fragments containing mouse actin DNA sequences, 5-7 could be assigned to
chromosome 3, three to chromosome 17 and one to chromosome 2 (17).
At high stringency wash, p749 hybridizes to a single mouse DNA
fragment which was assigned to mouse chromosome 3 (Figure 3B,
Table 1). The assignment of the skeletal muscle actins to chromosome
3 was confirmed using a probe derived from the 5' end of the rat
skeletal muscle actin (17).

4. CONCLUSIONS

There are several examples of genes involved in specific
pathways of differentiation which are clustered; the best known
are the silkworm and Drosophila chorion genes·(20, 21), the antennapedia gene complex (22) and the bithorax gene complex (23)
in Drosophila. The present investigation shows that this is not
the case for the genes coding for the major contractile proteins
expressed during terminal differentiation of muscle cells. Rather,
the results indicate that these genes are organized according to
their evolutionary relatedness.

All myosin heavy chain genes that could be detected with
our probes are located on a single chromosome. The obvious next
step is to find out if these genes are indeed clustered and whether
there is a correlation between their arrangement on the chromosome
and the pattern of their expression.

In situ hybridization experiments showed that in Drosophila,
the actin genes are not clustered (24). On the other hand, cloning
of sea urchin actin genes have shown that several actin genes are

linked (25). The present experiments suggest that in the mouse, five to seven actin genes (or pseudogenes) are clustered on a single chromosome while three are grouped on another; other actin DNA sequences are scattered on several chromosomes. The detailed study of these genes may provide useful information on their evolutionary relatedness. It is of obvious interest to identify the actin genes which are clustered with the skeletal muscle actin gene on chromosome 3, and to correlate these finding with studies on the evolution of these genes based on their structure and location of introns (26).

5. ACKNOWLEDGEMENTS

    This work was supported in part by the National Institutes of Health, Grant No. GM-22767, by a United States - Israel Binational Science Foundation Grant 3114/82; and by a grant from the Muscular Dystrophy Association. H.C. and P.E.B. were supported by a Postdoctoral Fellowship from the Muscular Dystrophy Association. U.N. is the encumbent of the A. and E. Blum Career Development Chair.

6. REFERENCES

1) BUCKINGHAM, M.E. (1977). The international review of biochemistry: Biochemistry of cell differentiation II, (ed., J. Paul), Vol. 15, p. 269, University Park Press, Baltimore.
2) YAFFE, D. (1968). Retention of differentiation potentialities during prolonged cultivation of myogenic cells. Proc. Natl. Acad. Sci. USA., 61, 477.

3) YAFFE, D. and SAXEL, O. (1977). A myogenic cell line with altered serum requirements for differentiation. Differentiation, 7, 159.

4) WEINTRAUB, H. and GROUDINE, M. (1976). Chromosomal subunits in active genes have an altered conformation. Science, 193, 848.

5) WEISBROD, S. (1982). Active chromatin. Nature, 297, 289.

6) STALDER, J., GROUDINE, M., DODGSON, J.B. ENGLE, J.D and WEINTRAUB, H. (1980). Hb switching in chickens. Cell, 19, 973.

7) GAZIT, B., CEDAR, H., LERER, I. and VOSS, R. (1982). Active genes are sensitive to deoxyribonuclease I during metaphase. Science, 217, 648.

8) KATCOFF, D., NUDEL, U., ZEVIN-SONKIN, D., CARMON, Y., SHANI, M., LEHRACH, H., FRISCHAUF, A.M. and YAFFE, D. (1980). Construction of recombinant plasmids containing rat muscle actin and myosin light chain DNA sequences. Proc. Natl. Acad. Sci. USA, 77, 960.

9) NUDEL, U., KATCOFF, D., CARMON, Y., ZEVIN-SONKIN, D., LEVY, Z., SHANI, M. and YAFFE, D. (1980). Identification of recombinant phages containing sequences for different rat myosin heavy chain genes. Nucl. Acids Res., 8, 2133.

10) NUDEL, U., KATCOFF, D., ZAKUT, R., SHANI, M., CARMON, Y., FINER, M., CZOSNEK, H., GINSBERG, I. and YAFFE, D. (1982). Isolation and characterization of rat skeletal muscle and cytoplasmic actin genes. Proc. Natl. Acad. Sci. USA, 79, 2763.

11) CARMON, Y., CZOSNEK, H., NUDEL, U., SHANI, M. and YAFFE, D. (1982). DNAase I sensitivity of genes expressed during myogenesis. Nucl. Acids Res., 10, 3085.

12) ELGIN, S.C.R. (1982). DNAase I - hypersensitive sites of chromatin. Cell, 27, 413.

13) RUDDLE, F.H. (1981). A new era in mammalian gene mapping: somatic cell genetics and recombinant DNA methodologies. Nature, 294, 115.

14) WHALEN, R.G., SELL, S.M., BUTLER-BROWNE, G.S., SCHWARTZ, K., BOUVERET, P. and PINSET-HÄRSTRÖM, I. (1981). Three major heavy chain isozymes appear sequentially in rat muscle development. Nature, 292, 805.

15) WHALEN, R.G., BUGAISKY, L.B., BUTLER-BROWNE, G.S., PINSET-HÄRSTRÖM, I., SCHWARTZ, K. and SELL, M. (1982). Characterization of myosin isoenzymes appearing during rat muscle development. C.S.H. Symposium on Muscle Development, in press

16) SHANI, M., ZEVIN-SONKIN, D., SAXEL, O., CARMON, Y., KATCOFF, D., NUDEL, U. and YAFFE, D. (1981). The correlation between the synthesis of skeletal actin, myosin heavy chain, and myosin light chain and the accumulation of corresponding mRNA sequences during myogenesis. Develop. Biol. 86, 483.

17) CZOSNEK,,H., NUDEL, U., SHANI, M., BARKER, P.E., PRAVTCHEVA, D.D., RUDDLE, F.H. and YAFFE, D. (1982). The genes coding for the muscle contractile proteins, myosin heavy chain, myosin light chain 2 and skeletal muscle actin are located on three different mouse chromosomes. EMBO J., in press

18) VANDERCKERKHOVE, J. and WEBER, K. (1979). The complete amino acid sequence of actins from bovine aorta, bovine heart, bovine fast skeletal muscle and rabbit slow skeletal muscle. Differentiation, 14, 123.

19) GARRELS, J.I. (1979). Changes in protein synthesis during myogenesis in a clonal cell line. Dev. Biol., 73, 134.

20) EICKBUSH, T.H. and KAFATOS, F.C. (1982). A walk in the chorion locus of Bombyx mori. Cell, 29, 633.

21) SPADLING, A.C. (1981). The organization and amplification of two chromosomal domains containing Drosophila chorion genes. Cell, 27, 193.

22) LEWIS, R.A., WAKMIMOTO, B.T., DENELL, R.E. and KAUFMAN, T.C. (1980). Genetic analysis of antennapedia gene complex. (ANT-C) and adjacent chromosomal regions of Drosophila melanogaster. II. Polytene chromosome segments 84A-84B1,2. Genetics, 95, 383.

23) HOGNESS, D.S., SAINT, R.B., AKAM, M.E., GOLDSCHMIDT-CLERMONT, M. and BEACHY, P. (1982). A molecular analysis of the bithorax complex in Drosophila. J. Cell Biochem. Supplement 6, p. 263.

24) FYBERG, E.A., KINDLE, K.L., DAVIDSON, N. and SODJA, A. (1980). The actin genes of Drosophila: A dispersed multigene family. Cell, 19, 365.

25) SCHELLER, R.H., McALLISTER, L.B., CRAIN, W.R., DURICA, D.S. POSAKONY, J.W., THOMAS, T.L., BRITTEN, R.J. and DAVIDSON, E.M. (1981). Organization and expression of multiple actin genes in the sea urchin. Mol. Cell Biol. 1, 609.

26) ZAKUT, R., SHANI, M., GIVOL, D., NEUMAN, S., YAFFE, D. and NUDEL, U. (1982). The nucleotide sequence of the rat skeletal muscle actin gene. Nature, 298, 857.

27) D'EUSTACHIO, P., PRAVTCHEVA, D., MARCU, K. and RUDDLE, F.H. (1980). Chromosomal location of the structural gene cluster encoding murine immunoglubulin heavy chains. J. Exp. Med. 151, 1545.

28) D'EUSTACHIO, P., BOTHWELL, S.L.M., TAKARO, T.K., BALTIMORE, D and RUDDLE, F.H. (1981). Chromosomal location of structural genes encoding immunoglobulin lambda light chains. J. Exp. Med. 153, 793.

APPROACHES TO THE BIOCHEMISTRY OF DIFFERENTIATION OF MOUSE EMBRYONAL CARCINOMA CELLS

M.J. Evans, R.H. Lovell-Badge*, D. Latchman#,
A. Stacey, and H. Brzeski‡

Department of Genetics
University of Cambridge
Downing Street
Cambridge, UK

1. INTRODUCTION

A complete interrelationship has now been demonstrated between mouse embryonal carcinoma (EC) cells and normal mouse embryos (1, Figure 1). EC cells are probably homologous with 5.5 day old mouse embryo epiblast (2,3).

Pure populations of EC cells may be induced to differentiate in a synchronous manner by allowing them to form aggregates which are then kept in suspension. The first differentiation observed is the formation of a primary embryonic endodermal layer on the outside of the aggregate thus giving rise to a structure known as a simple embryoid body (Figure 2).

It might be supposed that this differentiation is the result of re-programming of gene transcription and the synthesis of a

---
Present address: MRC Mammalian Development Unit, Univ. Coll., London.
#*Present address: Eukaryotic and Molecular Genetics Res. Group,*
    Dept of Biochemistry, Imperial College, London.
‡*Present address: Dept of Biochemistry, University of Strathclyde.*

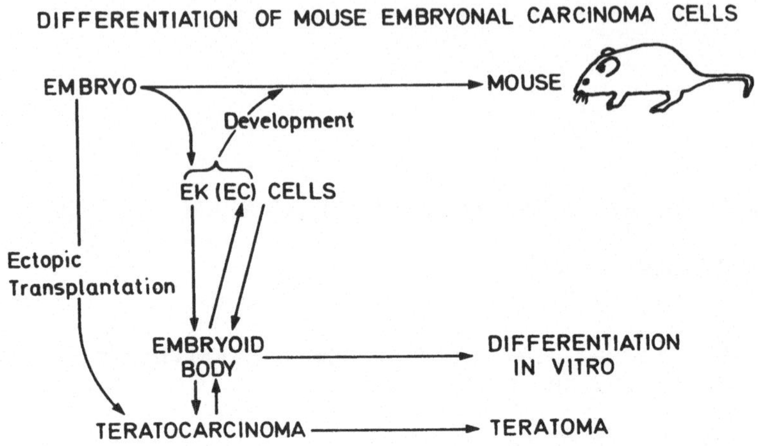

DIFFERENTIATION OF MOUSE EMBRYONAL CARCINOMA CELLS

Fig. 1. *Interrelationships between the embryo and EC cells*

changed spectrum of proteins. We have attempted to examine the range of change involved by the use of two dimensional gel electrophoresis and looked in more detail at the differential control of one major protein - alphafoetoprotein (AFP).

2. POLYPEPTIDES CHANGES IN DIFFERENTIATING EC CELLS

About 40% of the newly synthesized proteins in a simple embryoid body labelled by a 2 hr pulse of $[^{35}S]$-methionine are found in the outer endodermal cells. Two dimensional gel electrophoresis of samples taken from a time course of differentiation and each labelled over a 2 hr period provided the basis for a comparison of the changes in protein synthesis (4). About 1000 polypeptides were observed and of approximately 600 which could be compared with confidence, 53 were found to vary and of these

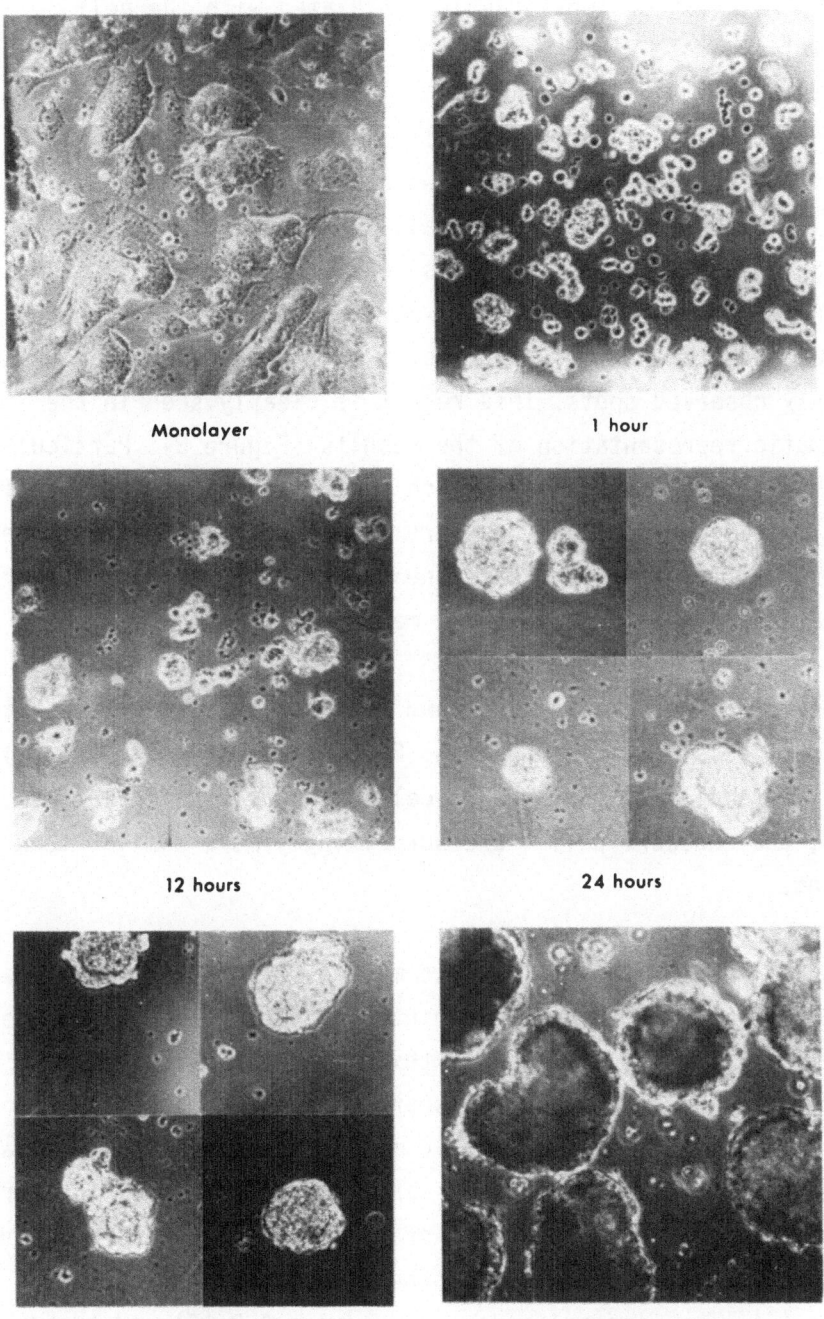

Fig. 2. *Embryoid body formation from PSMB EC cells.*

33 varied in a manner consistently correlated with the cell
differentiation.

Considering the completely different cell morphology and cell
behaviour of the endodermal cells which are proceeding towards
terminal differentiation such a small range of differences is at
first sight surprising. As EC cells persist throughout the time
course of this experiment it is to be expected that most changes
are appearances of new polypeptides rather than disappearances of
previously observed spots. This result is clearly seen in the
diagrammatic representation of the results (Figure 3). Particularly
interesting are those proteins which appear transiently (spots 1,
2 and 3). It might be that these are involved in transient processes
of cell committment but at least one (spot 1) is possibly AFP which
is secreted and mainly lost to the medium.

The main conclusion to be drawn from a comparison of such two
dimensional gels of differentiating EC cells is that the overall
changes are small and that the EC cells and their differentiated
product, the endoderm cells, are surprisingly more similar than
different.

It is known that there are two main types of primary endoderm
in the early mouse embryo, the parietal endoderm which lines the
trophoblast and which lays down a thick basement membrane known
as Reichert's membrane and the visceral endoderm which becomes
associated with the extra-embryonic mesoderm to form the yolk-sac.
Both types of endoderm synthesize large amounts of specific secreted
products. Differentiated products of the parietal endoderm include
Type IV collagen and laminin and other components of Reichert's
membrane (5). The visceral endoderm produces AFP (6) and transferrin (7).

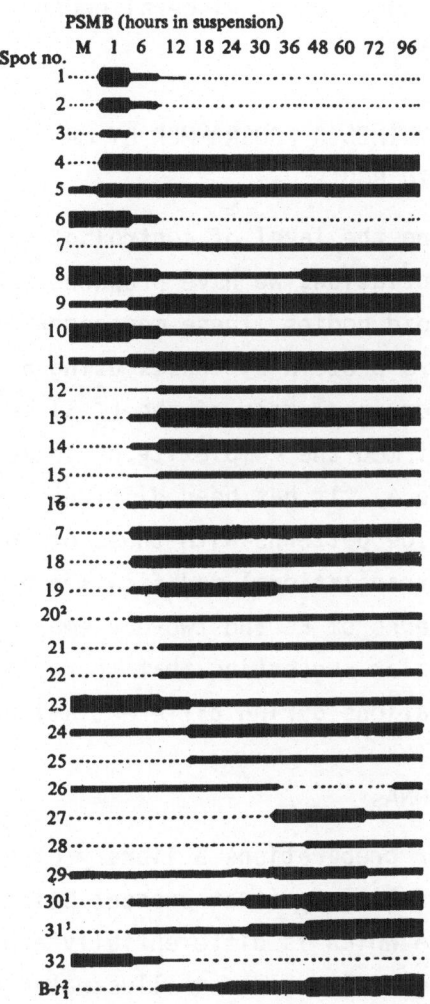

Fig. 3. *Diagram of changes in 33 spots seen in two dimensional gels of proteins labelled for 2 hr periods*. Differentiation of mouse EC cell line PSMB. From ref. 4.

Adamson *et al.* (8) have followed the production of AFP during embryonal carcinoma cell differentiation. The EC cells produced no AFP, but this was detectable as soon as recognizable endodermal cells appeared. It was also detected in cells of a particular clonal culture of endodermal cells derived from EC cell differentiation (PSA5E). Adamson (7) has recently reconfirmed that these cells do display the properties of visceral endodermal cells as they synthesize transferrin.

3. *IN VITRO* PROTEIN SYNTHESIS PROGRAMMED BY mRNAs FROM EC CELLS
   AND TWO-DAY EMBRYOID BODIES

In order to examine the level of control of protein synthesis during EC cell differentiation, we have prepared mRNA from EC cells and from two-day embryoid bodies. These messenger populations were used to program *in vitro* protein synthesis using a reticulocyte lysate system. Two dimensional gels of the [^{35}S]-methionine labelled *in vitro* products confirmed the complexity of intact messengers in the preparation (Figure 4). It has been difficult, however, to compare these polypeptide patterns with those of intact cells as the processes of post-translational modification are not the same. The translational patterns of EC and two-day embryoid body mRNA are, however, very similar suggesting that there is little change in the messenger populations during differentiation.

4. LIBRARY OF CLONED DNAs

From the messenger preparations a library of clones of cDNA has been prepared. This library is at present being screened for clones that detect mRNA which is differentially expressed in EC and endoderm cells (Figure 5). Clone 4 illustrated here, has been shown by Northern blotting to hybridize to an RNA species of approximately 4 Kb in length and that is present in total poly A RNA from undifferentiated PSMB cells (EC cells).

The function of these clones is at present unknown. Fortunately however, AFP is a major product of the visceral endoderm of the 16 day foetus and up to 25% of the mRNA is for AFP. We have cloned a segment of cDNA of the AFP message into the phage·M13mp7 (Latchman, Brzeski and Evans, unpublished). A *Mbo*-1 fragment of cDNA was in-

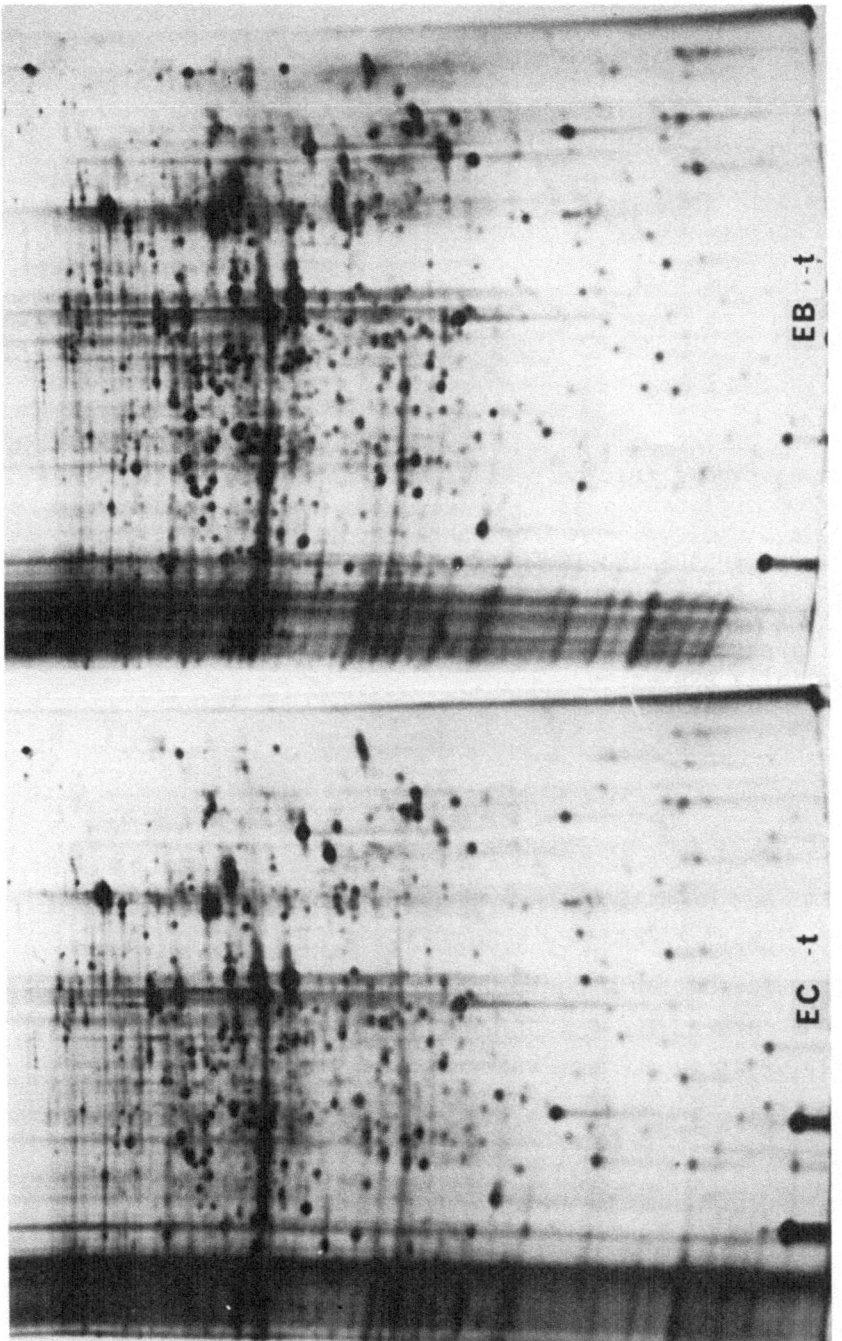

Fig. 4. *Two dimensional gels (IEF) of [^{35}S]-methionine labelled in vitro translation products of poly A+ mRNA from undifferentiated PSMB EC cells (EC-t) and embryoid bodies 48 hr after suspension of equivalent cells (EB-t).*

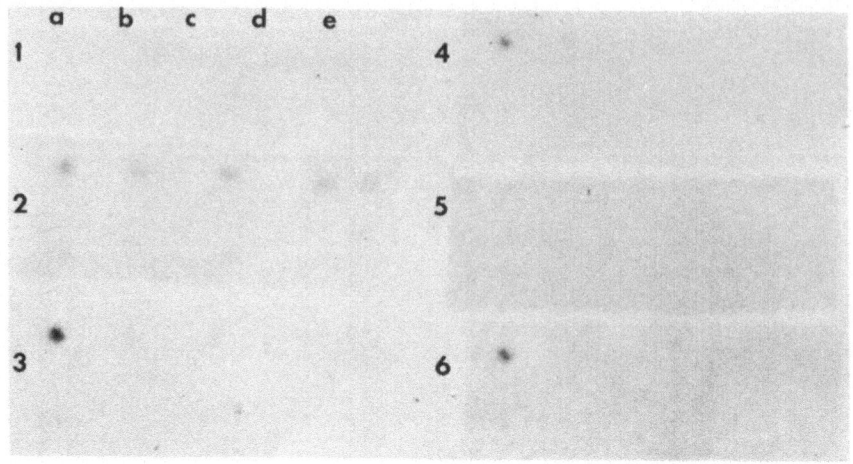

Fig. 5. *Clones of cDNA from embryoid bodies used as probes against poly-A+ RNA from different cell types.* Poly-A+ RNA was immobilised on ATP paper. a,b,c,d,e,f: RNA from PSMB (EC cells), STO (fibro-blast cells), PYS, PSA5E and ICME6, respectively. 1 and 5 non re-combinant vectors pAT 153 and M13mp7, respectively; 2 pAT 153 recombinant cDNA non-differentially expressed; 4 pAT 153 recombinant cDNA differentially expressed; 3 and 6 M13mp7 recombinant cDNA differentially expressed.

serted into the *Bam*-H1 site of the vector and the resulting population of clones screened by hybrid arrest of translation and positive message selection experiments to identify a clone contain-ing an AFP cDNA insert. The *Hae*-III and *Mbo*-1 cleavage of this insert identifies it on the known restriction site map of the mouse AFP gene (9).

The first question to be answered with this probe (AFP-1) was whether the transition to AFP production with differentiation of pure EC cells to embryoid bodies was associated with new synthesis of AFP message. The proportion of AFP message in total polyadeny-lated mRNA from undifferentiated EC cells and from embryoid bodies

was measured by binding 5 μg samples to diazobenzyloxymethyl paper
(DBM) (10), hybridizing with [^{32}P]-labelled AFP-1 clone DNA followed
by stringent washing and scintillation counting. A series of
standards was constructed by mixing yolk sac mRNA (20% AFP message
in this preparation) with brain mRNA (no AFP message) and applying
5 μg of each appropriate mixture to DBM paper. No stimulation of
binding was observed with EC cell mRNA whereas embryoid body mRNA
showed a stimulation of probe binding corresponding to a proportion
of 1.3×10^{-3} of AFP message. Only a proportion of cells in these
embryoid bodies are endodermal and this proportion is unlikely to be
above 40%. Using this figure, the proportion of AFP message in the
endodermal cells was calculated to be 3.25×10^{-3}. Immunofluorescence studies of similar embryoid bodies have previously shown (8)
that only a small proportion of cells may be actively synthesizing
AFP, so the message levels are probably severely underestimated for
particular cells. The level of counts were such that an upper limit
on the amount of AFP message present in the EC cell sample would
correspond to a proportion of about 5×10^{-5}. This does not, therefore, rule out its presence at low levels in the undifferentiated
cells but an increase of at least 60-fold accompanies differentiation. By the time that the visceral endoderm cells have fully differentiated in the yolk sac the AFP message level is at least 5000
times the relative concentration possibly present in EC cells.

Although these figures cannot rule out a low constitutive
rate of AFP message synthesis in the undifferentiated cells, the
simplest hypothesis is that transcription of the AFP gene is turned
on with differentiation. We have, therefore, examined the state of
the AFP gene in undifferentiated and differentiated cells.

There is now considerable evidence that expression of a
particular gene is associated with a conformation in the chromatin
which renders it preferentially sensitive to DNAse I digestion

(see ref. 11 for review). Nonexpressed genes are resistant to such digestion. This sensitivity is not related, however, to the level of gene expression but rather to the potential for such expression. This leads to the question of how such a difference arises during development. One possibility is that the majority of genes in a pluripotent cell could be in a closed state and that during differentiation there is an opening of particular genes. The alternative is that the genes could be open with closure occurring in cells as they become restricted in their developmental potentiality. We have used the AFP-1 cDNA probe which hybridizes with a single *Eco* R1 fragment of mouse genomic DNA, to investigate the state of the AFP gene in undifferentiated EC cells and in a line of endodermal cells (PSA5E) which are derived from them. It is not possible to perform this experiment directly with embryoid bodies as the two cell types are not separated. There was a high level of endogenous nuclease activity in the chromatin preparations from EC cells but its effect was minimised by conducting the DNAse-1 digestions at $4^{o}C$ with a high level of exogenous nuclease (2.5 μg/ml). Under these conditions the AFP gene was digested in chromatin from both foetal and adult liver but not from brain. This is the expected distribution, respectively, for an AFP-producing tissue, a tissue which is capable of being induced to produce AFP, and one which has no potentiality for AFP production. Similarly, the gene was sensitiv to DNAse-I digestion in PSA5E cell chromatin but not in EC cell chromati (Latchman, Brzeski, Lovell-Badge and Evans, submitted for publication). This result suggests that the unexpressed genes in EC cells are closed despite being potentially expressible in the various diversely differentiated progeny of these cells. Cell determination results in the "opening" of new genes during embryogenesis.

5. CONCLUSIONS

Embryonal carcinoma cells in culture allow access to large

numbers of cells which are equivalent to early mouse embryo cells. These cells should facilitate biochemical investigations into the molecular biology of early mammalian development. The most striking result from the two dimensional gel analyses of cells undergoing differentiation is the relatively small number of qualitative changes observed. This may be compared with the small number of primary changes observed in cell transformation (see for example article by Bravo *et al.* in this volume) despite in both cases an apparently major change in cell morphology, structure and behaviour.

One interpretation for these results may be that only a small proportion of the total spectrum of polypeptides which are synthesized appear in sufficient quantity to be detected in the two dimensional gels. The spots described are approximately an order of magnitude less in number than the estimated complexity of the message population (12), and thus many diffe.ences in polypeptide pattern might be below the limits of detection. This may, however, overestimate the discrepancy and a downward re-estimation of complexity figures for mRNA coupled with a determined upward estimate of polypeptide spot numbers might make the two figures more comparable (13, 14).

If it could be supposed that many of the significant changes in protein synthesis should be apparent, then we must conclude that cell determination and differentiation may depend not so much on qualitative switching of genes but upon quantitative regulation of the balance of the gene products which are common to most cells. Particular producs of terminal differentiation - including AFP, the example used here - may well undergo specific qualitative switches but this may be neither the general case nor the basis of cellular development.

6. ACKNOWLEDGEMENTS

We are grateful to the SERC and Cancer Research Campaign for project grants which have supported this work. Some of this work was carried out whilst H.B. was a SERC postdoctoral research fellow. D.S.L. was supported by a MRC postgraduate studentship and A.S. by one from the SERC.

7. REFERENCES

1) EVANS, M.J. and KAUFMAN, M.H. (1981). Establishment in culture of pluripotential cells from mouse embryos. Nature, 292, 154.

2) EVANS, M.J. (1981). Origin of mouse embryonal carcinoma cells and the possibility of their direct isolation into tissue culture. J. Reprod. Fert. 62, 625.

3) EVANS, M.J. (1982). BSDB symposium, in preparation.

4) LOVELL-BADGE, R.H. and EVANS, M.J. (1980). Changes in protein synthesis during differentiation of embryonal carcinoma cells, and a comparison with embryo cells. J. Embryol. Exp. Morphol., 59, 187.

5) COOPER, A.R., KURKINEN, M., TAYLOR, A. and HOGAN, B.L.M. (1981). Studies on the biosynthesis of laminin by murine parietal endoderm cells. Eur. J. Biochem., 119, 189.

6) DZIADEK, M. and ADAMSON, E.D. (1978). Localization and synthesis of alpha-fetoprotein in post-implantation mouse embryos. J. Embryol. Exp. Morphol., 43, 289.

7) ADAMSON, E.D. (1982). The location and synthesis of transferrin in mouse embryos and teratocarcinoma cells. Developmental Biology, 91, 227.

8) ADAMSON, E.D., EVANS, M.J. and MAGRANE; G.G. (1977). Biochemical markers of the progress of differentiation in cloned teratocarcinoma cell lines. Eur. J. Biochem., 79, 607.

9) LATCHMAN, D.S. (1981). Control of alpha feto-protein gene expression in the mouse. Ph.D. Thesis, University of Cambridge.

10) ALWINE, J.C., KEMP, K.J., PARKER, B.A., REISER, J., REINERT, J., STACK, G.R. and WAHL, M. (1979). Detection of specific RNAs or specific fragments of DNA by fractionation in gels and transfer to diazobenzyloxymethyl paper. Methods in Enzymology, 68, 220.

11) WEISBROD, S. (1982). Active Chromatin. Nature, 297, 289.

12) AFFARA, N.A., JACQUET, M., JAKOB, H., JACOB, F. and GROS, F. (1977). Comparison of polysomal polyadenylated RNA from embryonal carcinoma and committed myogenic and erythropoetic cell lines. Cell, 12, 509.

13) DUNCAN, R. and McCONCKEY, E.H. (1982). How many proteins are there in a typical mammalian cell?. Clin. Chem. 28, 749.

14) BRAVO, R. and CELIS, J.E. (1982). Up-dated catalogue of HeLa cell proteins: Percentages and characteristics of the major cell polypeptides labelled with a mixture of 16 [^{14}C]-amino acids. Clin. Chem., 28, 766.

CHANGES OF PROTEIN GLYCOSYLATION DURING DIFFERENTIATION OF MOUSE EMBRYONAL CARCINOMA CELLS

G. Cossu*# and L. Warren*

*The Wistar Institute, 36th Street at Spruce
  Philadelphia, PA 19104

#Istituto di Istologia ed Embriologia generale
  Universita' di Roma, Italy

1. INTRODUCTION

The search for a role of carbohydrates in mammalian development has been long hampered by the minute amounts of material available. Recently, the establishment of cell lines from mammalian teratocarcinomas has offered an alternative model for the study of early embryogenesis in mammals (1). Specifically, several stem cell lines, capable of either spontaneous or drug-induced differentiation, allow a biochemical study of certain aspects of early differentiation.

Surface carbohydrates, unique to stem cells and to early embryos, have been shown to be synthesized by teratocarcinoma cells (2) and to represent the structural determinants of several antigens, similarly unique to early embryonic cells (3, 4). The structural features of these carbohydrates have been partially elucidated: they are composed of linear or branched chains of lactosamine linked to a typical complex-type core and containing fucose and sialic acid (5). The term lactosaminoglycans has been recently used to describe these glycopeptides (6) whose size (5-7Kd) is markedly larger than that of typical asparagine-linked glycopeptides (2-3.5 Kd).

Presently, little is known about their distribution among
cellular glycoproteins, and it is unclear whether they might affect
the structural and functional characteristics of the proteins to
which they are linked. Furthermore, it remains to be answered
whether: 1) the reduced expression of lactosaminoglycans upon stem
cell differentiation reflects a reduced rate of synthesis of these
molecules or a change in relative proportion due to a concomitant
increased synthesis of small asparagine-linked glycopeptides; 2)
the changes in protein bound carbohydrates reflect appearance and/
or disappearance of specific glycoproteins or rather a differential
glycosylation of polypeptides common to both differentiated and undifferentiated cells.

In this article we provide evidence that such an increase in
asparagine-linked glycopeptide synthesis does occur during differentiation of embryonal carcinoma cells (F9) into primitive endodermal
cells (F9 AC C19). Furthermore we show the occurrence of similar
changes in the carbohydrate groups of fibronectin and explore the
possible consequences that such changes might determine on the
molecular characteristics of fibronectin.

2. CHANGES IN THE SYNTHESIS OF PROTEIN-BOUND CARBOHYDRATES DURING EMBRYONAL CARCINOMA CELL DIFFERENTIATION

F9 embryonal carcinoma cells, and their differentiated derivatives F9 AC C19 were metabolically labelled with [^{3}H]-glucosamine
(Clc-N), mannose, (Man), galactose (Gal), fucose (Fuc), and leucine
(Leu). Table 1 shows that while the incorporation of radiolabelled
precursors (Leu) into proteins is about the same between the two
cell types, incorporation of Glc-N, Man and Gal (but not Fuc) into
glycoproteins is increased 2-3 times in differentiated cells (F9
AC C19) as compared to undifferentiated cells (F9). This observation
allows a different interpretation of previously published data (3).
When the total pronase digested protein-bound carbohydrates from

undifferentiated and differentiated cells are compared by gel
filtration on a column of Sephadex G-50, the typical elution pattern
is as represented in Figure 1.

Table 1. *Incorporation of* [^{3}H] *precursors into trichloroacetic
acid (TCA)-insoluble material in teratocarcinoma cells*

Cell line	cpm incorporated per μg of protein				
	Leu	GlcN	Gal	Man	Fuc
F9	3,250	1,790	2,350	240	1,250
F9 AC C19	3,750	3,150	4,030	740	1,150

Cells were labelled for 6 hr with 10 μCi/ml of each precursor in
complete medium. Results are expressed as cpm incorporated per
μg of protein (lipid-extracted TCA-insoluble fraction of the cell
homogenate).

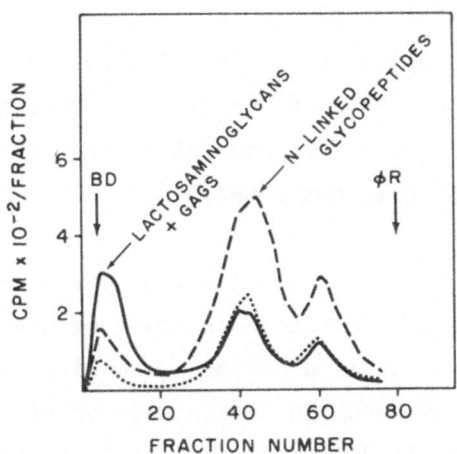

Fig. 1. *Gel filtration analysis on Sephadex G-50 column of pronase-
digested glycoconjugates from F9* (——) *and F9 AC 19* (--- *and* ...)
cells. About 5 x 10^3 cpm of [^{3}H]GlcN labelled glycopeptides from 2 x
10^6 F9 cells were applied to the column. The (···) line represents
the pattern obtained from F9 AC C19 cells when 5 x 10^3 cpm of [^{3}H]
GlcN labelled glycopeptides were applied to the column. The (---)
line represents the pattern obtained from F9 AC C19 cells when
material derived from 2 x 10^6 cells (10^4 cpm) was applied to the
column. The samples were run separately and their profiles superimposed on the basis of the positions of the marker dyes blue dex-
tran (BD) and phenol red (øR).

Glycopeptides from F9 undifferentiated cells are eluted from the column as a major peak of radioactivity partially excluded from the column (lactosaminoglycans). A minor peak corresponding to the small asparagine-linked glycopeptides is also detected among the glycopeptides from F9 cells. In contrast, the glyco-peptides from F9 AC C19 differentiated cells are eluted as a major peak corresponding in elution position to the small asparagine-linked glycopeptides while the lactosaminoglycans represent a minor component. By plotting data of Figure 1 so that glycopeptides from the same number of cells (rather than the same amount of radio-activity) are compared, it can be seen that, beside a decline in the synthesis of the lactosaminoglycans, embryonal carcinoma cell differentiation is accompanied by a dramatic increase in the syn-thesis of asparagine-linked small glycopeptides. Gas chromatographic analysis of alditolacetate derivatives of monosaccharides from large and small glycopeptides of differentiated and undifferentiated cells confirmed that the observed changes in radioactivity do correspond to changes in the real content of the carbohydrate molecules within cells (data not shown).

The increased synthesis of small glycopeptides in differenti-ated cells might reflect a selective increase of glycoprotein synthesis (as related to total proteins) or a differential glyco-sylation of the same glycoprotein species. In other words, the polypeptide synthesized in differentiated cells might bear an in-creased number of small glycopeptides as compared to the same poly-peptide synthesized in undifferentiated cells. If this is true, then it is necessary to postulate that the glycosylation machinery of undifferentiated cells fails to recognize glycosylation sites on the polypeptides which will be late glycosylated by differenti-ated cells. This would also imply the existence of more than one mechanism which allows the transfer of a carbohydrate moiety from the lipid donor to a specific asparagine on the protein.

3. CHANGES IN THE PROTEIN-BOUND CARBOHYDRATES OF FIBRONECTIN
    DURING DIFFERENTIATION OF EMBRYONAL CARCINOMA CELLS

    A possible experimental approach to the solution of such a
problem is that of isolating a glycoprotein which is synthesized
both before and after differentiation but bears different carbohydrate groups depending on the cells in which it is glycosi-
lated. A good candidate for such a study is a viral glycoprotein,
the only glycoprotein whose amino-acid sequence is certainly the
same in the two different cell types (all cellular glycoproteins
might in fact be isoforms with small variations in primary sequence).
Unfortunately the G protein of the Vesicular Stomatitis Virus does
not bear lactosaminoglycans when synthesized in F9 cells (7) and
cyanogen bromide (CNBr) cleaved peptides revealed the same glycosylation pattern, despite differences in sialylation (Cossu and
Warren, manuscript in preparation). As an alternative approach we
studied fibronectin from both differentiated and undifferentiated
cells.

    Several recent reports have indicated that fibronectins re-
present a heterogeneous family of molecules that differ in the
amount and the type of carbohydrates bound (8). Examples of such
differences are found in fibronectins from adult versus embryonic
fibroblasts (9), from normal versus transformed cells (10) and
from amniotic fluid versus plasma fibronectin (11). Furthermore,
it has been recently shown that human germ cell tumours synthesize
a fibronectin similar to that present in amniotic fluid (12) and
human teratocarcinoma cells synthesize fibronectin of a higher
molecular weight than that present on fibroblasts (P. Andrews,
manuscript submitted for publication).

    We isolated fibronectin from the culture medium of F9 and
F9 AC C19 cells by affinity chromatography on gelatin-Sepharose
(13), and from the surface of these cells by the urea extraction

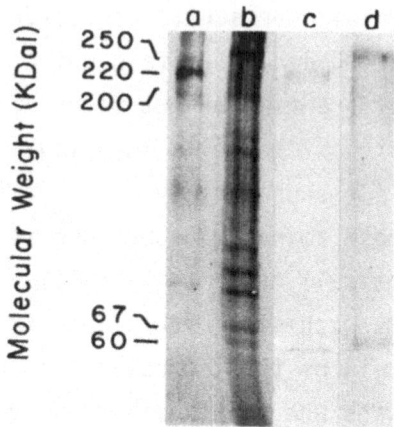

Fig. 2. *Electrophoresis on SDS-polyacrylamide gel (6%) of urea-extracted (a,b) and medium-released gelatin-bound (c,d) [^{35}S]-methionine labelled proteins from teratocarcinoma cell lines; F9 AC C19 (a,c), F9 (b,d).*

method (14). The molecular weight of such fibronectin, as estimated by sodium dodecyl sulphate-polyacrylamide gel electrophoresis (SDS-PAGE) differed: Figure 2 shows that fibronectin from F9 cells was apparently larger (250 Kd) than fibronectin from differentiated cells (220 Kd). Both molecules could be immunoprecipitated by an antiserum against fibronectin.

To determine whether bound carbohydrates might account for this higher molecule weight, F9 and F9 AC C19 cells were labelled metabolically with [^{14}C] GlcN, and fibronectin was purified, electrophoresed and then digested with pronase as described in detail elsewhere (14). The released carbohydrate groups were analyze. by gel filtration on a column of Sephadex G-50. The carbohydrate

groups released from both cellular and medium-derived fibronectin
of F9 AC C19 cells elute as a major peak (Figure 3) in a position
typical of the bi-antennary complex type glycopeptide described
in hamster fibroblasts (15). In contrast, when the glycopeptides
of fibronectin from F9 cells were run on a Sephadex G-50 column
(Figure 3b), most of the radioactivity eluted as a single peak
in the void volume of the column, consistent with molecular weight
of at least 6 Kd. In this sample, the peak representing typical
complex type glycopeptides (tubes 40-60) was barely detectable.
However, in the medium-released fibronectin (Figure 3a) two peaks

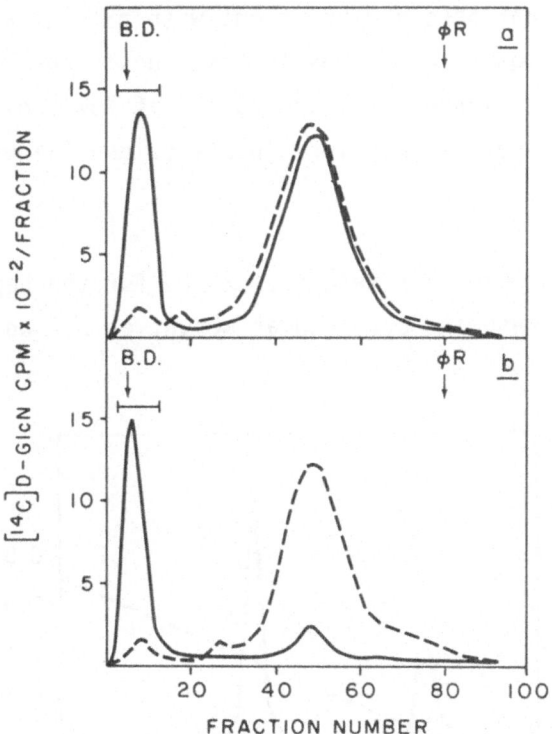

Fig. 3. *Gel filtration analysis on Sephadex G-50 column of pronase-
digested glycoconjugates from medium released (a) and from cell-
associated (b) fibronectins.* F9 (——) and FG AC C19 (---) derived
material were run separately on the same column and their elution
profiles superimposed on the basis of the positions of the marker
dyes BD and øR.

corresponding to both the typical complex type glycopeptides and
the large carbohydrate groups excluded from the column were detected
in comparable amounts.

Thus the increase of molecular weight in F9-derived fibronectin is due to high molecular weight carbohydrate groups which
are covalently linked to the 250 Kd fibronectin as demonstrated
by persistent association even after subjecting the material to
dissociative gradients of cesium chloride in 4 M guanidine HCl.
About 80% of these glycoconjugates consist of heparan sulphate as
analyzed by ion exhange chromatography (Figure 4, third peak),
digestion with heparinase and treatment with nitrous acid. The remaining glycoconjugates are large lactosaminoglycans, digestible
with endo-β-galactosidase (17), and eluted at low ionic strength
from a diethylaminoethyl (DEAE)-cellulose column (Figure 4, first
two peaks).

Estimating a molecular weight of 20 Kd for the heparan sulphate
chains (16) and presuming a molecular weight of 5-7 Kd for the

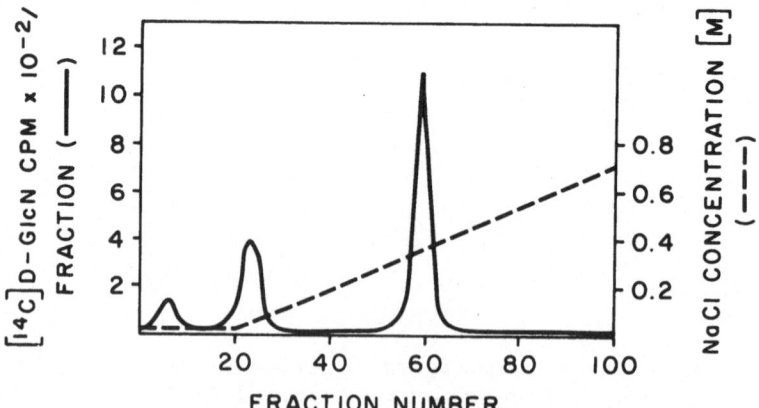

Fig. 4. *Ion exchange chromatography on DEAE-cellulose of material excluded from Sephadex G-50 column (Figure 3) from cell surface
fibronectin of F9 cells.* Glycoconjugates were eluted with a linear
salt gradient from 0 to 0.8 M NaCl in 5 mM Tris-HCl pH 7.

lactosaminoglycans (3), we suggest that one chain of each carbohydrate is present per fibronectin molecule which would result in
observed molecular weight increase of 25-30 Kd.

Interestingly, gel filtration analysis of carbohydrate groups
from F9 cells revealed that the cellular, but not the medium released, form of fibronectin lacks almost completely the small
glycopeptides typical of other fibronectins. Since both forms are
synthesized by the same cells, variations in cpm should reflect
actual variations in the amount of the molecule synthesized.
Because of this and the similarity in the radioactivity incorporated
into the large glycoconjugates from both cellular and medium-released fibronectins from F9 cells, the lack of the small glycopeptides in the cellular form probably does not reflect a variation
in relative proportions but an actual dramatic reduction in amount.
It is possible that some of the sites available for glycosylation
on cellular fibronectin are not recognized by the glycosylation
machinery of F9 cells, or, alternatively, that the available sites
are glycosylated with different carbohydrate chains (heparan sulphate
and/or lactosaminoglycans). On the other hand, it is possible that
the F9 fibronectin(s) might represent embryonic isoforms of the
molecule lacking sites for glycosylation and possessing additional
sites available for glycosaminoglycans. The answers to these
questions await the detailed structural characterization of different
fibronectins. It is most likely, however, that the developmentally
regulated presence of lactosaminoglycans on fibronectin reflects
changes in the glycosylation machinery since these glycopeptides
are known to be linked to asparagine as typical N-linked glycopeptides (5).

Taken together the evidence strongly suggests that glycosylation of fibronectin represents a cell-type specific function
and the same or a closely related polypeptide might be modified

by post-translational events specific to the cell in which it is
synthesized. Specifically, cellular fibronectin of F9 cells would
represent an example of a polypeptide glycosylated with lactosaminoglycans but with a reduced number of typical complex type glycopeptides. Thus, a programmed change in the glycosylation machinery
of early embryonic cells is likely to take place during the first
stage of differentiation.

4. EFFECTS OF LACTOSAMINOGLYCANS AND HEPARAN SULPHATE ON THE
     BIOCHEMICAL CHARACTERISTICS OF FIBRONECTIN

It was of interest to determine whether any of the known
activities of fibronectin might be modified when lactosaminoglycans
and heparan sulphate are covalently bound to the molecule.

We approached the problem by studying the interactions of
F9-derived fibronectin with extra-cellular matrix components. In
the experiments described below we compared the medium released
fibronectin (labelled with [^{35}S]-methionine) with human plasma
fibronectin (monitored by ultraviolet absorbance). Medium released
fibronectin was employed in these studies because it is soluble
in the absence of detergents or urea.

We first studied the affinity for gelatin of F9-derived fibronectin as compared to plasma fibronectin. After purification on a
coumn of gelatin-Sepharose, followed by gel filtration on a column
of Sephacryl-300 (to remove low molecular weight gelatin-binding
proteins) F9-fibronectin (20,000 cpm) was mixed with 0.5 mg of
human plasma fibronectin and applied again to a gelatin-Sepharose
column. Elution was carried out with a gradient from 0 to 6 M urea
in 50 mM Tris HCl (pH 7.2). The results obtained are shown in
Figure 5.

Human plasma fibronectin is eluted from the colums as discrete

peaks between 1 and 4 M urea. In contrast [^{35}S]-labelled F9 fibronectin is eluted by higher concentrations of urea with major peaks
eluting at 3, 4 and 5 M urea. Thus, it appears that heparan sulphate
(and lactosaminoglycans) increase the affinity of fibronectin for
gelatin, in agreement with previous data showing that soluble
heparin would increase the affinity of fibronectin for gelatin (8).

We then examined the characteristics of F9 fibronectin binding
to heparin, another macromolecule known to interact with fibronectin (17). F9-derived fibronectin binds to heparin and is co-
eluted (at 0.1 M NaCl) with plasma fibronectin when elution is
performed with a buffer without Ca^{2+} and Mg^{2+} ions (Figure 6).

However when the experiment was repeated with 10 mM MgCl$_2$ and
CaCl$_2$ added to the eluting buffer a different result was obtained.

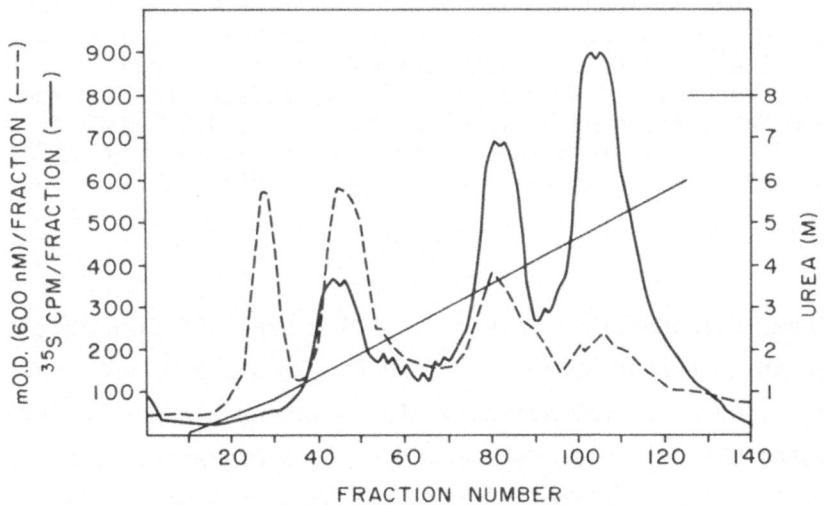

Fig. 5. *Binding of F9-derived and human plasma fibronectin to
gelatin.* [^{35}S]-methionine labelled fibronectin from the medium of
F9 cells was mixed with 0.5 mg of human plasma fibronectin and the
sample was applied to a gelatin Sepharose column. Elution was
carried out with (50 + 50 ml) a gradient from 0 to 6 M urea in
50 mM Tris HCl (pH 7.2). F9 fibronectin (——); human plasma fibronectin (---).

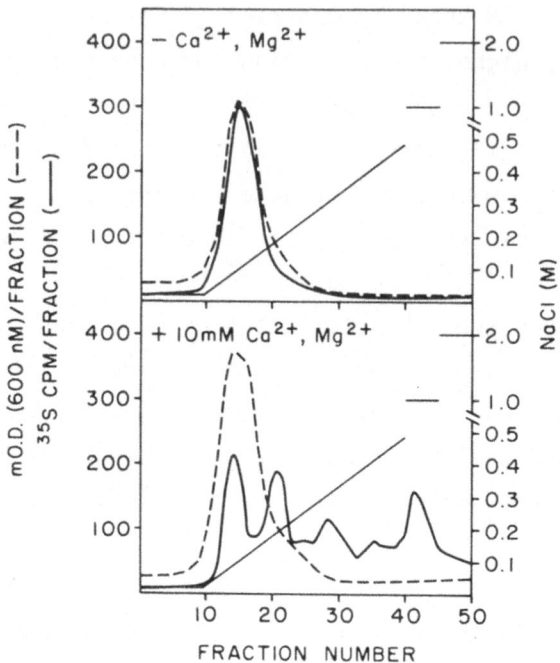

Fig. 6. *Binding of fibronectins derived from F9 medium and human plasma to heparin.* Samples, prepared as described in legend to Fig. 5, were applied to two identical columns of heparin-agarose. The samples were eluted with a (50 + 50 ml) linear gradient from 0 to 0.5 M NaCl, followed by 10 ml washed of 1 and 2 M NaCl in 50 mM Tris HCl (pH 7.2). In b the buffer also contained 10 mM $MgCl_2$ and $CaCl_2$. F9 fibronectin (——); human plasma fibronectin (---).

While plasma fibronectin is eluted at 0.1 M NaCl, F9 fibronectin is eluted as different peaks at 0.1, 0.2, 0.3 and 1 M NaCl, respectively. Therefore the affinity of F9 fibronectin for heparin is increased but only when Ca^{2+} and Mg^{2+} ions are present. A heterogeneity of the molecule (not revealed otherwise by SDS-PAGE or ion exchange chromatography studies) might be the cause of the elution from heparin column as discrete peaks at different salt concentrations, but the reason for the existence of such a heterogeneity is presently not understood.

5. CONCLUSIONS

The studies reported in this article demonstrate the existence of changes in the synthesis of protein-bound carbohydrates during early differentiation of mouse embryonal carcinoma cells. The differentiated cells do synthesize more asparagine-linked small glycopeptides and less lactosaminoglycans than the stem cells from which they are derived.

We propose that such an increased synthesis of small glycopep-tides may lead to a differential glycosylation of some polypeptides common to both undifferentiated and differentiated cells. Cellular fibronectin, derived from F9 stem cells might represent such as example, because of its lack of small glycopeptides. In addition, our studies revealed that lactosaminoglycans and heparan sulphate are covalently bound to both cellular and medium released fibro-nectins from undifferentiated cells. Such differential glycosylation of fibronectin results in an increased affinity of the molecule for gelatin and heparin (in the presence of Ca^{2+} and Mg^{2+} ions). Clearly more studies are needed in order to figure out the possible physiological consequences of such differential glycosylation of fibronectin.

It is interesting, however, to note that endodermal cells synthesize a complex basal lamina, consisting of type IV collagen, laminin and glycosaminoglycans (18). They also synthesize a membrane proteoglycan which is probably involved in cellular adhesion to basal lamina (19). Interestingly, the only glycosaminoglycan present on such proteoglycan is heparan sulphate. Stem cells, on the other hand, synthesize little if any of such components. It is possible that heparan sulphate containing fibronectin synthesized by stem cells, because of its increased affinity for extracellular matrix components, might facilitate the adhesion of cells to the little matrix present in the intercellular space, thus mimicking a function that will be later replaced by more specialized components.

6. ACKNOWLEDGEMENTS

This work was supported by grants CA-19130, PRP-28 and
CA-21069 from USPHS. G.C. was the recipient of the USPHS fellowship
TW-03012.

7. REFERENCES

1) MARTIN, G.R. (1980). Teratocarcinomas and mammalian embryo-
    genesis. Science, 209, 768.
2) MURAMATSU, T., GACHELIN, G., NICOLAS, J.F., CONDAMINE, H.,
    JAKOB, H. and JACOB, F. (1978). Carbohydrate structure and
    cell differentiation: unique properties of fucosylglycopeptides
    isolated from embryonal carcinoma cells. Proc. Natl. Acad. Sci.
    USA, 75, 2315.
3) MURAMATSU, T., GACHELINE, G., DAMONEVILLE, M., DELARBRE, C.
    and JACOB, F. (1979). Cell surface carbohydrates of embryonal
    carcinoma cells: polysaccharidic chains of F9 antigens and
    receptors for two lectins, FBP and PNA. Cell, 18, 183.
4) ANDREWS, P.W., KNOWLES, B.B., COSSU, G. and SOLTER, D. (1981).
    Teratocarcinoma and mouse embryo cell surface antigens:
    characterization of the molecule(s) carrying the SSEA-1 anti-
    genic determinant. In: Teratocarcinoma and the Cell Surface
    (eds, T. Muramatsu and Y. Ikawa), p. 103, Japan Sci. Soc.
    Press Tokyo.
5) MURAMATSU, T., GACHELIN, G. and JACOB, F. (1979). Characteriza-
    tion of glycopeptides isolated from membranes of F9 embryonal
    carcinoma cells. Biochem. Biophys. Acta, 587, 392.
6) MURAMATSU, H. and MURAMATSU, T. (1982). Decreased synthesis
    of large fucosylglycopeptides during differentiation of embry-
    onal carcinoma cells induced by retinoic acid and dibutyril
    cyclic AMP. Develop. Biol., 90, 441.

7) ETCHINSON, J.R., SUMMERS, D.F. and GEORGOPOULOS, C. (1981).
    Variations in the structure of radiolabelled glycopeptides
    from glycoprotein of Vescicular Stomatitis Virus grown in
    four mouse teratocarcinoma cell lines. J. Biol. Chem., 256,
    3366.

8) RUOSLAHTI, E., HAYMAN, E.G., PIERSCHBACHER, M. and ENGVALL,
    E. (1982). Fibronectin: purification, immunochemical pro-
    perties and biological activities. Meth. Enzymol., 82, 803.

9) YAMADA, K.M. and KENNEDY, D.W. (1979). Cell surface and plasma
    fibronectins are similar but not identical. J. Cell Biol.,
    80, 432.

10) WAGNER, D.D., IVATT, R., DESTREE, A.T. and HYNES, R.O. (1981).
    Similarities and differences between the fibronectins of
    normal and transformed cells. J. Biol. Chem., 256, 11708.

11) PANDE, H., CORKILL, J., SAILOR, R. and SHIVELY, J.E. (1981).
    Comparative structural studies on human plasma and amniotic
    fluid fibronectins. Biochem. Biophys. Res. Commun., 10., 265.

12) RUOSLAHTI, E., JALANKO, H., COMING, D.E., NEVILLE, A.M. and
    RAGHAVAN, D. (1981). Fibronectin from human germ-cell tumours
    resemble amniotic fluid fibronectin. Int. J. Cancer, 27, 763.

13) ENGVALL, E. and RUOSLAHTI, E. (1977). Binding of soluble
    form of fibroblast surface protein, fibronectin, to colla-
    gen. Int. J. Cancer, 20, 1.

14) YAMADA, K.M., SCHELESINGER, D.H., KENNEDY, D.W., and PASTAN,
    I. (1977). Characterization of a major fibroblast cell sur-
    face glycoprotein. Biochemistry, 16, 5550.

15) FUKUDA, M. and HAKOMORI, S. (1979). Carbohydrate structure
    of galactoprotein, a major transformation sensitive gly-
    coprotein released from hamster fibroblasts. J. Biol. Chem.,
    254, 5451.

16) COSSU, G. and WARREN, L. (1982). Lactosaminglycans and heparan
    sulfate are covalently bound to fibronectins synthesized by
    mouse stem teratocarcinoma cells. Submitted for publication.

17) YAMADA, K.M., KENNEDY, D.W., KIMATA, K. and PRATT, M. (1980).
     Characterization of fibronectin interactions with glycosamino-
     glycans and idenfication of active proteolytic fragments.
     J. Biol. Chem., 255, 6055.

18) OOHIRA, A., WIGHT, T.N.,MCPHERSON,J. and BORNSTEIN, P. (1982).
     Biochemical and ultrastructural studies of proteoheparan
     sulfates synthesized by PYS-2, a basement membrane-producing
     cell line. J. Cell Biol., 92, 357.

19) LEIVO, I., ALITALO, K., RISTELI, L., VAHERI, A., TIMPL, R.
     and WARTIOVAARA, J. (1982). Basal lamina glycoproteins lami-
     nin and type IV collagen are assembled into a fine-fibered
     matrix in cultures of a teratocarcinoma-derived endodermal
     cell lines. Exp. Cell Res., 137, 15.

FOCUSING ON A PARTICULAR MODEL OF CELL DIFFERENTIATION: THE VERTEBRATE EYE LENS

L. Simmonneau

Unité de Recherches Gérontologiques
INSERM U 118
29 Rue Vilhem
75016 Paris, France

1. INTRODUCTION

The vertebrate eye lens represents a suitable organ for the study of fundamental biological mechanisms. Its several features makes it a useful system for investigating, at a molecular level, the processes which control the different states of differentiation. Furthermore, such a study offers to contribute to our knowledge of ageing, a process still only partially understood.

This highly specialized organ is capable of proper refraction of incident light bringing visual acuity. It is completely devoid of blood vessels, connective tissues and nerve cells. Throughout life, this organ increases in volume and weight. The surrounding fluids, aqueous humour in the anteriour part, vitreous body in the posterior part, provide the necessary metabolites (Figure 1). A striking property of this organ is that it never sheds its cells. It is surrounded by a basal lamina - the capsule - that is synthesized by the epithelium. This capsule, to which the zonular fibers attach, is composed mainly of collagen (type IV) and proteoglycans (1).

117

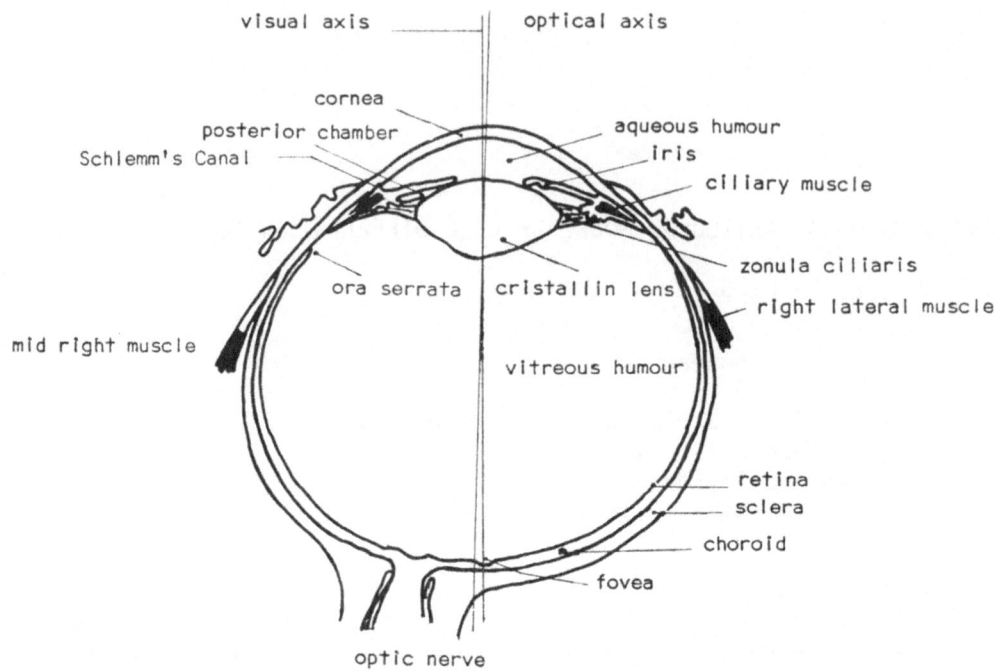

Fig. 1. *Diagram of a sagittal section of the human eye.*

The nucleus of the lens is the older part of the organ and contains molecules which have been synthesized during embryogenesis. Proteins in the centre do not turn over or do so exceedingly slowly and only the equatorial periphery contains newly synthesized proteins. In a broad sense, one is tempted to speculate that this organ offers a natural gradient of senescent biological components.

2. MORPHOGENESIS OF THE EYE LENS

During embryogenesis the lens is generated from ectoderm which is close to the optic vesicle, that is to say the presumptive retina (Figure 2). These two tissues are closely associated. The first morphological differentiation corresponds to the formation

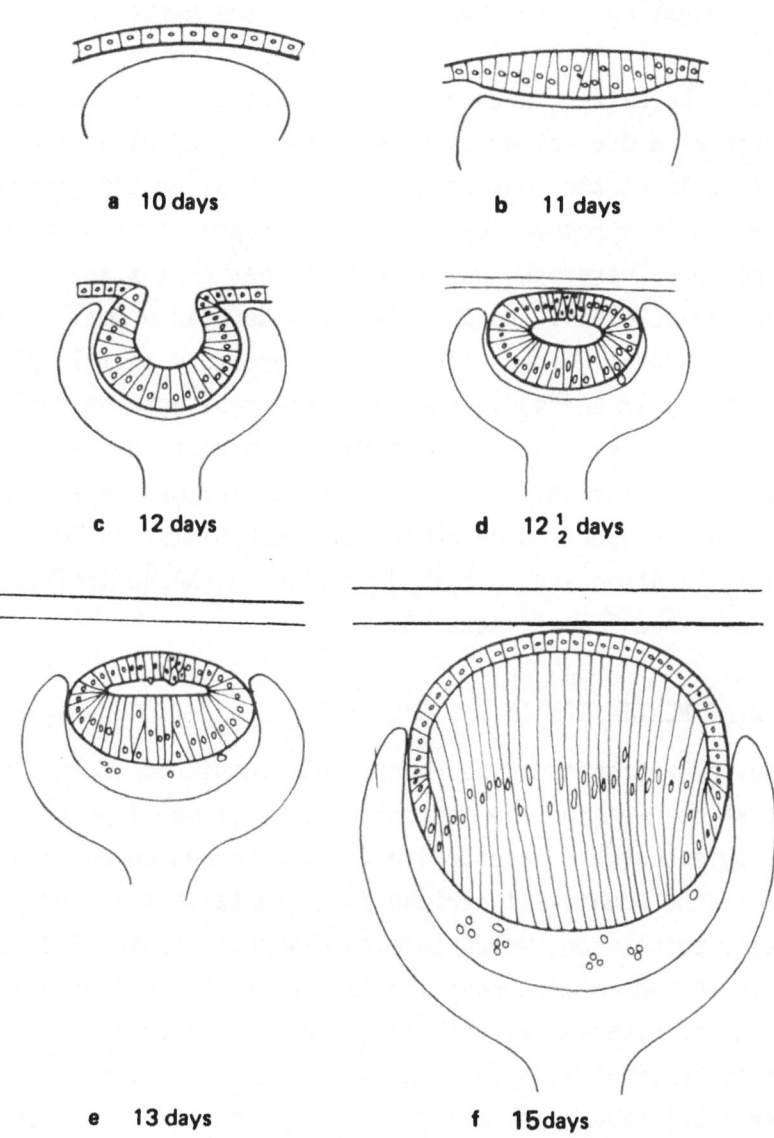

Fig. 2. *Sagittal sections of various stages of embryogenesis in rat lens development.* Diagrams from Mac Avoy (21).

of the lens placode (Figure 2b) which invaginates to form the lens
pit (Figure 2c) and subsequently the lens vesicle (Figure 2d). The
primitive elongation of the cells occurs in the posterior part of
the developing lens, filling the lumen and forming the primary
fiber cells. Thereafter, secondary fibers are added throughout
life, though at a decreasing rate, by division and elongation of
epithelial cells at the lens equator. Shortly after elongation
begins, the nucleus becomes pycnotic and disappears as do other
cell organelles. Therefore, cells in the inner cortex and
nucleus are featureless. The DNA is broken down into small pieces
and in the embryonic chicken lens these correspond to multiples of
about 180 base pair units, suggesting a cleavage between nucleosomes
(2). The lens fibers, however, contain the complete protein synthesizing machinery, including long life stabilized messenger RNAs.
It is of great interest to determine the mechanisms that contribute
to such stabilization and in that direction an RNAase inhibitor
has been described in calf lens (3, 4).

3. CELL DIFFERENTIATION IN THE EYE LENS

Three compartments can be roughly distinguished in this model.
In the first one we find the central epithelium monolayer whose
cells are quiescent and arrested in G_1. The second compartment is
localized at the equatorial periphery and contains the slowly
cycling stem population, which in mice divides between 17-20 days
(5). The third compartment contains the terminally differentiated
progeny, the anucleated cell fibers. In mice, the entire process
from mitosis to final differentiation takes about 4 months. Rafferty
and Rafferty (5) have calculated that in 8 weeks old CF_1 mice the
epithelium cells produce about 200 new lens fibers a day.

Molecular studies of differentiation or ontogenesis are based
on the concept that differentiated cells could be distinguished
by their own tissue-specific proteins. In the case of lens fiber

differentiation, authors have described an important increase in membrane (composed mainly of gap junctions) as well as in the specific lens proteins, the crystallins.

(i) *Membrane proteins (MP 26)*

MP 26 is the major polypeptide (M_r = 26 Kd) present in gap junctions and it can be considered as a marker of differentiation since it has not been detected in the lens epithelium plasma membrane (6). The messenger RNA encoding for MP 26 is localized in a polyribosome fraction that is bound to the cytoskeleton-plasma membrane complex (7).

In spite of increasing studies devoted to this polypeptide (8, 9), the main studies on lenses have been concentrated on the crystallin proteins. These proteins constitute the main bulk of the lens proteins and ensure a good refractive index for the transmission of light to the retina.

(ii) *Crystallins*

The lens contains about 35% by weight of structural proteins. In 1830, Berzelius described in this organ a protein-like material that he termed "Krystallin" (10). Crystallins contributed more than 90% of the total lens proteins and in general they are highly conserved within the whole vertebrate subphylum. Manski estimated that human or calf lens crystallins, for instance, still contain "fossil" antigenic determinants coming from the primitive Agnatha (jawless fish), which originated the vertebrates some 450 million years ago (11). Animals are unable to develop an immunological tolerance to their own crystallins as δ autologous lens proteins are not recognized as self. So, endophtalmitis phacoanaphylactica is a severe allergic reaction in the eye of individual sensitized to lens proteins.

Various isolation procedures and a wide survey of properties
of crystallins have been summarized in excellent reviews and books
(12-14). In mammals, four classes of crystallins have been separated
by gel filtration on Sephadex G 200 or Ultrogel ACA 34 and these
correspond to α, β_H, β_L and γ crystallin.

The α crystallins have the highest molecular weight (average
of 800 Kd). These aggregates can be of higher molecular weight and
reach more than 50.000 Kd in the case of cataracts and/or ageing.
Four polypeptides compose the α aggregates: two primary gene
products: αB_2 (basic subunit) and αA_2 (acidic subunit). Furthermore some αA_2 and αB_2 molecules yield αA_1 and αB_1 polypeptides
as a result of a post-synthetic deamidation process. In addition,
degradation by cleavage from the C-terminal end takes place
during ageing.

The β crystallins are separated into two fractions having
molecular weights of about 200 Kd (β_H) and 50 Kd (β_L), respectively.
The major polypeptide βB_p (principal basic) is shared by both
fractions.

γ crystallins are eluted after the β_L fraction from the gel
filtration. This fraction in the calf lens contains five components.
The primary structure of two of them has been determined (15).
Figure 3 shows an electrophoresis pattern of bovine water soluble
crystallins.

Lenses of reptiles and birds possess another crystallin which
is different from α, β and γ. Its name is δ crystallin and represents 50% of the total protein in adult lens and 75% in embryo-
nic lens. Like α and β, it is involved in polymeric structures.
Delta crystallin is the first lens polypeptide whose corresponding
cDNA has been prepared (16). In the calf lens system, αA_2, $\beta B_1 a$

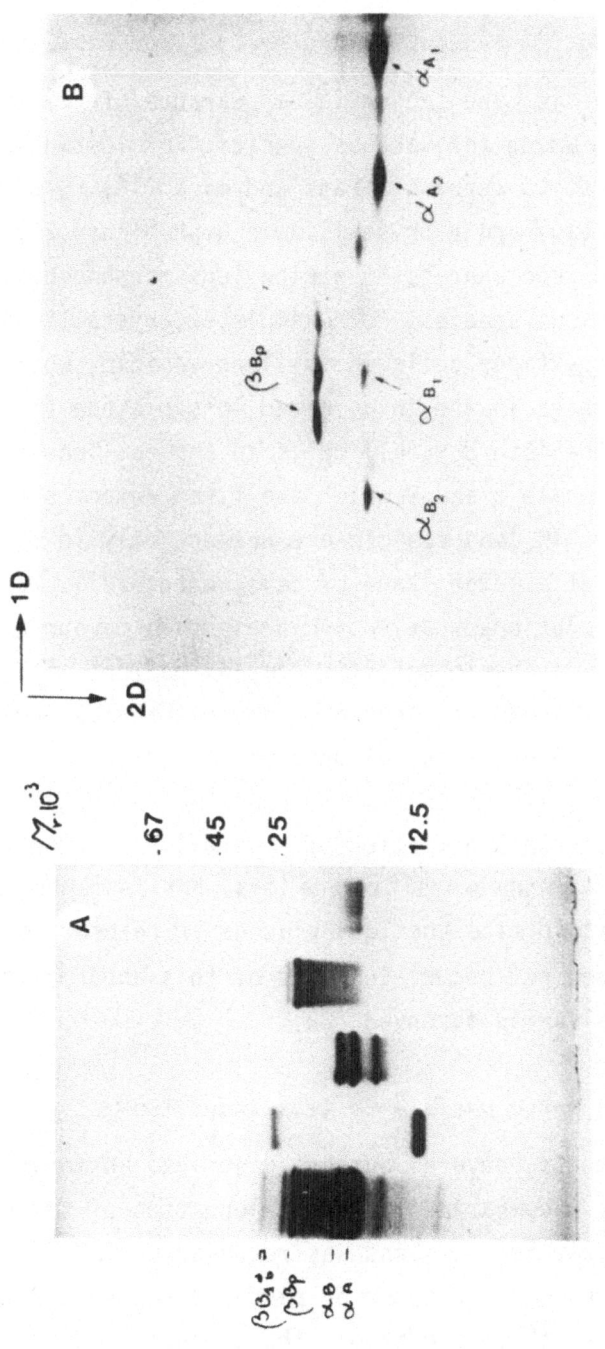

Fig. 3. *Electrophoretic pattern of bovine crystallins.* (A) 18% SDS PAGE of fractions separated by gel filtration: (1) total crystallins, (2) MW protein markers, (3) α-crystallins, (4) β-crystallins, (5) γ-crystallins. (B) Two dimensional PAGE of major crystallin polypeptides: 1D (one dimensional) means 7 M urea at pH 7.0; 2D (two dimensional) means 18% SDS PAGE.

and γcDNA have also been subjected to molecular cloning (17).

(iii) *Temporal and spatial sequential synthesis of crystallins*

During embryogenesis, the sequential appearance of crystallin
subunits is different among the various species. For instance, α
crystallin is the first to appear in rats and mice (18), γ crystallin
the first in chicken (19) and β crystallin in amphibians (20). The
pattern of crystallin gene expression during lens morphogenesis is
also different in various species. For example, β crystallin is
detected in the primary fiber cells in the lens vesicle, while in
rats α crystallin appears in the invaginated lens placode (12 days
of development). By the 15th day, all cells in the rat lenses
contain α crystallin while β and γ which are first detected at $12\frac{1}{2}$
days in primary fibers in lens vesicle are present only in fibers
and elongating cells at a later stage of development (21). We could
postulate a strong evolutionary selective pressure in favour of
mechanisms which allow a specie-specific complex pattern of aggre-
gated crystallin subunits during lens ontogeny but the determinism
of these different patterns is as yet unknown.

We must point out that initiation of crystallins synthesis
does not require normal morphogenesis of the lens. For instance, in the
hereditary anophtalmia in mice the lens placode is reabsorbed and
normal development does not occur. In spite of this abnormality
synthesis of α crystallin is observed (22).

(iv) *Influence of the optic vesicle in lens induction*

In early experiments, several authors described an inductive
effect of the optic cup compartment on the elongation of the fiber
cells (23, 24). For example, the lens was reversed by 180^0 so
that the fibers were now in the anterior part and the epithelium
faced the retina. Under these conditions the epithelial cells
elongated forming new fiber cells and a new epithelial monolayer

was formed several days later. The formation of the vertebrate lens
during embryogenesis is also related to the development of the
retina. Speman (25) working with the frog *Rana fusca* was the first
one to show an inductive effect of the optic vesicle on the head
ectoderm leading to lens formation. Furthermore, when explants of
newborn rat lens epithelium were cultured *in vitro* only a small
amount of crystallins were synthesized. The same tissue, however,
when cultured on a piece of neural retina increased in size and
synthesized large amounts of α, β and γ crystallins (26). Replacement
of neural retina by a corneal stroma, however, yielded similar
results (26).

Is this cell-cell interaction due to a diffusable signal?
Consistent with this idea McAvoy (26) has shown that a neural
retina conditioned medium is able to stimulate elongation and
synthesis of crystallins. Also, Van der Starre (27) demonstrated
the transfer of a substance from the optic cup, through an agar
slice, towards the ectoderm. This piece of agar was able to differentiate pieces of ectodermic tissue laid on it (27). Other authors
have described a transfer of RNA containing particles from the
optic cup to the ectoderm (28), while a factor in the vitreous
humour, named lentropin has been shown to promote cell elongation
and δ crystallin synthesis in primary explants of embryonic chicken
lens (29). There is, however, evidence indicating that direct cell
contact is also required. For example, at 11 days of embryonic rat
development a close association - i.e. cytoplasmic processes inter-
connection - between presumptive lens ectoderm and optic vesicle
has been described (30).

From the above data, it would seem that morphological and
biochemical features of lens differentiation appear following induction by optic vesicles. However, Karkinen-Jääskeläinen has shown
that presumptive lens ectoderm of chicks or mice, without any

contact with optic vesicles, is able to form lentoids during *in vitro* cultivation (31). Also Baravanov *et al.* (32) have demonstrated that in chick embryos (at least until stage 10-13) the presumptive lens ectoderm as well as other areas of the head ectoderm (lateral and stomodal areas), which are not involved in lens formation during morphogenesis, are able to form crystallins synthesizing lentoid bodies long before the beginning of induction by the optic vesicle.

(v) *Lentoid bodies and transdifferentiation* in vitro

Dissociated cells of lens epithelia from different species: chick embryo or hatched chickens (33, 34), adult rats (35), normal and cataractous mice (36), are able to form *in vitro* small clusters of elongated cells named lentoid bodies. These bodies synthesize crystallins but their type or amount vary from tissue to tissue as well as with the age and genotype of the starting tissue (37).

An interesting biological observation is that ocular tissues other than lens epithelium can escape from their stable differentiated state to transdifferentiate to another commitment pathway. In this respect, the vertebrates eye tissue are a marvelous exception to the concept of competence. Normally as development proceeds, competence to be committed to another differentiation pathway is gradually restricted and finally lost. The Wolffian regeneration in newts is now classical: after lentectomy, cells of pigmented epithelium of the dorsal and ventral iris epithelium regenerate a lens (38). Likewise, regeneration of a lens from cornea after lentectomy in larval Xenopus has been shown (39). In the latter case implantation of retina tissue into the anterior chamber (aqueous humour), even in the presence of an intact lens, transdifferentiates the iris into lens cells (40).

These phenomena also occur *in vitro*. For example, crystallins containing lentoids have been shown to arise from several eye tissues, such as the iris pigmented epithelium of the newt (41), the neural retina of chick embryos (42), quail embryos (43), and human foetus (44) as well as from pigmented retina of adult newts (45) or chick embryos (46). An excellent study carried out by Clayton *et al.* (47) on transdifferentiation of chick neural retina cells *in vitro*, demonstrated the importance of the embryonic stage. Thus, from a very early embryonic stage (3½ days old) to the late development stages (18 days old), the sequence of crystallins synthesis (δ in early stages, α and β in later stages) mimics the sequence of events occurring during lens ontogenesis.

(vi) In vitro *cellular differentiation*

Lentoid bodies are considered in the literature as densely packed lens epithelial cells in which terminal differentiation has occurred. In mammals, γ crystallin antiserum has been used to follow differentiation using immunofluorescence, however, one must be careful when studying elongation of cultured cells. Glässer *et al.* (48) have analyzed different enzyme activities (leucine amino peptidase, glucose 6 phosphate dehydrogenase, thiol protein disulfide oxidoreductase, lactate dehydrogenase) and found that neither the fiber like structures nor the aggregates represented a true fiber formation *in vitro*. Moreover, Creighton *et al.* (49) have shown that monolayers of rat lens epithelial cells cultured *in vitro* stain positively with specific rat γ crystallin antiserum. Also Miller *et al.* (50) observed α and β crystallins but not γ crystallin staining in one day old rat lens epithelium transformed with Rous Sarcoma virus. These cells exhibited a fibroblastic morphology. Weinstein *et al.* (51) described an α and γ crystallin containing bovine lens cell lines with morphological appearance of epithelial cells but with no evidence of elongation on aggregation (as fibers or lentoid bodies). Transformation with SV40 did not influence the synthesis of crystallins.

Vermorken and Bloemendal (52) have shown, however, that bovine lens cell lines growing on the capsule synthesize mainly αB_2 crystallin while cells that grew off the capsule onto the plastic Petri dish did not. Based on these results they concluded that a dedifferentiation has taken place (53). We agree with these authors on the fact that, considering the α crystallin subunit synthesis in bovine epithelium lens cells (BEL) grown on the capsule, the stoichiometric ratio αB_2 : αA_2 is largely in favour of αB_2 as it is in the central epithelium *in situ* (54). In our hands, however, BEL cells cultivated on plastic still synthesize αB_2 (but not αA_2) and βBp polypeptides. At early passages, radioactive spots migrating on two dimensional gels in the position of αB_2 and βBp have been observed. At late passages the intensity of these spots were very low but nevertheless detectable. Even if these BEL cells are not committed to terminal differentiation *in vitro*, we think that they retain roughly a central epithelium differentiated state. Moreover, BEL cell shape modification by cultivation onto a basal lamina membrane or in a retinal extract conditioned medium (55) did not affect crystallin synthesis (56). Injection of BEL cells in athymic mice gave small cell clusters in which we still detected α crystallins (57). BEL cells at early passages synthesized mostly type IV collagen, while cells in late passages synthesized collagen of type I, III and IV (58).

All these different data, perhaps conflicting at a first observation, suggest that differentiated features (crystallins gene expression, metabolic activities, cell shape etc.) are regulated independently from one another in this model. Many experiments performed at different levels of gene expression support such an idea.

(ii) *Analysis of the modulation of gene expression in lens epithelium*

Piatigorski and collaborators have reported several interesting

findings. First they have found that RNA and protein synthesis may
be blocked without preventing cell elongation (59, 60), while
inhibition of elongation by colchicine did not prevent accumulation
of crystallin mRNA or δ crystallin synthesis (61). A decreased
proportion of δ crystallin synthesis was observed in lens epithelial
cells during development and serum starvation *in vitro*, but this
decrease was not associated with a reduction of the number of δ
crystallin mRNA sequences per cell (62). Serum stimulation of lens
cell division did not increase δ crystallin synthesis but promoted
the accumulation of δ crystallin mRNA (63). Furthermore, these
authors have shown that different elements of the differentiation
program of epithelium lens may be lost at different times during
embryogenesis. For instance, 6 days old chick embryo epithelial
lens cells grown in foetal calf serum supplemented medium elongated,
accumulated δ crystallin mRNAs and synthesized δ crystallins. In
the same medium the 19 days old epithelium divided rather than
elongated, accumulated δ crystallin mRNA but did not synthesize
a high relative proportion of δ peptide (63, 64).

At present it is not known whether the appearance and accumulation of crystallin mRNA is regulated at the level of transcription,
hnRNA processing in the nucleus, or cytoplasmic mRNA stabilization.
Hybridization experiments using a δ crystallin cDNA probe demonstrated an equal amount of δ crystallin mRNA in lens epithelium of
6 or 19 days old chicken embryos. However, δ crystallin synthesis
was higher in the former case (62). Cell-free translation of the
postpolysomal RNAs pool from one day old chicken lenses yielded
more β crystallin peptides than the polysomal fraction (64). These
experiments show a storage of specific messengers and a post-
transcriptional control. Such regulation of pre-existing mRNAs has
been described in other systems such as myeloid leukemic cells (65).
A reverse transcriptase activity has been found in the calf lens
(66) which could block crystallin mRNA translation. Alteration of

the ratio of intracellular Na^+/K^+ ions modify the synthesis of δ crystallin peptides in cultured cataracted mouse lenses (67). Also, the sensitivity to cap analogues (m7 GpppG) is different among the several lens mRNA molecules. For example, in calf lenses, γ crystallin mRNA is less sensitive than the β or α mRNA (68). In pathological states, some posttranslational processing defects could also take place, as in the experimental cataract induced by injection of bleomycin in newborn rats (69).

Different forms of iso accepting tRNVs have been reported during terminal differentiation. For instance a new tRNA^{Phe} has been described in the bovine cortical fibers which is not present in the epithelium (70). A correlation between presence of site specific hypomethylation of the crystallin genes in the embryonic lens DNA and δ crystallin gene expression has been found in the chicken (71).

4. CONCLUSION

The vertebrate eye lens offers an exciting model for investigating many biological problems. In this short article, I have not mentioned important research fields such as the study of the cytoskeleton and plasma membrane during lens cell differentiation, taxonomic studies of crystallin peptides and the evolutionary origin of the eye and lens, the many changes of crystallins occurring in lens diseases (cataract in particular) or during ageing, enzyme activities in lens etc. Excellent reviews on these topics have been published (12, 14, 72-75).

In vitro experiments offer excellent possibilities for analyzing tissue interactions or the effect of several factors (dexamethaxone, insulin, theophylline, cyclic AMP, growth factors, etc) on epithelial cells. However, one must be careful when dealing with differentiation of cells in culture. Lens epithelial cells are already differentiated cells with precise properties and the entry to their commitment pathway to form fibers results in the appearance of many new characteristics. The shift from one compartment to another needs analysis of both compartments. In that direction, a double stranded endonuclease activity which appears in fiber cells in 11 days old chick embryos could help to define these compartments (75).

Our understanding of lens development is progressing with growing experimental data on the embryogenesis and/or differentiation at the physiological or molecular biology levels. It would seem that many controls must operate since morphological and biochemical parameters seem to change in an unrelated manner during lens fiber differentiation. Concerning the specific problem of crystallin synthesis and regulation it would be necessary to carry out detailed studies of their messenger RNAs as well as of the whole protein synthesizing machinery. Modifications of some elements of this machinery, as rRNA modifications found by Vandenhaute (76) in ageing mouse liver cells, could disturb the accuracy and/or efficiency of the protein synthesis process.

5. ACKNOWLEDGEMENTS

The author would like to thank Dr. Bernadette Herve for valuable discussions.

6. REFERENCES

1) KEFALIDES, M.A. (1971). Structure and biosynthesis of basement membranes. Int. Rev. Exp. Pathol., 10, 1.

2) APPLEBY, D.W. and MODAK, S.P. (1977). DNA degradation in terminally differentiating lens fiber cells from chick embryos. Proc. Natl. Acad. Sci. USA, 74, 5579.

3) ORTWERTH, B.J. and BYRNES, R.J. (1972). Further studies on
    the purification and properties of a ribonuclease inhibitor
    from lens cortex. Exp. Eye Res., 14, 114.

4) VAN DER BROEK, W.G.M., KOOPMANS, M.A.G. and BLOEMENDAL, H.
    (1974). Electrophoretic identity and rapid detection on
    gel slices of the alkaline RNAse inhibitor from several organs.
    Mol. Biol. Rep., 1, 295.

5) RAFFERTY, N.S. and RAFFERTY, K.A. (1981). Cell population
    kinetics of the mouse lens epithelium. J. of Cell Physiol.,
    107, 309.

6) BLOEMENDAL, H. (1979). Lens proteins as markers of terminal
    differentiation. Ophthal. Res., 11, 243.

7) RAMAEKERS, F.C.S., SELTEN-VERSTEEGEN, A.M.E., BENEDETTI, E.L.,
    DUNIA, I. and BLOEMENDAL, H. (1980). *In vitro* synthesis of the
    major lens membrane protein. Proc. Natl. Acad. Sci. USA, 77,
    725.

8) BROEKHUYSE, R.M., KUHLMAN, E.D. and STOLS, A.L.H. (1976).
    Lens membranes. VII. MIP is an immunologically specific
    component of lens fiber membranes and is identical with 26K
    band protein. Exp. Eye Res., 23, 365.

9) KIBBELAAR, M.A. and BLOEMENDAL, H. (1979). Fractionation of
    the water insoluble proteins from calf lens. Exp. Eye Res.,
    29, 679.

10) BERZELIUS, J.J. (1830). Lärobok, Kicmen, Deel 2, 512.

11) MANSKI, W. and MALINOWSKI, K. (1978). The evolutionary
    sequence and quantities of different antigenic determinants
    of calf lens alpha crystallins. Immunochemistry, 15, 781.

12) HARDING, J.J. and DILLEY, K.J. (1976). Structural proteins of
    the mammalians lens: a review with emphasis on changes in
    development, ageing and cataract. Exp. Eye Res., 22, 1.

13) PALMER, W.G. and PAPACONSTANTINOU, J. (1968). Biochemical
    properties of α crystallins during lens development. In:
    Symp. of Internat. Soc. Cell Biol. (ed. K.B. Warren), tome
    7, 165.

14) BLOEMENDAL, H. (1981). The lens proteins. In: Molecular and Cellular Biology of the Lens (ed. H. Bloemendal), p. 1, Wiley-Interscience Publication.

15) SLINGSBY, C. and CROFT, L.R. (1978). Structural studies on γ-crystallin fraction IV: A comparison of the cysteine-containing tryptic peptides with the corresponding amino-acid sequence of γ-crystallin fractionII. Exp. Eye Res., 26, 291.

16) BHAT, S.P. and PIATIGORSKY, J. (1979). Molecular cloning and partial characterization of δ-crystallin cDNA sequences in a bacterial plasmid. Proc. Natl. Acad. Sci. USA, 76, 3299.

17) DODEMONT, H.J., ANDREOLI, P.M., MOORMANN, R.J.M., RAMAEKERS, F.C.S., SCHOENMAKERS, J.G.S. and BLOEMENDAL, H. (1981). Molecular cloning of mRNA sequences encoding rat lens crystallins. Proc. Natl. Acad. Sci. USA, 78, 5320.

18) McAVOY, J.W. (1978). Cell division, cell elongation and distribution of α-, β- and γ-crystallins in the rat lens. J. Embryol. Exp. Morphol., 44, 149.

19) ZWAAN, J. and IKEDA, A. (1968). Macromolecular events during differentiation of the chicken lens. Exp. Eye Res., 7, 301.

20) McDEVITT, O.S. and BRAHMA, S.K. (1981). Ontogeny and localization of the α, β and γ-crystallins in newt eye lens development. Dev. Biol., 84, 449.

21) McAVOY, J.W. (1980). Induction of the eye lens. Differentiation, 17, 137.

22) KONYUKHOU, B.V., MALININA, N.A., PLATONOV, E.S. and YAKOVLEV, M.J. (1978). Immunohistochemical study of crystallin synthesis in mouse lens morphogenesis. Biol. Bull. Acad. Sci. USSR, 5, 397.

23) COULOMBRE, A.J. and COULOMBRE, J.L. (1963). Lens development fibre elongation and lens orientation. Science, 142, 1489.

24) GENIS-GALVALEZ, J.M., SANTOS-GUTIERREZ, L. and RIOS-GONZALES, A. (1967). Causal factors in corneal development: an

experimental analysis in the chick embryo. Exp. Eye Res.,
6, 48.

25) SPEMAN, M. (1901). Über Korrelationen in der Entwicklung des Auges. Verh. Anat. Ges., 15, 61.

26) McAVOY, J.W. (1980). β- and γ-crystallin synthesis in rat lens epithelium explanted with neural retina. Differentation, 17, 85.

27) VAN DER STARRE, H. (1977). Biochemical investigation of lens induction *in vitro*. II. Demonstration of the induction substance. Acta Morphol. Neerl. Scand., 16, 109.

28) HUNT, H.H. (1961). A study of the fine structure of the optic vesicle and lens placode of the chick embryo during induction. Dev. Biol., 3, 175.

29) BEEBE, D.C., FEAGANS, D.E. and JEBENS, H.A.H. (1980). Lentropin: a factor in vitreous humor which promotes lens fiber cell differentiation. Proc. Natl. Acad. Sci. USA, 77, 490.

30) McAVOY, J.W. (1980). Cytoplasmic processes interconnect lens placode and optic vesicle during eye morphogenesis. Exp. Eye Res., 31, 527.

31) KARKINEN-JAASKELAINEN, M. (1978). Permissive and directive interactions in lens induction. J. Embryol. Exp. Morphol., 44, 167.

32) BARABANOV, V.M. and FEDTSOVA, N.G. (1982). The distribution of lens differentiation capacity in the head ectoderm of chick embryos. Differentiation, 21, 183.

33) OKADA, T.S., EGUCHI, G. and TAKEICHI, M. (1973). The retention of differentiated properties of lens epithelial cells in clonal cell culture. Dev. Biol., 34, 321

34) DE POMERAI, D.I., PRITCHARD, D.J. and CLAYTON, R.M. (1977). Biochemical and immunological studies of lentoid formation in cultures of embryonic chick neural retina of day-old chick lens epithelium. Dev. Biol., 60, 416.

35) HAMADA, Y. and OKADA, T.S. (1977). The differentiating ability
     of rat lens epithelium cells in cell culture. Dev. Growth and
     Differ., 19, 265.
36) RUSSEL, P., FUKUI, H.N., TSUNEMATSU, Y., HUANG, F.L. and
     KINOSHITA, J.M. (1977). Tissue culture of lens epithelial
     cells from normal and Nakano mice. Invest. Ophtalmol. Visual.
     Sci., 16, 243.
37) DE POMERAI, D.W., CLAYTON, R.M. and PRITCHARD, D.J. (1978).
     Delta crystallin accumulation in chick lens epithelial
     cultures: dependence on age and genotype. Exp. Eye Res., 27,
     365.
38) YAMADA, T. (1977). Control mechanisms in cell-type conversion
     in newt lens regeneration. In: Monographs in Developmental
     Biology (ed. A. Wolsky), Vol. 13, Karger, Basel, New York.
39) CAMPBELL, J.C. (1965). An immunofluorescent study of lens
     regeneration in larval Xenopus laevis. J. Embryol. Exp.
     Morphol., 13, 171.
40) WILLIAMS, L.A. (1970). The effect of a normal lens on lens
     regeneration in Notophthalmus viridescens viridescens. Am.
     Zool., 10, 322.
41) EGUCHI, G., ABE, S. and WATANABE, K. (1974). Differentiation
     of lens like structures from newt iris epithelial cells in
     vitro. Proc. Natl. Acad. Sci. USA, 71, 5052.
42) OKADA, T.S., ITOH, Y., WATANABE, K. and EGUCHI, G. (1975).
     Differentiation of lens in culture of neural retinal cells
     of chick embryos. Dev. Biol., 45, 318.
43) MASASUKE, A. and OKADA, T.S. (1978). Effects of culture
     media on the "foreign" differentiation of lens and pigment
     cells from neural retina in vitro. Dev. Growth and Differ.,
     20, 71.
44) OKADA, T.S., YASUDA, K., HAYASHI, M., HAMADA, Y. and
     EGUCHI, G. (1977). Lens differentiation in cultures of
     neural retina cells of human foetuses. Dev. Biol., 60, 305.

45) EGUCHI, G. (1976). "Transdifferentiation" in vertebrate cells
     in cell culture. In: Embryogenesis in mammals. Ciba Foundation
     Symposium, 40, 241.

46) EGUCHI, G. and OKADA, T.S. (1973). Differentiation of lens
     tissue from the progeny of chick retinal pigment cells cultured
     in vitro: a demonstration of a switch of cell type in clonal
     cell culture. Proc. Natl. Acad. Sci. USA, 70, 1495.

47) CLAYTON, R.M., THOMSON, I. and DE POMERAI, D.I. (1979).
     Relationship between crystallin mRNA expression in retina
     cells and their capacity to re-differentiate into lens
     cells. Nature, 282, 628.

48) GLASSER, D., IWIG, M., ANSORGE, S. and FISCHER, C. (1976).
     Investigations of the differentiation of bovine lens epithe-
     lial cells in vitro. Ophthal. Res., 8, 283.

49) CREIGHTON, M.D., MOUSA, G.Y. and TREVITHICK, J.R. (1976).
     Differentiation of rat lens epithelial cells in tissue
     culture. (I) Effects of cell density, medium and embryonic
     age. Differentiation, 6, 155.

50) MILLER, G.G., BLAIR, D.G., HUNTER, E., MOUSA, G.Y. and
     TREVITHICK, J.R. (1979). Differentiation of rat lens epi-
     thelial cells in tissue culture (III). Functions in vitro of
     a transformed rat lens epithelial cell line. Develop.
     Growth and Differ., 21, 19.

51) WEINSTEIN, B.I., SCHWARTZ, J., LONIAL, H., MARCIA OCHOA
     DOMINGUEZ, GORDON, G., HOCHSTADT, J., SOUTHERN, D.B.,
     DUN, M. and SOUTHERN, L. (1982). Normal and conditionally
     transformed bovine lens epithelial cell lines containing
     alpha and gamma crystallins. Exp. Eye. Res., 34, 71.

52) VERMORKEN, A.J.M. and BLOEMENDAL, H. (1978). α-crystallin
     polypeptides as markers of lens cell differentiation.
     Nature, 271, 779.

53) VERMORKEN, A.J.M., GROENEVELD, A.A., HILDERINK, J.M.H.C., DE WAAL, R. and BLOEMENDAL, H. (1977). Dedifferentiation of lens epithelial cells in tissue culture. Mol. Biol. Reports, 3, 371.

54) SIMMONNEAU, L., HERVE, B., JAZQUEMIN, E. and COURTOIS, Y. (1982). State of differentitation of bovine epithelial lens cells *in vitro*. Modulation of the synthesis and of the poly-merisation of specific proteins (crystallins) and non specific proteins in relation to cell divisions, submitted for publication.

55) ARRUTI, C. and COURTOIS, Y. (1978). Morphological changes and growth stimulation of bovine epithelial lens cells by a retinal extract *in vitro*. Exp. Cell Res., 117, 283.

56) SIMMONNEAU, L., HERVE, B., JACQUEMIN, E. and COURTOIS, Y. (1982). State of differentiation of bovine epithelial lens cells *in vitro*. Relationship between the variation of the cell shape and the synthesis of crystallins, submitted for publication.

57) COURTOIS, Y., SIMONNEAU, L., TASSIN, J., LAURENT, M.V. and MALAISE, E. (1978). Spontanous transformation of bovine lens epithelial cells. Kinetics analysis and differ-entiation in monolayers and in nude mice. Differentiation, 10, 23.

58) LAURENT, M., KERN, P., COURTOIS, Y. and REGNAULT, F. (1981). Synthesis of types I, III and IV collagen by bovine lens epithelial cells in long term cultures. Exp. Cell Research, 134, 23.

59) CRAIG, S.P. and PIATIGORSKY, J. (1973). Cell elongation and δ-crystallin synthesis without RNA synthesis in cultured early embryonic chick lens epithelium. Biochem. Biophys. Acta, 299, 642.

60) PIATIGORSKY, J., WEBSTER, H. de F. and and CRAIG, S.P. (1972). Protein synthesis and ultrastructure during the formation of embryonic chick lens fibers *in vivo* and *in vitro*. Dev. Biol., 27, 176.

61) PIATIGORSKY, J., WEBSTER, H. de F. and WOLLBERG, M. (1972). Cell elongation in the cultured embryonic chick lens epithelium with and without protein synthesis. J. Cell Biol., 55, 82.

62) BEEBE, D.C. and PIATIGORSKY, J. (1977). The control of δ-crystallin gene expression during lens cell development: dissociation of cell elongation, cell division, δ-crystallin synthesis, and δ-crystallin mRNA accumulation. Dev. Biol., 59, 174.

63) MILSTONE, L.M., ZELENKA, P. and PIATIGORSKY, J. (1976). δ-crystallin mRNA in chick lens cell: mRNA accumulates during differential stimulation of δ-crystallin synthesis in cultured cells. Dev. Biol., 48, 197.

64) THOMSON, I., WILKINSON, C.E., JACKSON, J.F., DE POMERAI, D.I., CLAYTON, R.M., TRUMAN, D.E.S. and WILLIAMSON; R. (1978). Isolation and cell free translation of chick lens crystallin mRNA during normal development and transdifferentiation of neural retina. Dev. Biol., 65, 372.

65) HOFFMAN-LIEBERMAN, B., LIEBERMAN, D. and SACHS, L. (1981). Control mechanisms regulating gene expression during normal differentiation of myeloid leukemic cells: differentiation defective mutants blocked in mRNA production and mRNA translation.

66) CHEN, J.H. and SPECTOR, A. (1974). The presence of a poly-adenylic acid-dependent DNA polymerase in calf lens. Exp. Eye Res., 19, 389.

67) SHINOHARA, T. and PIATIGORSKY, J. (1980). Persistance of crystallin messenger RNAs with reduced translation in hereditary cataract in mice. Science, 210, 914.

68) ASSELBERGS, F.A.M., PETERS, W.H.M., VAN VENROOIJ, W.J.
    and BLOEMENDAL, H. (1978). Inhibition of translation of lens
    mRNAs in a messenger dependent reticulocyte lysate by cap
    analogues. Biochem., Biophys. Acta, 520, 577.

69) WEILL, J.C., LECAT, G., VINCENT, A., CIVELLI, O. and
    POULIQUEN, Y. (1980). Messenger RNAs for cataractous lens
    are also present on normal lens polyribosomes. Eur. J.
    Biochem., 11, 593.

70) ORTWERTH, B.J., HORWITZ, J. and CHU-DER, O.M.Y. (1982).
    The induction of tRNA^{phe} in mammalian lens cortex: a possible
    control point in the synthesis of a α-crystallin. Exp. Eye Res.,
    34, 767.

71) JONES, R.E., DE FEO, D. and PIATIGORSKY, J. (1981). Transcrip-
    tion and site-specific hypomethylation of the δ-crystallin
    genes in the embryonic chickens lens. The Journal of Biological
    Chemistry, 256, 8172.

72) CLAYTON, R.M. (1978). Divergence and convergence in lens cell
    differentiation: regulation of the formation and specific
    content of lens fibre cells. In: Stem cells and Tissue Homeo-
    stasis (eds, B.I. Lord C.S. Potten and R.J. Cole), p. 115,
    Cambridge University Press.

73) BLOEMENDAL, H. (1977). The vertebrate eye lens. A useful
    system for the study of fundamental biological process on a
    molecular level. Science, 197, 127.

74) PIATIGORSKY, J. (1981). Lens differentiation in vertebrates.
    A review of cellular and molecular features. Differentiation,
    19, 134.

75) HERVE, B., JACQUEMIN, E. and LESCURE, B. (1982). Endogenous
    nuclease activity in chick embryo lens cells. Cell Diff., in
    press.

76) VANDENHAUTE, J. (1981). Influence de l'âge sur les caractéri-
    stiques structurales et les propriétés fonctionelles de la
    machinerie de synthèse protéique dans le fois de souris. Thesis.
    Facultés Universitaires Notre Dame de la Paix (Namur-Belgique).

THE NEOPLASTIC CELL AND ITS ANALYSIS BY CELL HYBRIDIZATION:
1. THE NATURE OF THE TRANSFORMATION PROCESS AND ITS MARKERS
2. ANALYSIS OF TRANSFORMATION BY CELL HYBRIDIZATION

L.M. Franks

Imperial Cancer Research Fund
Lincoln's Inn Fields
London, WC2A 3PX
U.K.

1. THE NATURE OF THE TRANSFORMATION PROCESS AND ITS MARKERS

There is no absolute marker for neoplastic transformation, particularly for cells *in vitro*. To understand the reasons for this it is necessary to understand the series of changes which takes place during the process of neoplasia (see 1 for extensive general review).

(i) *Stages in carcinogenesis*

Carcinogenesis is a multistage process. The application of a cancer producing agent (carcinogen) does not lead to the immediate production of a tumour. There are a series of changes after the initial initiation step induced by a carcinogen. The subsequent stages - tumour promotion - may be produced by the carcinogen or by other substances (promoting agents) which do not themselves produce tumours. Initiation, which is the primary and essential step in the process, is very rapid but once the initial change has taken place the initiated cells may persist for a considerable time, perhaps the lifespan of the individual. The most likely site for the primary event is in the genetic material (DNA),

141

although there are other possibilities. The carcinogen is thought
to damage or destroy specific genes probably in the DNA of cells
from the stem population of the tissue involved (see 2 for review).

"Initiated" cells remain latent until acted upon by promoting
agents. Many of these transformed cells or their progeny may not
grow at all or grow very slowly. It is at the post-initiation
stage that the influence of growth appears. Promoting agents are
not carcinogenic in themselves but they do induce cell proliferation. Many other substances will induce cell growth, so that
although cell growth is necessary for tumour development there
must also be other factors involved. The suggestion is that promoting agents may interfere with the process of differentiation
which normally takes place when cells move from the dividing stem
cell population into functioning and usually non-dividing cells.
Even though the cells are being acted on by growth promoting
agents, the cells may still be sensitive to the normal growth
inhibiting factors in the body so that the final outcome and the
rate at which changes occur depends on the balance between the host
factors and the extent of the changes which have been induced in
the initiated cells. This explains why morphologically preneoplastic or apparently fully transformed cells can be found in high
frequency but which do not appear to be growing and sometimes even
regress (3).

Another major and unexplained area is concerned with the time
scale of carcinogenesis. The latent period between initiation and
the appearance of tumours is one of the least understood aspects
of tumour development. After exposure to industrial carcinogens
it may take over 20 years before tumours develop. Even in animals
given massive doses it may take up to a quarter or more of the

total lifespan before tumours appear. Yet another unexplained fact is that only a very small number of cells "initiated" by a carcinogen will eventually produce tumours - perhaps only one or two from many millions of treated cells.

Most chemical and physical cancer inducing agents are very highly reactive and when they react with DNA in the affected cell they usually damage many other sites as well as the relatively few which are thought to control neoplastic transformation. Thus, the same agent may produce tumours in a given organ which differ greatly from each other, depending on the specific genes which have been altered or lost. At one extreme if only the "transforming" sites have been altered the resulting tumour cells will still retain much of the normal differentiated structure and function of the cell from which they have arisen. In the skin for example it will still resemble a skin cell and may still produce normal skin products and will still be responsive to some normal growth controlling factors. If the genes responsible for normal structure are more severely damaged, the resulting tumour cells have fewer normal properties; at the other extreme the cells may have lost almost all the normal properties of the cell from which it has arisen. The loss of normal characteristics is known as dedifferentiation or anaplasia. The pathologists can grade tumours by making an approximate assessment of the degree of structural dedifferentiation by examining sections of tumours under the microscope. As a rule, there is an approximate correlation between the tumour grade and growth rate. The pathologist's assessment of tumour grade is based only on alterations in structure and these are not invariably related to changes in function. Some cells may have lost their structural characters but still retain differentiated biochemical characters and others may still appear structurally differentiated but have lost any normal functional attributes.

(ii) *Tumour heterogeneity*

Another practical problem in the assessment of tumours is
that tumours are not homogeneous and some may contain areas with
more than one tumour grade. It used to be thought that tumours
arose from a single altered cell, i.e. were clonal in origin, but
there is now some doubt about this. But even if it were true
there is no doubt that by the time a tumour is detectable clini-
cally, whether it has arisen from one or many cells, it has been
present for a long time and the cells have had to go through a
large number of cell division so that the opportunity for variation and selection of different cell populations has occurred.
A tumour about 0.5 cm in diameter, which is just detectable may
contain over 500 million cells. The developed tumour usually
consists of a mixed population of cells, which may differ in
structure, function, growth potential, resistance to drugs or X-
rays and ability to invade and metastasize. Many of these characters may not be stable and may be influenced by the host response
or by treatment. An obvious example is the destruction of X-ray
sensitive cells by X-ray treatment. If the tumour also contains
X-ray resistant cells the cancer cells which are left after
treatment will be X-ray resistant. Any individual character
may vary independently.

So that when we are considering transformed cells *in vitro*
we must decide whether we are trying to identify transformed cells
in the early stages i.e. in the transition from normal to fully
transformed, or whether we are looking for alterations in behaviour of fully transformed cells. In the early stages of carcinogenesis where the transformation frequency is low we may have one
transformed cell in a population of 10^7 or 10^8 unaltered cells.
These transformed cells may not be growing at all. At the other
end of the scale we have mass populations of fully transformed
rapidly growing cells.

(iii) *Normal and tumour cells in culture*

Another complication is the degree of selection which occurs when normal or tumour cells are put into culture (see 4-6 for discussion). In explants of normal tissue most of the differentiated cells die out and the cultures are repopulated by a stem cell population within the first few days. Tumour tissue behaves in the same way. It is probably true to say that although some normal and tumour cells can be maintained in primary cultures for some months no truly normal cells will survive as transferable cell lines. Cells derived from normal human tissues invariably die out after a variable number of population doublings but a number of cell lines have been established from rodent mesenchymal and some epithelial cells. Almost all of these are chromosomally abnormal and many are tumour producing in the right conditions. Only a small proportion of tumours can be established in culture - about one in 20 in our hands. The stages in the establishment of tumour lines have been described elsewhere (6) but the significant point is that there is selection *in vitro* from a heterogeneous starting population. One human bladder tumour line which we have studied had several different clonal populations early in culture but in later cultures were uniform (7). The major selection pressures are concerned with the ability of cells to survive in the artificial, standardized conditions of substrate, nutrition gas phase etc. The only cells which will survive *in vitro* are those which can adapt to these conditions so that cells in culture may have more in common with each other than with the tissues from which they were derived. The practical importance is that these populations of tumour and "normal" cells are highly selected and in behaviour and metabolism may be very far removed from the situation in the body.

Bearing these caveats in mind, I shall consider some of the markers which occur in the early stages of transformation and

those which can be looked for in the later stages, in mass populations. There are many reviews on markers for transformation
(eg. 8-10). I have discussed the ultrastructural and other changes
which occur in some detail elsewhere (4, 11). Here I shall only
comment on a few points of special interest in the early and later
stages of transformation. Any distinction between early and late
in a continuing process must be arbitrary and the particular
character I have used is concerned with cell adhesion. In the
early stages cells cannot easily be transferred. In the later stages
successful cell transfer is possible.

(iv) *Early changes in transformation*

The systems which we have established to study the early
stages use rodent salivary gland (12, 13) and bladder (14). In
the salivary gland we treat cells with a carcinogen for 24 hours
very early in the culture period on day 3-4, when we expected the
cells to still be normal, although we now know that this is probably not so. The epithelial cells are also in their normal
relationship to their stroma, so that if the stroma plays a role
in carcinogenesis, it is able to do so in this system. The cultures
pass through four well-defined stages before lines of tumour
producing cells can be established. In Stage I (10-30 days) there
is an outgrowth of epithelium, in Stage II (30-70 days) ductal
differentiation occurs in some epithelium in Stage III (70-100
days) small slowly proliferating foci develop either from the ducts
or from flat epithelial areas. In Stage IV (100-200 days) the
proliferation rate in a very few foci increases and the cells become
more irregular. The cells cannot be transferred easily until about
150 days, after which time they are tumour producing in syngeneic
mice. The tumours are epithelial and similar in morphology to
salivary gland tumours which occur in man. The biological behaviour
and time scale suggests that a process of initiation and promotion
similar to that found *in vivo* is taking place. Using this system
we can look at the early and late stages more clearly.

One of the earliest recognisable changes is the alteration
in morphology of small groups of cells. The standard pathologist's
criteria for the diagnosis of neoplasia applies equally well to
cell culture - increased cellularity, loss of polarity, pleomorphism,
lack of cohesion of cells, variation in shape and size and increased
staining intensity of nuclei, increased mititic activity and
abnormal mitosis. Unfortunately, as *in vivo*, none of these criteria
are absolute. Some are due to tissue culture conditions. In the
early stages although the cells are morphologically "neoplastic"
they do not as a rule behave as fully transformed cells and in
particular they do not grow rapidly. They correspond to the carcinomas *in situ* found in man in the cervix for example, where
apparently transformed cells are not growing and frequently regress. For those who do not possess the visual discrimination of
the experienced pathologist, many of the morphological features
can be measured, although the results are usually less accurate.
Machines are available which will measure cell size or total DNA.
Increased DNA content is a common feature of transformed cells.
We have a microfluorimeter (15) which measures DNA content in
whole single cells *in situ*. This is particularly useful in
studying altered foci since the growth rate at this stage is very
low and it is difficult to obtain a sufficient number of satisfactory chromosome preparations. Chromosomal changes during neoplasia
have been reviewed by Cowell (16) and others.

Light microscope morphology is usually easier to interpret
than transmission EM except for the detection of abnormal mitochondria which are a common feature of transformed cells (4 and
others). Reduced uptake and retention of a mitochondria specific
membrane potential probe, Rhodamine 123, has been reported in a
virus transformed cell line (17). It may be possible to detect
small numbers of transformed cells *in situ* using this probe. Scanning
EM is useful to demonstrate alterations in the pattern and form of

microvilli, often found in transformed cells but as with other
markers these changes are not absolute. Alterations in surface
proteins and enzymes have been described but are not consistent (4).

A particular character of transformed epithelial cells, in
the early stages of the process is that like their normal counterparts they are very firmly adherent to the substrate and cannot
be successfully transferred. This must reflect changes in cell
surface and sub-membrane proteins such as laminin, fibronectin,
vinculin etc., it should be possible to study these changes in
single transformed cell.

(v) *Later changes during in vitro transformation*

The major feature which indicates a transition from the
"early" stages is the ability of cells to grow after transfer.
This does not imply that the cells are necessarily tumour producing - a number of so-called preneoplastic cell lines have been
established from salivary gland foci (18), skin (19), bladder (20),
liver (21) and trachea (22), but it seems to be a necessary preliminary to the development of fully transformed cells. The ability
to survive after transfer is usually associated with an increase
in growth rate and the appearance of some or all the other characters listed in Table 1. These are the characters used as markers
for malignancy in established tumour cell lines but again none are
absolute. Some of the changes in the early stages of course persist,
although some may change in character, eg. changes in cell surface
responsible for reduced adhesion. These changes are responsible
for results obtained in some tests for transformation such as
alterations in permeability or aggluinability either spontaneously
or after treatment with lectins (8, 9). Other structural and
functional changes are discussed elsewhere (4). Here I shall discuss some of the commonly used markers.

Since most of the commonly used markers for transformation *in vitro* involve measurement of growth this is the first stage at which they can be used. How accurately do these markers reflect neoplastic transformation in epithelium? Most of the markers have been established using mesenchymal cells, usually transformed by viruses. Marshall *et al.* (23) looked at seven different markers including density dependent inhibition of growth, fibrinolysin production, growth in agar, presence of LETS protein (fibronectin) and growth in nude mice, of six different human tumour cell lines all shown not to be HeLa cells. The only consistent markers were lack of density dependent inhibition of growth and absence of LETS protein. In a more detailed study of LETS protein as well as other characters, including microfilament bundle distribution and growth in agar (24, 25) it was found that in hybrids between normal rat embryo mesenchymal cells and cells from a mouse mammary gland epithelial tumour there was a gradual transition in properties between tumour producing and non-tumour producing hybrids. The non-tumour producing hybrids had high levels of LETS protein, a normal microfilament distribution and did not grow in agar suspension, features which were absent from the tumour producing hybrids. As well as cells at the two extremes there were hybrids in which all the changes appeared to be quantitative rather than all or none effects. A further complication, when dealing with epithelial systems is that the normal distribution of LETS protein is not known in any detail (26). In the hybrids used by Marshall and Dave (25) the normal rat embryo mesenchymal cell gene products could be regarded as the source of the LETS protein. Wigley and Summerhayes (27) showed that although most normal and preneoplastic bladder and salivary gland cells, and tumours from cells transformed *in vivo* by direct implantation of a carcinogen, or transformed *in vitro*, did not have significant LETS protein, two transformed bladder cell lines did have appreciable amounts.

Growth in agar is another commonly used marker for transformation, established for mesenchymal cell systems. Marshall *et*
al. (24) showed that in epithelial/mesenchymal tumour hybrids
there was a fair degree of correlation between colony forming
efficiency in agar and malignancy. A high colony forming efficiency was associated with tumour production but low efficiency
did not exclude tumour development. One hybrid line with a very
low colony forming efficiency did produce tumours but only after
a long latent period. With human epithelial tumours (23) the
correlation was even less marked. High colony forming efficiency
did not always correlate with tumorigenicity in nude mice. One
line which did not grow in agar was tumorigenic and two which
grew in agar were not tumorigenic. Knowles and Franks (28) looked
at this type of correlation in more detail using cell lines from
adult mouse salivary gland transformed *in vitro*. We used five
different cell lines. All were tumour producing and all grew in
agar but the colony forming efficiency of the parent lines varied
from 0.1% to 4.4%. Agar clones from two separate lines also varied
in efficiency (0.7-11%), showing that the cells of the primary
tumours were heterogeneous in this particular character. We also
confirmed Marshall's observations that the ability of epithelial
cells to grow in agar, as with other markers, parallels transformation only in some closely controlled conditions, and with some
tumours. Another marker is multinucleation induced by *Cytochalasin*
B. Although some consider this a useful marker for transformation
in epithelial and virus transformed cells (eg. 29, 30) others do
not (31).

(iv) *The production of new and altered proteins by tumour cells*

Many of the markers I have discussed are of course mediated
by proteins so that direct investigations of proteins in normal and
tumour cells is logical (see also other articles in this volume).

These fall into three groups: (1) expression of normal proteins, perhaps in abnormal amounts or situations; (2) expression of proteins not normally present in the cell of origin;(3) new tumour specific proteins. Those in the first group are the biochemical expression of structural changes eg. alterations in cytoskeletal elements, cell surface proteins, loss of specific junctional complexes and so on. The second group include the production of foetal proteins, eg. CEA, or of ectopic hormones. These groups both reflect disturbances in differentiation. The third group is more interesting since this group contains true tumour markers. Some may be detected immunologically as tumour specific antigens (eg. 32 and many others), or biochemically using one or two dimension gel electrophoresis (33-36). Although two dimensional gels offer greater resolution we have found that a computer assisted analysis of one dimensional gels can identify single band differences between clones of one human tumour cell line. Recently Harris and his colleagues (37) described a glycoprotein, the Ca antigen, recognisable by a monoclonal antibody, which was present in many malignant tumours (38) and in tumour producing hybrid cells. This should be of value in diagnosis. Another important group of proteins are the DNA associated proteins since it is likely that these control gene expression. A number of workers have now reported specific differences in DNA associated proteins in transformed cells (39-41). Until we can detect directly minor alterations in the genome which are responsible for neoplastic transformation the expression of tumour specific proteins are likely to be the most reliable marker available.

(vii) *Conclusions*

   Neoplastic transformation is a multistage process and detectable alterations in structure and behaviour do not appear until a very late stage. Although there are many differences between normal

and tumour cells, most are probably the result of random damage
to the genome during the process of tumour initiation and development and not directly related to the changes responsible for
"malignancy". The random changes are probably responsible for the
variations in structure and behaviour between different tumours
or between cell populations in the same tumour. Most of the commonly
used *in vitro* tests for transformation can only be applied to mass
populations of rapidly growing cells and cannot be used to study
the early stages of the process when only a small number of non-
proliferating cells may be involved. Most of these tests have been
established using mesenchymal cells but the majority of tumours
are derived from epithelial cells. Cell lines can be established
from only a small number of tumours and there are few, if any,
normal cell lines which can be used as controls. Many so-called
"normal" cell lines are partially transformed and some (eg. 3T3)
are already tumour producing in the right conditions. Cells which
can be established *in vitro* and used for investigations on neoplasia
are therefore highly selected. The common features shared by those
tumour cells which can be maintained in culture may reflect their
capacity to adapt to the *in vitro* environment rather than their
"malignancy". Even growth *in vivo* is not an infallible test for
transformation since there are many conditions in which undoubted
tumour cells will not grow, even in the individual in which the
tumour has arisen. Until we have methods for identifying minor
departures from the normal in the genome we must depend on alterations in complex patterns of cell behaviour, most of which may be
random accompaniments of the transformation process. The random
nature of these changes also explains the discordant results when
different tests for transformation are applied to the same tumour
cells. In the analysis of neoplastic transformation the techniques
of somatic cell genetics can be used to investigate the alterations
in gene expression which accompany the process, but there are many
problems which are discussed in the next section.

Table I. *Early and later changes accompanying transformation*

Early changes in transformation (cells not transferable)

1. Morphological changes
2. Increased DNA content
3. Chromosome changes
4. Substrate dependence
5. Alterations in surface proteins
6. Alterations in surface enzymes

Etc.

Later changes in transformation (cells transferable)

1. Alterations in growth rate
2. Reduced serum requirement
3. Reduced substrate dependence
4. Changes in cytoskeleton
5. Chromosome changes
6. Alterations in DNA associated proteins
7. Production of "new" proteins
8. Altered production of normal proteins
9. Multinucleation after *Cytochalasin B.*
10. Tumour production

Etc.

2. ANALYSIS OF TRANSFORMATION BY CELL HYBRIDIZATION

Research in many areas in biology produces confusing results but nowhere is the confusion greater than in the answers
to apparently simple questions in the field of transformation
and ageing studied by cell hybridization. I shall deal with each
separately but the principles involved are the same. On the face
of it the experiments are simple. For example, if the genes

responsible for tumour production are recessive it should be
possible to repress their activity by cell hybridization with
genes from normal cells. If the reverse is true i.e. the char-
acter(s) are dominant, hybridization should not suppress their
expression. Even a brief survey of published work (42-45) shows
the complexity of the results obtained. One school (46, 47) claims
suppression of "malignancy" by normal cells, a second claims lack
of suppression (42, 47) and a third (45) claims that both are
true. In this paper I shall make no attempt to review the whole
field but only try to consider some of the discrepancies.

(i) *Some divergent results*

     The following tables illustrate the results of some experi-
ments using tumour and non-tumour cells. Table 2 shows <u>selected</u>
results from papers by Harris and his group (48, 49) compared
with results reported by Kucherlapati and Shin (47). One source
of confusion concerns the cell line A9. Harris *et al.* (48) used
these cells as the low tumour parent (described in the abstract
as non-tumour but in fact tumour producing). Kucherlapati and
Shin (47) used cells with the same designation as a high tumour
line giving tumours in all mice inoculated with a dose as low
as 10^4 cells although tumours were produced with as few as 50
cells. The cells used by both groups were developed from a cell
line originally derived by Littlefield (50) from L cells. The
differences in behaviour can be explained by selection of variants
from a heterogeneous population or by the development of stable
variants so that this apparent anomaly, although confusing, is
explicable. The results reported by Harris *et al.* (48) suggest
that intraspecific hybridization of highly malignant Ehrlich
mammary tumour cells with cells of low malignancy, even if not
"normal", will suppress the expression of malignancy i.e. tumorigenicity. In a second series of experiments interspecific hybrids
were made with a second highly malignant mouse tumour - melanoma

Table 2. *Tumour suppression in hybrids*

Parental cells		Tumorigenicity		
Non tumour	tumour	Non tumour	Tumour	Hybrids
1 Mouse fibroblast (A9)	Mouse Ehrlich mammary	3/15	100%	5/50
1 Human fibroblast (MRC5; Strain 375)	Mouse melanoma (PG19)	0	100%	0/3;0/4;0/17
1 Human lymphocytes	Mouse melanoma	0	100%	2/3;1/3;1/4
2 Human fibroblast (DUB)	Mouse Sarcoma (A9)	0	100%	33/33
2 Human fibroblast (BP)	Mouse melanoma (PG19)	0	100%	13/14
2 Human fibroblast (GM1429)	Mouse melanoma (PG19)	0	100%	16/16

1. From Jonasson and Harris (49).
2. From Kucherlapati and Shin (47).

PG 19, and human diploid cells - two cell strains cultured from
human embryos (?) and peripheral blood lymphocytes from a normal
(?) male donor. These experiments showed that if 2×10^6 cells
were inoculated into *nude* mice, PG19 x human fibroblast hybrids
were non-tumorigenic and that hybrids between PG19 and diploid
human lymphocytes showed a reduced tumorigenicity. I shall consider other aspects of this experiment later. Kucherlapati and
Shin (47) repeated these experiments using A9 cells and PG19
cells as the mouse tumour lines and three lines of diploid human
fibroblasts, DUB, BP and GM1429. They isolated 19 different hybrid
clones and showed no suppression of tumorigenicity in any of their
hybrids, using 2×10^6 cells or in one group, 4×10^6 cells. There
are many other reports, some reporting suppression, some non-
suppression by hybridization of tumour and non-tumour cells (see
for example 42, 43 and 45 for reviews and references). The production of monoclonal antibodies using hybridomas is of course an
example of non-suppression, in use in very many laboratories.

Another group of confusing experiment concerns the effects
of hybridising tumour cells with other tumour cells. Harris (43)
states that in general a malignant cell fused with another malignant
cells remains malignant, but others (45) and some of our own
results described later show that this is not always the case. How
can these divergent results be explained?

(ii) *Factors influencing the expression of "malignancy" in hybrid*
     cells

Some of these are listed in Table 3 and clearly illustrate
the source of the confusion since many of the factors cannot be
clearly defined (11). Obviously some tumours are more "malignant"
than others. Some grow more rapidly and spread more readily than
others. Some are slow growing and may not spread at all. The
whole question of "malignancy" and methods for assessing it is

Table 3. *Factors influencing expression of malignancy*

Basic "malignancy" of tumour cells

    Cell number required to produce tumour

    Latent period

    Host reaction (variability of hosts)

    Nature of the transforming agent

Growth capacity of hybrids

Genetic stability of hybrids

    Selection of tumorigenic variants

    Random loss or retention of "suppressor" genes

    Gene dosage

Role of cytoplasm

Basic "suppressor" activity of non tumour cells

---

discussed earlier but it seems reasonable to suppose that less
"malignant" tumours may be more easily suppressed than more
malignant. It is worth remembering at this stage that any of the
cells which can be used for these experiments are very highly
selected as already discussed and almost all tumour cells which
can be maintained in culture are derived from highly malignant
tumours. However, there are two factors in tumorigenicity experiments which can be measured and which do give a guide to
malignancy - the minimum cell number necessary to produce a tumour,
and the latent period before a tumour can be detected. The third
factor in this particular equation - the host response - cannot
yet be measured but there is a considerable variation even between
animals in the same group eg. the one animal which fails to produce a tumour when all others in the group do, let alone variations between animal stocks in different laboratories. This
applies particularly to interspecies hybrids tested in *nude* or
immunosuppressed mice which may vary in the degree of immune

suppression, a situation further complicated by variations in the
degree of tumour-specific antigenicity in hybrid clones derived
from the same parental cells (51). The difficulties involved in
making direct comparisons is shown in Table 4 again comparing
results from Jonasson and Harris (49) and Kucherlapati and Shin
(47). One could postulate that the PG19 line in the latter laboratory is more malignant than that used by Harris's group since it
produces tumours more quickly and with a smaller cell inoculum but
there are discrepancies even here, eg. 3×10^2 cells produces
tumours more quickly than 3×10^3. A wise man should reserve
judgement at this stage. The only thing that is clear from this
table is that rather more hybrid cells are needed to produce a
tumour than are the parental cells, but the number of experiments
are small.

In one specific instance the nature of the transforming agent
which has induced the tumour is of importance. Experiments using
SV40 transformed cells can be expected to differ from those using
spontaneously or chemically induced tumours since the SV40 genome
is known to be incorporated in the tumour cell DNA and the ability
to form tumours appears to be dominant in these cells (52).

The growth capacity of the hybrids also influences the results obtained. The experimental procedure obviously selects for
hybrids which will grow *in vitro* but we know that many tumours
which grow rapidly *in vivo* will not grow *in vitro* and many cells,
even those derived from tumours, which grow *in vitro* will not
produce tumours after mouse inoculation. The factors involved are
not understood but it is likely that the ability of cultured cells
to grow in mice is not directly related to "malignancy",

The genetic stability of the hybrids is also of major importance. The general statement "hybrids were selected by cloning

Table 4. *Cell dose and latent period*

	Cell type	Cell number	Tumours	Latent period (days)
1	Mouse melanoma (PG19)	5×10^4	4/4	20 (18-24)
		5×10^5	5/5	
2	Mouse melanoma (PG19)	3×10^1	1/3	20
		3×10^2	2/3	13
		3×10^3	3/3	24
		3×10^4	2/3	8
		3×10^5	3/3	7
1	Mouse melanoma (PG19) x fibro-blasts	2×10^6 (3 clones)	0/24	-
		2×10^6 (4 clones)	12/12	14 (11-14)
2	Mouse melanoma (PG19) x fibro-blasts	3×10^5	2/2	28
		3×10^4	1/2	
		3×10^3	1/2	
		4×10^6 (5 clones)	13/14	5
		3×10^4	2/2	28
		3×10^3	1/2	

1. From Jonasson and Harris (49).
2. From Kucherlapati and Shin (47).

from a mass population after fusion" covers a very complex series
of events. Firstly, hybrids not able to grow at clonal densities
are eliminated. Chen (53) described changes in uncloned populations
of diploid human fibroblasts hybridised with heteroploid mouse
RAG cells. Heterokaryons containing complete genomes of both
parental cells randomly lost chromosomes from both species. The
majority of cells in early growth stages, however, still possessed
a nearly complete human genome. The rate of human chromosome loss
in subsequent growth periods was not uniform, being gradual in some
and rapid in others. The initially predominant 2n human-1s (1h:1m)
type was soon replaced by a less frequent 2n human 2s mouse (1h:2m)
type. Over an increased period of time in mass culture, the number
of stemlines decreased. One stemline, often a (1h:2m) type with
a greatly reduced human complement, outgrew the others and occupied
the entire culture. Therefore, the usual process of clonal isolation
may confer a negative selection bias against cell hybrids retaining
a large number of human chromosomes. Hybrid stemlines with stable
karyotype were established before 36 days after fusion had elapsed.
The stage in the process at which cloning is carried out may thus
influence the type of cell which is selected. Once selected, some
clones may remain stable but others may continue to lose chromosomes
and show changes in their behaviour. Stanbridge *et al.* (45) for
example found that most HeLa x human fibroblast hybrids were non-
tumorigenic but after extended culture, tumour producing segregants
appeared. This was a rare event, although Harris (43) believes
that it may occur with quite high frequency. It is difficult to
establish definite figures for the rate at which these changes
occur, since they are influenced by the specific cells involved.
In intraspecies hybrids most chromosomes are retained but in
interspecies hybrids there is usually rapid chromosomes loss,
mostly in one direction but not always. For example, in human x
mouse hybrids there is usually a rapid loss of human chromosomes
but the human fibrosarcoma HT1080 hybridised with mouse peritoneal

Table 5. *Factors influencing basic "supressor" activity*

Nature of the suppressor cell

Interspecific v intraspecific hybrids

Effects of tissue type

    Epithelium x mesenchyme
    Epithelium x epithelium
    Mesenchyme x mesenchyme

Tumour x tumour hybrids

---

macrophages, spleen cells or primary fibroblasts segregated mouse
chromosomes (42). The type of segregation is influenced by the
nature of the parental cells. This is discussed later. The general
pattern seems to be that where the chromosomes of the non-malignant
parent are retained preferentially the cells tend to be non tumorigenic; if these chromosomes are eliminated tumorigenicity is pre-
served, but again these findings are not absolute (43). In general,
no specific chromosomes or chromosome group have been found to be
associated with tumorigenicity or its suppression in rodent or
human tumour hybrids (54-56). It has been suggested that gene dosage
may play a role (57). Jonasson *et al.* (58) have suggested a relationship (see original paper for details) but their most interesting
finding was that tumour suppression occurred even when no human
chromosomes were detectable. They suggested that an extrachromoso-
mal element was responsible.

The role of the cytoplasm in tumour suppression has been investigated by Howell and Sager (59). They reported suppression or
reduction in tumorigenicity by fusing cytoplasm of non tumour cells
and intact tumour cells but the results are confusing and the numbers
are small. These experiments have been criticised by Croce (42)
and must be considered inconclusive (60).

Table 6. *Tumour suppression in hybrids*[1]

Cell type	Cell number			Latent period (range)
	5×10^4	5×10^5	5×10^6	
Mouse melanoma (PG19)	4/4	5/5	-	20 (18-24)
Mouse melanoma (PG19) x fibroblasts	0/1	-	0/3	
	0/3	-	0/4	
	0/3	-	0/17	
Mouse melanoma (PG19) x fibroblasts	-	-	5/40	50 (42-55)
	-	-	16/19	45 (28-64)
	-	-	4/5	58 (51-62)
Mouse melanoma (PG19) x lymphocytes	0/2	2/23	2/3	52 (50-54)
	0/4	4/14	1/3	52 (46-54)
	1/4	0/4	4/4	31 (28-40)

1. From Jonasson and Harris (49).

The last factor influencing the expression of "malignancy" in hybrids is the basic suppressor activity of the non-tumour parent.

(iii) *Factors influencing basic "suppressor" activity*

Although there is obvious disagreement on the incidence of tumour suppression in hybrids, there is no doubt that it often does occur. Some of the variability in the results lies in characters associated with the tumour cell but others are due to the non-tumour parent. Some of these factors are listed in Table 5.

The exact nature of the cell has a considerable effect. An important point is that cells with the same name are not always the same, as we have already seen with A9 cells. Even cells from the same primary culture may show considerable deviation during the establishment of the culture and particularly when transferred to different laboratories, even if the possibility of contamination with other cell types is excluded. All human fibroblasts are not the same. Some are established from adults, some from embryos and some from individuals with genetic defects. Although all arise from the same multipotential mesenchymal cell type, different cells may develop into stable lines with different differentiated characters (see 4 for review). The MRC 5 strain used by Jonasson and Harris (49) may have completely different characters from the DUB or PB strains used by Kucherlapati and Shin (47). The human fibroblasts are invariably non-tumour producing and eventually die out in culture, however rodent (especially mouse) mesenchymal cell lines almost invariably transform spontaneously. Many of the commonly used so called normal cell lines like 3T3 or BHK are already tumour-producing. Most show chromosomal abnormalities. It is probably true to say that there are no truly normal epithelial or mesenchymal tissue culture cell lines available (see 4 and 11 for discussion). This has an important bearing on the results.

Hybridisation with tissue culture cells may produce different
results from hybridisation with normal cells. Thus Jonasson and
Harris (49) reported complete suppression of PG19 melanoma crossed
with human fibroblast cell strains but only partial suppression
when crossed with human lymphocytes. Croce (61) hybridised human
HT1080 cells with normal mouse peritoneal macrophages, spleen
cells or primary fibroblast cultures. There was no tumour
suppression but the hybrids segregated mouse, not human chromosomes.
If the HT1080 cells were hybridised with phenotypically normal
mouse cell lines the hybrids segregated human chromosomes and
produced phenotypically normal and malignant cells. These examples
are sufficient to illustrate the complexity of the problems.

Both inter- and intra-specific hybrids have been produced
and there seems to be little doubt that the exact role of species
specificity in tumour suppression is not understood. Both suppression
and non-suppression have been reported with each (see 42 for review).

The effects of tissue type are another aspect which has not
been considered to any extent. Since most so called normal cell
lines are mesenchymal these "fibroblasts" have been used as the
normal parent in most tumour suppression experiments. These cells
are expressing genes which differ greatly from those expressed by
epithelial cells so that hybridisation across somatic cell type
barriers may interfere with mechanisms involving growth and differentiation rather than with the same tumour gene(s) themselves.
Peehl and Stanbridge (62) fused normal skin epithelium with HeLa
cells derived from an epithelial (but not squamous) carcinoma and
found that this induced squamous differentiation of the tumour but
not tumour suppression. In our own work (Cowell and Franks, in
preparation) fusion of an *in vitro* transformed mouse bladder
tumour cell line with normal bladder epithelium did not always
suppress tumorigenicity. Interestingly enough, fusion of the tumour

cells with adult mouse fibroblasts did not suppress tumorigenicity
but some tumours were mesenchymal in type and others intermediate
with some epithelial characters. In experiments in which bladder
tumour cells were fused with normal epithelium from other (i.e.
non bladder) organs, preliminary results suggest that there may
be complete suppression. This raises many problems to be discussed
elsewhere, but it seems likely that experiments using like
parents i.e. epithelium x epithelial tumour of the same tissue
and mesenchymal tumour x mesenchyme may give a better understanding
of the control of tumorigenicity without complicating factors
associated with the control of differentiation.

Tumour cells have also been fused with tumour cells in an
attempt to study complementation. Harris (43) suggested that if
a single gene is responsible for malignancy then all hybrids between two tumour cells should be malignant but if more than one
site was involved some hybrids would be non-tumorigenic. Stanbridge
et al. (45) found that hybrids between different carcinoma lines
(type not specified) or between carcinoma and lymphoma lines were
not suppressed but there was stable suppression in carcinoma x
sarcoma and carcinoma x melanoma crosses. They conclude that there
is a common genetic locus governing tumorigenicity in carcinoma
cells therefore complementation does not occur but different loci
govern tumorigenicity in sarcomas and melanomas. This is an
attractive hypothesis but unfortunately tumour x tumour crosses
using clones from a single human bladder carcinoma also produced
some non tumorigenic clones (64; Hastings and Franks, in preparation).
The tumour x tumour clones also illustrate the variability of the
whole process. Firstly, hybrids selected from the same parental
population have different morphological characters and show differences in malignancy as measured by tumour growth rate. Some grow
more slowly than the parents, some more rapidly (Cowell and Franks,
in preparation). These differences can be more easily explained

as a consequence of random loss of different characters.

(iv) *Cell ageing and hybridization*

Normal diploid cells from most species have a finite lifespan
in culture (65) and it has been suggested that the process is
similar to normal ageing, although this seems unlikely (66). After
a variable number of population doublings, most cultures die out
but in some species, especially the mouse, some cells survive,
transform spontaneously and develop into tumour producing transferable cell lines. A great deal of work has been done comparing
cells in the early (young) and late (old) stages of culture. In
hybridisation experiments young and old cells have been fused with
each other, with separated cytoplasm and nuclei or with continuous
(transformed) cell lines. The results show that fusion of human
diploid cells (67) at whatever stage or differentiated rhesus cells
(melanocytes or lymphocytes) (68) with continuous cell lines do
not suppress growth of the continuous line i.e. "the senescence
factor" is not dominant but since there is random chromosome loss
the experiments are not satisfactory. A more satisfactory experiment
is that of Norwood *et al.* (69) who found that the senescent phenotype was dominant when "young" and "old" cells were fused. Hayflick
and his colleagues (70, 71) showed that the dominant factor was
nuclear by hybridization of young and old cells with separated cytoplasm or nuclei. Young or old cells could not be made to proliferate
continuously by infection with SV40 virus or by fusion with lethally
irradiated, SV40 transformed or HeLa cells, although young cells
passed through a transient proliferation phase (72). This suggested
that an intact nucleus from a transformed cell line was necessary
for continuous proliferation although fusion of human diploid cells
with cytoplasm from continuous line cells did lead to a moderate
increase in DNA synthesis but no continuous cell lines (73).

(v) *The use of cells without selectable markers*

A major disadvantage of hybridisation experiments is that until very recently it was necessary to use cell lines with built-in genetic markers which can be used to select against homokaryons and unfused parental cells. This problem can now be bypassed by treating each parent with different irreversible biochemical inhibitors so that only heterokaryons will survive (74). This technique can also be used in combination, in fusions where one parent has a selectable marker, or in hybridisations using nuclear preparations, minicells and so on. There are still many technical problems to be solved. Whichever method is used the yield of hybrids is low, the number of hybrids capable of growing is even lower, and the retention of specific pieces of information is random. In spite of this, fusion techniques now allow us to ask some basic questions about the control of the expression of the transformed phenotype.

(vi) *A speculation*

Each cell may be expected to contain groups of active genes with specific functions and specific "strengths" (penetrance), together with inactive genes, some of which may be reactivated, given the appropriate stimulus, and some of which may be irreversibly switched off during differentiation. This latter group must include those responsible for determining somatic cell types since apart from the teratomas which are a special case there are no recorded examples (as far as I am aware) of cells crossing somatic cell barriers in postnatal life. The active genes control differentiated cell functions as well as the ordinary "housekeeping" functions. Another group must control the machinery necessary for cell growth and division. This group may remain active in some cells eg. stem cells, reactivatable if necessary in others and irreversibly switched off in terminally differentiated cells eg. neurons. The control of cell division is thus genetically controlled

by genes with different degrees of penetrance, depending on the
cell type. The transforming genes may be associated with this
group or may be independent. A carcinogen must activate the
transforming site, by definition, but may affect other groups at
random, as already discussed.

If these assumptions are correct, it should be possible to
predict the effects of cell hybridisation as function of the
balance between the degrees of penetrance of the different gene
groups retained by the hybrids. For example, fusion of tumour
cells with terminally differentiated cells, or with cells of
another somatic cell group would be expected to be more likely to
show tumour suppression than fusion with cells of the same cell
type. Unfortunately this simple picture is complicated by the
random loss of chromosomes, the unknown extent to which the
mechanisms which control gene expression may be altered by the
carcinogen and the practical problems of determining the exact
nature and state of differentiation of the non-tumour parental
cell.

(vii) *Conclusions*

The only rational conclusion to be drawn from cell hybridization experiments is that the variations in gene expression and
in phenotypic behaviour are due to random loss or retention of
genes or factors controlling gene expression. From the reported
experiments, general trends can be detected eg. non tumour, mesenchymal cells tend to suppress the expression of tumorigenicity in
heterotypic hybrids (i.e. when fused to a cell of different somatic
type) but not always. Tumour x tumour crosses also tend to be
suppressed if the parental cells are derived from different somatic
cells - but not always. An accurate analysis will only be possible
when single genes can be isolated. The principles involved in
studying phenotypic expression in ageing and in tumour cells are

the same as those involved in studying any other type of phenotypic expression. The problems which arise are due to the complexity
of the processed of ageing and of carcinogenesis and the difficulties involved in isolating specific characters which can be
identified and measured. A further complication is the fact that
almost all the techniques used must be applied to cells in culture.
Since only a very small proportion of tumour cells and very few,
if any, normal cells can be maintained in culture, experiments
must be done with highly selected, non-representative cell populations. The degree of selection is exaggerated in most experiments
by the need to use cells with selectable markers. As a consequence
of this, results must be interpreted with caution since they are
based on experiments with very marked limitations. Many of the far
reaching conclusions on the nature and control of ageing and neoplasia based on cell hybridization cannot be extrapolated directly
to the situation in the intact individual. The simplest explanation
of the results so far obtained is one based on the principle of
controlled random selection i.e. a random event occurs and the
selection pressures are exerted by the results hoped for or
expected.

3.   REFERENCES

1)   FOULDS, L. (1969). Neoplastic Development 1. Academic Press,
     London.

2)   CAIRNS, J. (1975). Mutation selection and natural history of
     cancer. Nature, 255, 197.

3)   FRANKS, L.M. (1956). Latency and progression in human tumours.
     Lancet, 2, 1037.

4)   FRANKS, L.M. and WILSON, P.D. (1977). Origin and ultrastructure
     of cells *in vitro*. Int. Rev. Cytol., 48, 55.

5)   FRANKS, L.M. (1980) Primary cultures of human prostate. In:
     Methods in Cell Biology (ed. D. Prescot) Vol. 21B, p. 153,
     Academic Press, London.

6)   FRANKS, L.M. (1982). Tissue culture and transplantation of
     bladder. In: The Pathology of Bladder Cancer (eds. G.T. Bryan
     and S.M. Cohen). CRC Press Inc. USA, in press

7)   HASTINGS, R.J. and FRANKS, L.M. (1982). Cellular heterogeneity
     in a tissue culture cell line derived from a human bladder
     carcinoma. Int. J. Cancer, in press

8)   CAMERON, I.L. and POOL, T.B. (eds)(1981). The Transformed
     Cell. Academic Press, London.

9)   SELL, S. (ed) (1980). Cancer Markers: Diagnostic and
     Developmental Significance. Clifton NJ: Humana Press.

10) BUSCH, H. and YEOMAN, L.C. (eds)(1982). Methods in Cancer
     Research. Vol. 19. Tumour Markers. Academic Press, New York.

11) FRANKS, L.M. (1979). What is a cancer cell? Phenotypic markers
     in *in vitro* carcinogenecis in epithelial systems. In: Neo-
     plastic Transformation in Differentiated Epithelial Cell
     Systems *in vitro* (eds. L.M. Franks and C.B. Wigley), p. 287,
     Academic Press, London.

12) WIGLEY, C.B. and FRANKS, L.M. (1976). Salivary epithelial cells
     in primary culture. Characterisation of their growth and
     functional properties. J. Cell Sci., 20, 149.

13) KNOWLES, M.A. and FRANKS, L.M. (1977). Stages in neoplastic
     transformation and adult epithelial cells by 7,12-dimethyl-
     benz(a)anthracene *in vitro*. Cancer Res., 37, 3917.

14) SUMMERHAYES, I.C. and FRANKS, L.M. (1979). Effects of donor
     age on neoplastic transformation of adult mouse bladder
     epithelium *in vitro*. J. Natl. Cancer Inst., 62, 1017.

15) COWELL, J.K. and FRANKS, L.M. (1980). A rapid method for
     accurate DNA measurements *in situ* using a simple micro-
     fluorimeter and Hoechst 33258 as a quantitative fluorochrome.
     J. Histochem. Cytochem., 28, 206.

16) COWELL, J.K. (1979). Chromosome changes associated with

epithelial cell transformation with special reference to
in vitro systems. <u>In</u>: Neoplastic Transformation in Differentiated Epithelial Cell Systems *in vitro*. (eds. L.M. Franks
and C.B. Wigley), p. 159, Academic Press, London.

17) JOHNSON, L.V., SUMMERHAYES, I.C. and CHEN, Lan Bo (1982).
Decreased uptake and retention of Rhodamine 123 by mitochondria
in feline sarcoma virus-transformed mink cells. <u>Cell</u>, <u>28</u>, 7.

18) WIGLEY, C.B. (1979). Transformation *in vitro* of adult mouse
salivary gland epthelium: A system for studies on mechanisms
of initiation and promotion. <u>In</u>: Neoplastic Transformation
in Differentiated Epithelial Cell Systems *in vitro* (eds. L.M.
Franks and C.B. Wigley), p. 3, Academic Press, London.

19) COLBURN, N. (1979). The use of tumour promoter responsive
epidermal cell lines to study preneoplastic progression. <u>In</u>:
Neoplastic Transformation in Differentiated Epithelial Cell
Systems *in vitro* (eds. L.M. Franks and C.B. Wigley), p. 113,
Academic Press, London.

20) HASHIMOTO, T. and KITAGAWA, H.S. (1974). *In vitro* transformation of epithelial cells of rat urinary by nitrosamine.
<u>Nature</u>, <u>252</u>, 497.

21) GERSCHENSEN, L:E. and THOMPSON, E.B. (eds)(1975). Gene
Expression and Carcinogenesis in Cultured Liver. Academic
Press, New York.

22) STEELE, V.A., MARCHOK, A.C. and NETTESHEIM, P. (1977).
Transformation of tracheal epithelium exposed *in vitro* to
N-methyl-N'-nitrosoguanidine (MNNG). <u>Int. J. Cancer</u>, <u>20</u>, 234.

23) MARSHALL, C.J., FRANKS, L.M. and CARBONELL, A.W. (1977).
Markers of neoplastic transformation in epithelial cell lines
derived from human carcinomas. <u>J. Natl. Cancer Inst</u>., 58, 1743.

24) MARSHALL, C.J., HUMPHRYES, K.C. and POLLACK, R.E. (1978).
Microfilament bundles, LETS protein and growth control in
somatic cell hybrids. <u>J. Cell Sci</u>., <u>33</u>, 191.

25) MARSHALL, C.J. and DAVE, H. (1978). Suppression of the
     transformed phenotype in somatic cell hybrids. J. Cell Sci.
     33, 171.

26) QUARONI, A., ISSELBACHER, K.J. and RUOSLAHTI, E. (1978).
     Fibronectin synthesis by epithelial crypt cells of rat small
     intestine. Proc. Natl. Acad. Sci. USA, 75, 5548.

27) WIGLEYS, C.B. and SUMMERHAYES, I.C. (1979). Loss of LETS
     protein is not a marker for salivary glands or bladder epi-
     thelial cell transformation. Exp. Cell Res. 118, 394.

28) KNOWLES, M.A. and FRANKS, L.M. (1978). Ultrastructure and
     biological markers of neoplastic change in adult mouse epi-
     thelial cells transformed in vitro. Brit. J. Cancer, 37, 603.

29) MEDINA, D., OBORN, C.J. and ASCH, B.B. (1980). Distinction
     between preneoplastic and neoplastic mammary cell populations
     in vitro by cytochalasin B-induced multinucleation. Cancer
     Res., 40, 329.

30) SOMERS, K.D. and MUEPHEY, M.M. (1980). Cytochalasin B-induced
     multinucleation of human tumour and normal cell cultures.
     Cell Biol. Int. Rep., 4, 487.

31) O'NEILL, F.J. (1980). Cytochalasin B response of cultured
     human tumours. Cell Biol. Int. Rep., 4, 1143.

32) BOONE, C.W., VEMBU, D., WHITE, B.J., TAKECHI, N. and PARANJPE,
     M. (1979). Karyotypic, antigenic, and kidney invasive properties
     of cell lines from fibrosarcomas arising in CsH/10T½ cells
     implanted subcutaneously attached to plastic plates. Cancer
     Res., 39, 2172.

33) GARRELS, J.I. (1979). Two-dimensional gel electrophoresis
     and cumputer analysis of proteins synthesized by clonal cell
     lines. J. Biol. Chem., 254, 7961.

34) McCONKEY, E.H. (1980). Identification of human gene products
     from hybrid cells: A new approach. Somatic Cell Genet.,
     6, 139.

35) BRAVO, R., BELLATIN, J. and CELIS. (1981) [^{35}S]-methionine labelled polypeptides from HeLa cells. Coordinates and percentage of some major polypeptides. Cell Biol. Int. Rep., 5, 93.

36) LEAVITT, J., GOLDMAN, D., MERRIL, C. and KAKUNAGA, T. (1982). Changes in gene expression accompanying chemically-induced malignant transformation of human fibroblasts. Carcinogenesis, 3, 61.

37) ASHALL, F., BRAMWELL, M.E. and HARRIS, H. (1982). A new marker for human cancer cells. 1. The Ca antigen and the Cal antibody. Lancet, ii, 1.

38) McGEE, J.O'D., WOODS, J.C., ASHALL, F., BRAMWELL, M.F. and HARRIS, H. (1982). A new marker for human cancer cells. 2. Immunohistochemical detection of the Ca antigen in human tissues with the Cal antibody. Lancet, ii, 7.

39) DUHL, D.M., BANJAR, Z., BRIGGS, R.C., PAGE, D.L. and HNILICA, L.S. (1982). Tumor-associated chromatin antigens of human colon adenocarcinoma cell lines HT-29 and LoVo. Cancer Res., 42, 594.

40) HNILICA, L.S. and BRIGGS, R.C. (1980). Nonhistone protein antigens. In: Cancer Markers (ed. S. Sell), p. 463. Clifton NJ: Humana Press.

41) SMITH, G.J. (1981). Chromatin-associated proteins as markers of neoplastic transformation in mammalian cells. Cell Biol. Int. Rep., 5, 635.

42) CROCE, C.M. (1980). Cancer genes in cell hybrids. Biochim. Biophys, Acta, 605, 411.

43) HARRIS, H. (1979). Some thoughts about genetics, differenti-ation and malignancy. Somatic Cell Genet., 5, 923.

44) SABIN, A.B. (1981). Suppression of malignancy in human cancer cells: Issues and challenges. Proc. Natl. Acad. Sci. USA,78, 7129.

45) STANBRIDGE, E.J., DER, C.J., DOERSEN, C-J., NISHIMI, R.Y.,
    PEEHL, D.M., WEISSMAN, B.E. and WILKINSON, J.E. (1982).
    Human cell hybrids: Analysis of transformation and tumori-
    genicity. Science, 215, 252.

46) KLINGER, H.P. (1980). Suppression of tumorigenicity in som-
    atic cell hybrids. Cytogenet. Cell Genet., 27, 254.

47) KUCHERLAPATI, R., and SHIN, S-I. (1979). Genetic control of
    tumorigenicity in interspecific mammalian cell hybrids. Cell,
    16, 639.

48) HARRIS, H., MILLER, O.J., KLEIN, G., WORST, P. and TACHIBANA,
    T. (1969). Suppression of malignancy by cell fusion. Nature,
    223, 363.

49) JONASSON, J. and HARRIS, H. (1977). The analysis of malig-
    cancy by cell fusion. VIII. Evidence for the intervention of
    an extra-chromosomal element. J. Cell Sci., 24, 255.

50) LITTLEFIELD, J.W. (1964). Selection of hybrids from mating
    of fibroblasts in vitro and their presumed recombinations.
    Science, 145, 709.

51) KIM, B.S., LIANG, W. and COHEN, E.P. (1979). Tumor-specific
    immunity induced by somatic hybrids. J. Immunol., 123, 733.

52) CROCE, C.M. and KOPROWSKI, H. (1978). The genetics of human
    cancer. Sci. Am., 238, 117.

53) CHEN, T.R. (1979). Cytogenetics of somatic cell hybrids. 1.
    Progression of stemlines in continuous uncloned cultures of
    man-mouse cell hybrids. Cytogenet. Cell Genet., 23, 221.

54) SHAFER, R., DOEHMER, J., DRUGE, P.M., RADEMACHER, I. and
    WILLECKE, K. (1981). Genetic analysis of transformed and
    malignant phenotypes in somatic cell hybrids between tumori-
    genic Chinese hamster cell and diploid mouse fibroblasts.
    Cancer Res., 41, 1214.

55) KLINGER, H.P. (1981). In: Chromosomes Today (eds. C.E. Ford,
    M.D. Bennett, M. Babrow and G. Hewitt), Vol. 7, p. 220, Allen
    & Unwin, London.

56) AVILES, D., RITZ, E. and JAMI, J. (1980). Chromosomes in
     tumours derived from mouse tumour x diploid cell hybrids
     obtained *in vitro*. Somatic Cell Genet., 6, 171.

57) CARNEY, D.N., EDGELL, C.J., GAZDAR, A.F. and MINNA, J.D.
     (1979). Suppression of malignancy in human lung cancer (A549/8)
     x mouse fibroblast (3T3-4E) somatic cell hybrids. J. Natl.
     Cancer Inst., 62, 411.

58) JONASSON, J., POVEY, S. and HARRIS, H. (1977). The analysis
     of malignancy by cell fusion. VII. Cytogenetic analysis of
     hybrids between malignant and diploid cells and of tumours
     derived from them. J. Cell Sci., 24, 217.

59) HOWELL, N. and SAGER, R. (1981). Genetic analysis of tumori-
     genesis. VIII. Suppression of SV40 transformation in cell
     hybrids and cytoplasmic transferants. Cytogenet. Cell Genet.
     31, 214.

60) ZIEGLER, M.L. (1978). Phenotypic expression of malignancy
     in hybrid and cybrid mouse cells. Somatic Cell Genet., 4, 477.

61) CROCE, C.M. BARRICK, J., LINNENBACH, A. and KOPROWSKI, H.
     (1979). Expression of malignancy in hybrids between normal
     and malignant cells. J. Cell Physiol., 99, 279.

62) PEEHL, M. and STANBRIDGE, E.J. (1981). Characterization of
     human keratinocyte x HeLa somatic cell hybrids. Int. J. Cancer,
     27, 625.

63) WIENER, F., KLEIN, G. and HARRIS, H. (1974). The analysis of
     malignancy by cell fusion. VI. Hybrids between different
     tumour cells. J. Cell Sci., 16, 189.

64) HASTINGS, R.J. (1981). The relationship between karyotype
     and markers of transformation in cell lines and clones
     derived from human bladder carcinomas. Ph.D. thesis, University
     of London.

65) HAYFLICK, L. (1980). Recent advances in the cell biology of
     ageing. Mech. Ageing and Develop., 14, 59.

66) FRANKS, L.M. (1974). Ageing in differentiated cells.
    Gerontologia, 20, 51.

67) GOLDSTEIN, S. and LIN, C.C. (1971). Rescue of senescent human
    fibroblasts by hybridization with hamster cells *in vitro*.
    Exptl. Cell. Res., 70, 436.

68) HU, F., PASZTOR, L.M. and TERAMURA, D.J. (1977). Somatic cell
    hybrids derived from terminally differentiated rhesus cells
    and established mouse cell lines. Mech. Ageing and Develop.,
    6, 305.

69) NORWOOD, T.H., PEDNERGRASS, W.R., SPRAGUE, C.A. and MARTIN,
    G.M. (1974). Dominance of the senescent phenotype in hetero-
    karyons between replicative and post-replicative human fibro-
    blast-like cells. Proc. Natl. Acad. Sci. USA., 71, 2231.

70) WRIGHT, W.E. and HAYFLICK, L. (1975). Nuclear control of
    cellular ageing demonstrated by hybridization of anucleate
    and whole cultured normal human fibroblasts. Exptl. Cell Res.,
    96, 113.

71) MUGGLETON-HARRIS, A.L. and HAYFLICK, L. (1976). Cellular
    ageing studied by the reconstruction of replicating cells from
    nuclei and cytoplasms isolated from normal human diploid
    cells. Exptl. Cell. Res., 103, 321.

72) MATSUMURA, T., PFENDT, E.A., ZERRUDO, Z. and HAYFLICK, L.
    (1980). Senescent human diploid cells (WI-38). Exptl. Cell
    Res., 125, 453.

73) NETTE, E.G., SIT, H.L. and KING, D.W. (1982). Reactivation
    of DNA synthesis in ageing deploid human skin fibroblasts
    by fusion with mouse L karyoplasts, cytoplasts and whole L
    cells. Mech. Ageing and Develop., 18, 75.

74) WRIGHT, W.E. (1978). The isolation of heterokaryons and
    hybrids by a selective system using irreversible biochemical
    inhibitors. Exptl. Cell. Res., 112, 395.

DNA AND TIME IN CARCINOGENESIS

M. Radman*, R. Wagner and P. Jeggo#

*Département de Biologie Moléculaire, Université libre
de Bruxelles, 1640 Rhode-St-Genèse, Belgium and *
Institut Jacques Monod, C.N.R.S., Université Paris 7
Tour 43, 2, Pl. Jussieu, 75251, Paris, Cedex 05
France*

1. INTRODUCTION

Epidemiological and experimental studies of carcinogenesis
have shown that time and DNA are important factors in the process
of carcinogenesis (see ref. 1-3 for review). Whereas the importance
of DNA in carcinogenesis has been revealed in its multiple aspects
and is incorporated in all current models of carcinogenesis, few
models deal with a crucial fact of carcinogenesis, its relationship with time. This short article presents a summary of time and
DNA factors involved in carcinogenesis and is structured around
a simple model which predicts that specific genes exist which
control the activity of cellular *oncogenes* (or cancer genes) and
suggests strategies of how to look for such onco-regulator genes
and for their products.

The basic suggestion is that, given the limitations to the
stability of DNA in somatic cells, it is possible to avoid (or to

#*Present address: Genetics Division, National Institute for Medical
Research, Mill Hill, London NW7 IAA, UK.*

delay in a programmed way) the deleterious effect of some genetic (or epigenetic) chromosomal alterations by delaying the "pheno-typic expression" of such "mutations", for example by synthesizing large quantities of a stable gene product.

2. THE CURRENT PARADIGM OF CARCINOGENESIS

The current paradigm of carcinogenesis is based on the recent rapprochement of viral oncology and chemical carcinogenesis when it was realized that, at least for one class of oncogenic viruses (e.g. retroviruses) and possibly for all oncogenic viruses, the virus is a vector, an activator and/or an amplifier of genuine cellular genes: *oncogenes* or "cancer genes" (reviews 4-6; see also other articles in this volume). The activation of *oncogenes* seems to be sufficient to cause malignant transformation, although their diversification may in theory also influence the efficiency of transformation. Activation of cellular *oncogenes* in non-viral carcinogenesis is supposed to occur by some kind of DNA alteration such as point mutation, chromosomal rearrangement or DNA methylation (see ref. 3 for review). There are at least twenty forms of cellular *oncogenes* and for a given tissue there is a preferential form of cellular *oncogene* which upon activation, causes malignant cell transformation (reviews 4-6). Thus, according to this paradigm, the process of carcinogenesis can be arbitrarily divided into two stages: (1) the process of activation of *oncogene(s)* and (2) the complex sequence of events following *oncogene* activation that leads to a fully malignant cell. If we consider that the second stage process starts when a virus carrying an already active cellular *oncogene* (e.g. a retrovirus) infects a normal cell (or an animal), then this complex process is rather short: a few weeks either in cell cultures or in infected susceptible animals. So, it is likely that the long latency period in chemical and "spontaneous" carcino-genesis lies in the first stage, i.e. the process of activation of the *oncogene(s)*. Unlike the second stage, which is under intense

research by viral oncologists, the first stage is still an open
area for research.

3. DNA IN CARCINOGENESIS

The direct proof that DNA is a target in carcinogenesis and
the carrier of the malignant phenotype came from DNA transfection
experiments which have shown that the malignant phenotype can be
generated by DNA transfer from one non-malignant cell to another
(7) and from a malignant cell into a non-malignant receptor cell
(reviews 4, 5).

The correlation between unremoved DNA lesions and carcinogenesis (8, 9) suggested that in radiation and chemically induced
carcinogenesis, a DNA alteration may be the initial causal event.
A remarkable correlation between the mutagenic and carcinogenic
capacity of chemicals supported the hypothesis that carcinogenesis
may be caused by somatic mutations. To avoid repetitition, we
suggest that the reader consult our recent comprehensive review
and critical evaluation of the role of DNA alterations in carcinogenesis (3).

4. TIME IN CARCINOGENESIS

The incidence of various classes of carcinomas (which account
for about 90% of all human cancers) rises with a high power with age
(usually fifth to seventh) whereas the relationship with the dose
of a carcinogen is usually (although not always) linear. In fact,
carcinogenesis rises with a high power of exposure time to carcinogen and with a low power of the dose of carcinogens (1, 2). Thus,
age appears to be the most potent carcinogen, and on the other hand,
the life span of the animal determines the susceptibility of its
target cells (i.e. stem cells) to carcinogenesis (carcinomas). It
is a "rule of thumb" that the latency period in carcinogenesis is

equivalent to 1/4 to 1/3 of the life span, i.e. about 20 years in
man (1). Most intriguing is that a "memory" of this rule is carried
by the cells when grown in cultures: the life span of the animal
species from which the cells are derived determines the susceptibility of the cells in culture to transformation by chemical carcinogens and radiation. Peto estimated that the stem cells of the
mouse are about 10^9-fold more suseptible to carcinogenesis than
human stem cells (10). This is in agreement with the fact that
mouse fibroblasts are readily transformed in cell cultures by a
variety of chemical mutagens, whereas human fibroblasts are (with
few exceptions) practically not transformable. It is unlikely that
this difference is due to a much higher susceptibility of mouse
chromosomes to either mutation or chromosomal aberration (11), or
to the shorter period between the activation of *oncogene(s)* and
the appearance of malignant phenotypes. When infected with the same
oncogenic virus both human and mouse cells are transformed with
similar efficiencies and within a comparable number of cell generations. Therefore, the period required for the activation of *oncogenes* seems the most likely stage of the carcinogenic process which
could account for the large difference in the "expression time"
between mouse and man in the respective carcinogenic processes.

Time in carcinogenesis is not chronological time but biological
time corresponding to, or proportional to, the number of cell
divisions (12, 21 and refs therein). Agents which shorten the latency
period in carcinogenesis are called tumour promoters; agents which
lengthen the latency period are antipromoters. Therefore, tumour
promoters are carcinogenic only for "initiated" cells, i.e. cells
which have been modified by exposure to some initiating carcinogen,
and are non-carcinogenic for intact cells. If initiation corresponds
to a latent DNA alteration occurring with comparable frequencies in
human and rodent cells, then human cells would appear as promotion-
resistant; in other words, promotion may be the predominant rate-

limiting step in human carcinogenesis. Therefore, preventing promotion, i.e. increasing the latency period, may be a promising
strategy towards cancer prevention.

5. STEPS IN CARCINOGENESIS

For a complete understanding of carcinogenesis it is important
to determine the total number of steps involved and to investigate the
nature of the individual steps. Available epidemiological data do
not allow a direct estimate of the number of steps in carcinogenesis
because of a disparity between the power relationship of dose and
tumour incidence compared with the exposure time and tumour incidence. Tumour incidence is usually related to dose by a power of
1 or 2 and to exposure time by a power of 4-7 (2). One possible
way to explain this disparity is to postulate that carcinogenesis
involves one or more purely "passive" steps. These may correspond
to time or cell division dependent steps such as dilution or inactivation of a negative control factor or build-up of a positive
control factor. Steps of this type would occur only after some
change in gene expression in the cells involved, and for a given
change, the expression time should be relatively constant. Thus,
if a number of cells have undergone a simultaneous change in gene
expression, the fraction which reveals the alteration phenotypically
at a given time should increase with time to some maximum at the
mean expression time for that event and then decline.

If the long exposure time analyzed in epidemiological studies
(involving large human populations) are equivalent to a series of
individual short time exposures, each capable of generating at
least one potential tumour cell in some subpopulation of the individuals exposed, then the number of individuals which reveal tumours
at a given time will be the sum of the numbers of individuals in
each of the subpopulations which have developed tumours at that

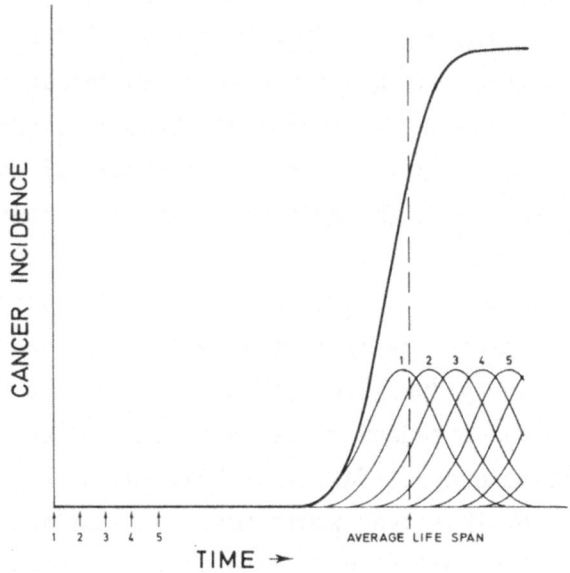

Fig. 1. *Theoretical cancer incidence as a function of time in a time delay model of carcinogenesis.* 1 to 5 are initiation events occurring in some short time interval as rare events in a large population. Small curves represent the Poisson distributions of the appearance of tumours resulting from initiation events 1 to 5. Heavy line indicate the cumulative incidence obtained from the sum of the individual distributions.

time. In such a case, if the total population is large and the
tumour incidence relatively low, a graph of tumour incidence versus
exposure time would be expected to rise exponentially and plateu
at a level corresponding to the number of individuals affected by
each exposure (see Figure 1 and its legend). However, if the average
life span of the individuals being studied is such that no individuals are living at the time when the plateau would normally be
reached, then the curve would appear to be exponential and exposure
time could appear to be related to incidence by a power not reflecting the number of steps involved in the process. The epidemiological cancer studies with humans so reveal a tendency towards
a plateu at advanced age (2 and refs therein).

   Dose-response data from studies of *in vitro* cell transformation
may provide some indication of the number of steps involved in carcinogenesis. Data from studies of X-ray irradiation of mouse C3H
10T1/2 cells in the dose range of 50-400 rads reveal a nearly
quadratic dose-response curve for transformation (data replotted
by us from ref. 13, Fig. 4), suggesting that these X-ray doses
induce two steps. However, it was discovered that even lower doses
of X-rays appear to increase the likelihood of transformation for
cells exposed (12), suggesting that at least three steps can be
induced by X-rays in this system. An alternative possibility is
that the dose-response curve might appear to be quadratic for some
reason other than the existence of two distinct steps, such as a
saturable repair process. Studies in the same system with the potent
tumour promoter 12-O-tetradecanoylphorbol-13-acetate (TPA) indicate
that the dose-squared curve is not an artefact. Although no effect
of TPA on cellular repair processes has been detected when cells
are treated with TPA after irradiation, the dose-response curve
for X-ray induced transformation is linear if the cells are treated
with TPA after irradiation. The slope of the log/log plot of transformation as a function of X-ray dose changes from two without TPA

to one with TPA (data from Figure 4 of ref. 13), indicating that
TPA is not simply reducing the expression time of some X-ray
induced event, an effect which would have shifted the curve but
not altered its slope.

Data from experiments on mouse skin carcinogenesis suggest
that TPA may act on two distinct steps, one occurring after a
single application and another requiring repeated applications.
The single application effect of TPA is inhibited by protease
inhibitors (14), which also inhibit the TPA effect on cell transformation *in vitro* (15, 16), suggesting that the effect of TPA on
cells in culture may correspond to the first promoter induced step
in mouse skin carcinogenesis. It may be that the second step correspond to some effect on the time dependent processes discussed
above and later in this article.

If the mouse skin and *in vitro* transformation systems are
comparable and representative of a general process of carcinogenesis, then it appears that there may be as many as four steps
in the process: two steps that can be induced only by carcinogens, one step that can be induced by either carcinogens or promoters, and a late, perhaps "dilution" or "expression time" step,
that may be facilitated by promoter action. It may be that there
are still other steps that have not been detected, perhaps because
they occur earlier or at lower doses than the steps studied, or
because they already occurred in the course of establishment of the
cell line studied.

6. A TIME-DELAY MODEL FOR CARCINOGENESIS AND ITS PREDICTIONS

The following speculation summarizes our considerations of DNA
and time in carcinogenesis. It is a version of the Comings' model for
carcinogenesis (17) adapted specifically to account for the species-
specific long latency period in carcinogenesis and to offer several

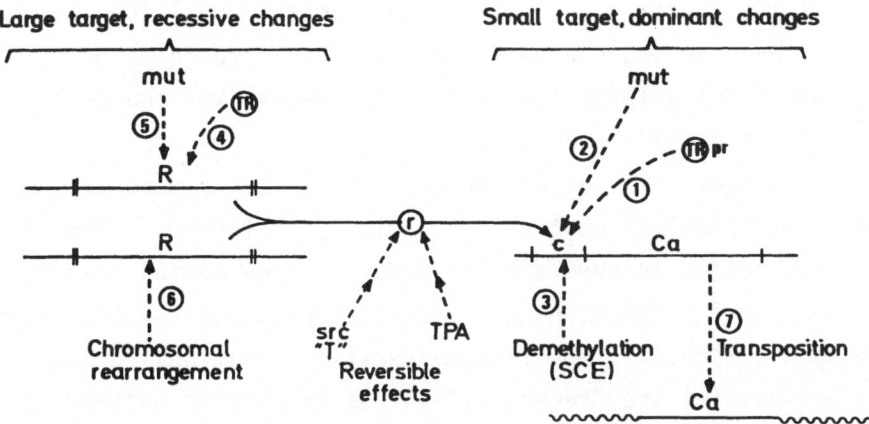

Fig. 2. *Scheme for the regulation of* oncogenes: *one way to summarize time and DNA factors in carcinogenesis.* The *oncogene* or cancer gene (Ca) is negatively controlled by an onco-regulator *r*-protein, encoded by the onco-regulator (R) gene (here in a functionally diploid state), via the control region (c) proximal to the Ca-gene. mut stands for point mutations including frameshift; TR for transposons; Pr for transcritional promoter; SCE for sister chromatin exchange and TPA for tumour promoting agent (such as tetradecanoyl-phorbol-acetate). Carcinogenesis can be induced either by inactivating the R-genes (pathways 4, 5, 6) (or by in-activating the *r*-protein) or by direct activation of the Ca-gene (pathways 1, 2, 3 and 7). Human cancer prone syndroms, Bloom's and Fonconi's anemia, could cause cancer through pathways 3, 6 and 1, 4, 7, respectively, whereas chemicals and radiations could, in theory, induce all seven pathways (see refs. 3, 22 and 23 and this text for further explanations.

testable predictions. Figure 2 is just a didactic scheme to illustrate our speculation and to visualize the experimental pre-dictions. The essence of our suggestion is that there may exist cellular onco-regulator genes (R) coding for the onco-regulator proteins (*r*). The amount and intrinsic stability of the *r*-protein(s) is supposed to determine the latency period ("expression time") in one pathway of carcinogenesis and is expected to be species and organ specific. For example, to account for the discussed difference in cancer susceptibility between man and mouse (see section 4), mouse cells are expected to have less and/or less stable *r*-protein than human cells. Furthermore, if the *r*-protein(s) inhibits the

activity of the specific *oncogene(s)* (or cancer genes) (Ca in
Figure 2), then one would expect two principal genetic targets
in carcinogenesis: a large target for the "recessive" changes
(i.e. inactivation of the R-genes) and a small target for the
"dominant" changes (i.e. direct activation of the Ca-gene by the
changes in its proximal control (c) region. The former "latency"
pathway is expected to have a much longer latency period than the
latter "dominant" pathway. Hence, the model proposes two mechanisms
for the species-specific and tissue-specific variations in the
latency period: (1) the "recessive" versus "dominant" pathway on
the level of DNA alterations and (2) in the "recessive" pathway,
variations in the amount and intrinsic stability of the r-protein.
These two mechanisms are testable and the predictions suggested
below may help us to understand the species-specific and the tissue-
specific differences (e.g. leukemias versus carcinomas) in the
latency period.

Let us now review the arguments which encouraged these specula-
tions. The first complete report on transformation of human cells
in culture has shown that the key factor in the transformation of
human primary fibroblasts by the potent mutagens N-methyl-N'-nitro-
N-nitrosoguanidine and 4-nitrosoquinoline-1-oxide is the long
expression period (21). Only if at least five to seven passages
(40 to 50 days of continuous growth in culture corresponding to at
least 13 cell generations) were allowed to occur after the treatment by the carcinogen, did any significant transformation occur
(morphological, agar-growth and tumorigenicity). After similar
mutagenic treatments mouse fibroblasts acquire the same transformed
phenotypes following only four to five cell generations.

The existence of *oncogenes* with control regions is suggested
by the results of the transfection experiments, in which the requirement for fragmenting the DNA from normal cells may reflect the need to

separate the cancer gene from its control region (7). In cells
transformed with the fragmented DNA, the cancer gene is presumably
associated with actively transcribed DNA, as the DNA of these
cells is able to transform without being fragmented (7).

Chromosomal rearrangements of the non-homologous recombination
class may cause activation of cancer genes either by transposing
the cancer gene to a location with a transcription control region
or by inserting such a control region into the cancer gene control
region or between the cancer gene (Ca) and its control region (c).
If DNA methylation is involved in the control of cancer genes,
then chromosomal rearrangements of the homologous recombination
class may affect expression by altering the pattern of methylation
(ref. 3 for review). It may also be possible to activate cancer
genes by means of sequence alteration in the control region, which
could be accomplished either by chromosomal rearrangement or by
base substitution or frameshift. However, if the control region
is necessary for transcription of the cancer gene, i.e. contains
a promoter, sequence alteration may not be possible or else may
need to be so specific as to make its occurrence unlikely. Any
activation of cancer genes involving chromosomal rearrangements
may require the induction of cellular recombination systems. Such
induction would represent additional steps in the process of
carcinogenesis.

The DNA from most transformed cells that have been examined so
far cannot cause transformation without being fragmented (reviews
4, 5), indicating that there must be a pathway of carcinogenesis
not involving the linking of cancer genes with transcription control
regions. It may be that the control region is a binding site for
a "repressor" (a binding protein or a specific methylase or de-
methylase). If so, there must be "regulator" (R) genes in the cell.
In such a case, an alternative, and clearly multistep pathway of

carcinogenesis would be the inactivation of the R-genes. Inactivation of a gene could presumably be accomplished by base substitution,
frameshift, integration of a virus, or any of the non-homologous
recombination class of chromosomal rearrangements (3). If one
copy of a repressor gene is inactivated, homologous recombination
or ploidy change should be able to make the cell homo- or hemizy-
gous for the mutant allele and, after the dilution or inactivation
of any "r-protein" molecules remaining in the cell, allow activation
of the *oncogene*.

It may be that one of the effects of tumour promoters (e.g.
TPA) and viral gene products necessary for the maintenance of
the transformed state (e.g. RSV *src* and SV40 and polyoma "T"
antigen proteins) is to act as antagonists of the *r*-protein, either
by inactivating it more rapidly than it can be synthesized or by
repressing the R-genes. Such action could account for the reversible mimicking of transformed phenotypes by TPA and for the
reversible nature of the effects of some viral transforming
proteins (19, 20). In the presence of a functional R-gene(s) the
effects of promoters would be reversible and in the absence of
functional R-genes ("initiated" target cells) the effects on *r*-
protein would not be reversible, but would hasten the appearance
of the stable malignant state.

If *oncogenes* (Ca) and onco-regulator genes (R) are involved
in carcinogenesis, then: (1) it may be possible to isolate onco-
regulator (*r*) protein(s) from cell nuclei, perhaps as a protein
binding TPA or modified (e.g. cleaved or phosphorylated) as a
result of TPA or *src* protein action, or as a protein present in
nuclei of normal cells but missing in transformed cells. The *r*-
protein may be present in the nuclei or those "dominant" tumour
cells whose DNA can transform without fragmentation; (2) it may
be possible to identify onco-regulator (R) genes by using DNA from

normal cells to transfect tumour cells (of those "recessive" majority tumours whose DNA cannot transform without first being fragmented) and then selecting or screening for reversion of the transformed phenotype and; (3) it may be possible to test in DNA transfection experiments whether tumours with a known history of short and long latency (e.g. leukemias and carcinomas) yield DNA which is different (e.g. "dominant" versus "recessive") in its transforming activity.

The most exciting prediction of the model is the existence of the r-protein(s), whose amount and stability should be species-specific and related to the life span and which also may vary among different tissues of the same organism.

7.   ACKNOWLEDGEMENTS

This work was performed under Euratom contract BIO-E-420B. P. Jeggo was an EMBO post-doctoral fellow. M. Radman was supported by a visiting professorship of the Université de Paris-Sud, Centre d'Orsay, Laboratoire de Génétique. Dr. Shoshana Wodak assisted in the construction of Figure 1. M. Radman acknowledges useful discussions with Dr. L.M. Franks and other participants during the International Summer School "Gene expression in normal and transformed cells", Sintra-Estoril, 1982.

8.   REFERENCES

1)   CAIRNS, J. (1978). Cancer, Science and Society. W.H. Freeman and Co., San Francisco.

2)   PETO, R. (1977). Epidemiology, multistage models and short-term mutagenicity tests. In: Origins of human cancer (Eds, H.H. Hiatt, J.D. Watson and J.A. Winsten), p. 1403, Cold Spring Harbor Laboratory Press, New York.

3) RADMAN, M., JEGGO, P. and WAGNER, R. (1982). Chromosomal rearrangement and carcinogenesis. Mutation Res., 98, 249.

4) WEINBERG, R.A. (1982). Fewer and fewer *oncogenes*. Cell, 30, 3.

5) COOPER, G.M. (1982). Cellular transforming genes. Science, 218, 801.

6) BISHOP, J.M. (1981). Enemies within: the genesis of retro-virus *oncogenes*. Cell, 23, 5.

7) COOPER, G.M., OKENQUIST, S. and SILVERMAN, L. (1980). Trans-forming activity of DNA of chemically transformed and normal cells, Nature, 284, 418.

8) RAJEWSKY, M.F., AUGENLICHT, L.H., BIESSMANN, H., GOTH, R., HULSER, D.F., LAERUM, O.D. and LOMAKINA, L., Ya (1977). Ner-vous-system-specific carcinogenesis by ethylnitrosourea in the rat: molecular and cellular aspects. In: Origins of Human cancer (Eds, H.H. Hiatt, J.D. Watson and J.A. Winsten), p. 709, Cold Spring Harbor Laboratory Press, New York.

9) SETLOW, R.B. (1978). Repair deficient human disorders and cancer. Nature, 271, 713.

10) PETO, R. (1979). Detection of risk of cancer to man. Proc. R. Soc. Lond. B 205, 111.

11) DE BOER, P., VAN BUUL, P.P.W., VAN BEEK, R., VAN DER HOVEN, F.A. and NATARAJAN, A.T. (1977). Chromosomal radiosensitivity and karyotype in mice using cultured peripheral blood lympho-cytes and comparison with system in man. Mutation Res., 42, 379.

12) KENNEDY, A.R., FOX, M.S., MURPHY, G. and LITTLE, J.B. (1980). Relationship between X-ray exposure and malignant transforma-tion in C3H 10 T 1/2 cells. Proc. Natl. Acad. Sci. USA, 77, 7267.

13) LITTLE, J.B. (1981). Influence of non-carcinogenic secondary factors on radiation carcinogenetic. Radiation Res., 87, 240.

14) SLAGA, T.J., FISHER, S.M., NELSON, K. and GLEASON, G.L. (1981). Studies on the mechanism of skin tumour promotion: evidence for several stages in promotion. Proc. Natl. Acad. Sci. USA, 77, 3659.

15) KENNEDY, A.R. and LITTLE, J.B. (1978). Protease inhibitors suppress radiation-induced malignant transformation *in vitro*. Nature, 276, 825.

16) KUROKI, T. and DREVON, C. (1979). Inhibition of chemical transformation in C3H/10 T 1/2 cells by protease inhibitors. Cancer Res. 39, 2755.

17) COMINGS, D.E. (1973). A general theory of carcinogenesis. Proc. Natl. Acad. Sci. USA, 70, 3324.

18) MOOLGAVKAR, S.H. and KNUDSON, A.G. (1971). Mutation and cancer; a model for human carcinogenesis. J. Natl. Cancer Inst., 66, 1037.

19) WEINSTEN, I.B. (1980). In: Mechanisms of toxicity and hazard evaluation (eds, B. Holmstedt, R. Lauwereys, M. Mercier and M. Roberfroid), p. 149, Elsevier/North Holland Biomedical Press, Amsterdam.

20) BISSELL, M.J., HATIE, C. and CALVIN, M. (1978). Is the product of the *src* gene a promoter? Proc. Natl. Acad. Sci. USA, 76, 348.

21) KAKUNAGA, T. (1977). The transformation of human diploid cell by chemical carcinogens. In: Origins of human cancer (eds, H.H. Hiatt, J.D. Watson and J.A. Winsten), p. 1537. Cold Spring Harbor Laboratory Press, New York.

22) KINSELLA, A.R. and RADMAN, M. (1978). Tumour promoter induces sister chromatid exchange: relevance to mechanisms of carcinogenesis. Proc. Natl. Acad. Sci. USA, 75, 6149.

23) KINSELLA, A.R. and RADMAN, M. (1980). Inhibition of carcinogen-induced chromosomal aberrations by an anti-carcinogenic protease inhibitor. Proc. Natl. Acad. Sci. USA, 77, 3544.

THE MOLECULAR GENETICS OF AVIAN ERYTHROBLASTOSIS VIRUS

M.L. Privalsky*, L. Sealy#, B. Vennstrom≠,
and J.M. Bishop*

*Department of Microbiology and Immunology, University
of California, San Francisco, California, USA; #McArdle
Laboratory for Cancer Research, University of Wisconsin
Madison, Wisconsin, USA and ≠Department of Microbiology
University of Uppsala, Uppsala, Sweden

1. INTRODUCTION

The discovery that viruses could function as etiological agents
of cancer was first made early in this century (1). Since that
time a variety of oncogenic viruses have been isolated from many
different animal species. This brief review will be confined
to a discussion of avian erythroblastosis virus (AEV), a retrovirus
capable of inducing sarcomas and a rapidly progressive erythroblastosis in susceptible chickens (2, 3). AEV is a useful prototype
for several reasons. First, the tools brought to bear, and the
discoveries already made in the study of AEV serve as excellent
examples of how contemporary research on retroviruses is conducted.
Second, AEV is one of several different avian retroviruses that
transform fibroblasts; and understanding of the mechanism of
transformation by AEV may clarify whether different oncogenic
viruses transform by related or unique mechanisms. Finally, an
elucidation of the means by which AEV transforms two distinct
target cell types (erythoblasts and fibroblasts (3)) may provide

193

insight into the manner in which growth and differentiation are controlled in both normal and neoplastic cells.

2. MOLECULAR GENETICS OF AVIAN ERYTHROBLASTOSIS VIRUS

The genome of avian erythroblastosis is diploid with haploid subunits of single-stranded RNA. Each haploid subunit is approximately 5.4 Kb in length, and is of positive polarity (i.e. the same polarity as messenger RNA) (4-9). On entering a susceptible cell, the AEV RNA genome is copied by reverse transcriptase into a double stranded DNA intermediate which can subsequently integrate into the host cell genome (10). Transcription of the integrated provirus by host cell polymerases generates both new RNA genomes, and also mRNA for the synthesis of virally encoded proteins. Alone, the integrated AEV genome is capable of inducing cellular transformation but cannot produce new infectious virions (3, 11). Co-infection with a suitable helper virus, such as a Rous-associated virus (RAV), is necessary to complement this replication defect and complete the replicative cycle. Therefore, infectious stocks of avian erythroblastosis virus are actually mixtures of AEV and helper (3, 11).

AEV is a member of a class of retroviruses that contain within their genomes specific regions (*oncogenes*) which have been implicated in the ability of these viruses to induce neoplasms (12). Though different viruses may possess different *oncogenes*, the vast majority of the retroviral *oncogenes* thus far analysed appear to have been acquired from counterparts within the normal, uninfected cellular genome (12). Indeed, several of these cellular *oncogenes* can themselves induce neoplastic transformation when placed in certain milieus. For example, DNA constructs generated *in vitro* that link a strong viral promoter to a molecularly cloned cellular *oncogene* are capable of oncogenic transformation when reintroduced into cells (13). There is evidence that the induction of bursal

lymphomas by Rous associated viruses (retroviruses that do not
contain *oncogenes*) may be due in part to an activation of a cellular
oncogene (14).

3. CLONING AND CHARACTERIZATION OF THE AEV GENOME

An important aid to our analysis of the molecular biology
of avian erythroblastosis virus was the molecular cloning of the
AEV genome (10). To achieve this, we exploited one particular
characteristic of the replicative cycle of AEV: among the population
of unintegrated virus-related nucleic acid molecules in a cell
productively infected by AEV are closed circular DNA forms of both
the AEV and the helper genomes. The DNA form of the AEV genome
was purified from this mixture by cleaving the helper genome DNA
molecules with *Hpa* I. a restriction endonuclease that does not
cleave within the erythoblastosis virus molecule. Isolation of the
remaining closed circular DNA molecules yielded a population
greatly enriched for AEV. The AEV molecules were then linearized
at an *Eco* RI site and cloned into the bacterial phage vector λgt
WES · B. The resulting molecular clones represented circularly
permuted DNA forms of the AEV RNA genome.

We first employed the AEV molecular clones to map the viral
genome by use of restriction endonuclease digestions and hetero-
duplex mapping ((10) and unpublished data). Our analysis of the
molecularly cloned viral genome substantiated and refined the
results previously obtained by other investigators without the
aid of molecular cloning (15-17). As can be seen in Figure 1, the
AEV genome appears to have been derived from a RAV-like helper
viral genome by a deletion of several replicative genes and a substitution with sequences unique to the erythroblastosis virus.
The Rous-associated viruses are not acutely transforming; the AEV-
unique region (referred to as *v-erb* (18)) is therefore implicated
in the ability of AEV to transform cells and to induce erythro-

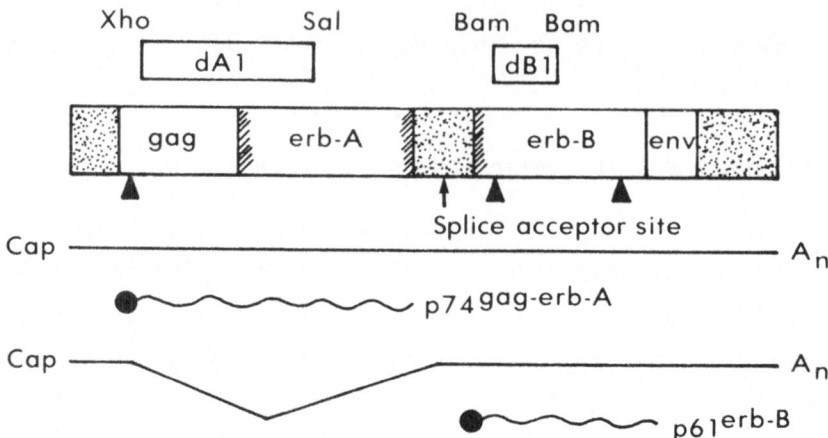

Fig. 1. *The molecular genetics of avian erythroblastosis virus.*
A map of the AEV genome is illustrated, as are the two messenger
RNA's known to be encoded by the virus (a 5.4 Kb RNA of genome
length and a 3.5 Kb subgenomic RNA). The polypeptides translated
from the AEV mRNA's are represented by wavy lines. The positions
of mutations created *in vitro* are also shown; the sites of frameshift mutations are indicated by triangles, and the limits of deletion mutations are indicated as boxed regions (dA1 and dB1)
above the genome map.

blastosis and sarcomas in birds. Analysis of the expression of the
AEV genome in transformed cells suggests the *v-erb oncogene* region
can be divided into two domains, a 5' proximal *v-erb-A* domain and
a 3' proximal *v-erb-B* domain (see Figure 1). Consistent with the
characterizations of other known retroviral *oncogenes*, both *v-erb-A*
and *v-erb-B* domains were found to be homologous to sequences found
in normal uninfected avian and mammalian cellular DNA (19, 20).
The viral *erb* domains were not detectably homologous to the *oncogene* of any other retrovirus analyzed.

Analysis of the map of the AEV genome provided an explanation
of the replication defectiveness of the erythroblastosis virus;
the AEV genome lacks all, or major, portions of, the genes necessary

for autonomous replication: *pol*, *gag* and *env* (genes for the reverse transcriptase and structural protein components of the virion). These functions can be complemented in trans, however, by co-infection with a suitable helper.

4. EXPRESSION OF THE AEV GENOME IN TRANSFORMED CELLS: RNA AND PROTEINS

The availability of specific cloned DNA restriction fragments representing various regions of the AEV genome greatly aided our analysis of viral RNA expression in AEV-transformed cells. Two viral RNA's were detected in both AEV-transformed fibroblasts and erythroblasts: a 5.4 Kb RNA similar or identical to the AEV genome itself, and a 3.5 Kb subgenomic RNA, (5, 6). The 3.5 Kb AEV RNA contains a portion of the 5' end of the AEV genome spliced onto the 3' proximal portion of the *erb* region (see Figure 1). These results suggested that the *erb* region might be functionally divided into two domains: *erb-A* (the 5' proximal portion that is not expressed in the subgenomic RNA) and *erb-B* (the 3' proximal portion contained within the 3.5 Kd subgenomic message). This hypothesis was confirmed by analysis of the polypeptides encoded by the AEV RNA's and by genetic dissection of the *erb* region (discussed below).

Proteins had been implicated as the effectors of oncogenic transformation by other viruses; the identification of mutants of AEV temperature-sensitive for the transformed phenotype implied a protein, or proteins, were also probably involved in oncogenesis by the erythroblastosis virus (21, 22). Several laboratories, including our own, therefore began a search for *erb* encoded polypeptides. Two different approaches were pursued: identification of the *in vitro* translation products of the AEV *erb*-related RNA's discussed above, and immunoprecipitation analysis of AEV-transformed cell proteins.

By use of *in vitro* translation, several investigators independently identified an approximately 75 Kd polypeptide as the
translation product of the 5.4 Kb AEV RNA (6-9). Peptide mapping and
immunological analysis demonstrated that this p75 protein is a
gag-erb fusion protein, presumably initiating in the helper-related
gag sequences on the AEV genome and reading through into *erb-A*
sequences (Figure 1). We were able to map the p75 protein coding
region on the AEV genome by use of specific cloned AEV DNA probes
in a hybridization-arrested translation procedure; the hybridization
arrested translation technique confirmed previous indications that
p75 was encoded by both *gag* and *erb-A* related sequences (23).

Identification of the p75 protein left the expression of
approximately 2 Kb of potential *erb* coding sequence, including
the entire *B* domain, unaccounted for. Work performed in other
laboratories demonstrated that AEV RNA the size of the subgenomic
erb-B message could be translated *in vitro*, yielding an approximately 41 Kd protein as the major polypeptide product (6-9). We
noticed in our experiments that in addition to the 41 Kd species
(p41), a 61 Kd protein (p61) was also translated from RNA the
size of the AEV subgenomic RNA (23). The peptide maps of the p41
and p61 proteins were closely related to one another. We were
able to demonstrate by the hybridization-arrested translation
technique that both p61 and p41 were translated from within *erb-B*
related sequences (23). The synthesis of p61 was inhibited by
hybridization of DNA representing the 5' leader region of intact
AEV subgenomic RNA, whereas the translation of p41 was not,
suggesting that only p61 was translated from message with an intact
5' end (23). The p61 protein is therefore likely to be the authentic
translation product of the *erb-B* domain, whereas p41 appears to be
an artifact of the *in vitro* translation technique - the product
of translation of RNA fragments. A similar artifact was also seen
in the early attempts to identify the protein product of the Rous

sarcoma virus *src oncogene* (24-26).

The AEV *erb-A* and *B* domains therefore appear to be expressed
separately: the *A* domain in the form of a p75 *gag-erb-A* fusion
protein translated from the 5.4 Kb genome length RNA, and the *B*
domain as a p61 polypeptide translated from the 3.5 Kb subgenomic
RNA species. Characterization of the *erb* proteins in the AEV-
transformed cells was of great interest. Identification of the viral
oncogene polypeptides amidst the background of host protein synthesis
was achieved by application of immunoprecipitation techniques. The
first AEV protein to be so identified in virally transformed cells
was p75, by virtue of its *gag* antigenic determinants (27-29). More
recently, antisera to the *erb-A* region of the protein has also been
obtained (30). Immunoprecipitation analysis with either serum
demonstrated that the p75 protein synthesized in transformed cells
appears to be identical to the 75 Kd *in vitro* translated product.
Little is known about the exact subcellular localization or possible
enzymatic activities of this protein.

Antisera to the *erb-B* protein has been successfully obtained
by two very different approaches: isolation of serum from neonatal
rats bearing tumours induced by AEV (Michael Hayman and colleagues,
personal communication), and expression of a portion of the *erb-B*
polypeptide in bacteria and immunization of rabbits with this
bacterial *erb-B* related polypeptide (M.L.P., J.M.B., J.P. McGrath,
and A.D. Levinson, manuscript in preparation). The bacterial
expression vector employed is diagrammed in Figure 2. The sera
raised in tumour-bearing animals have the advantage of possessing
activity against both p75 and p61 proteins, and are probably capable
of recognizing antigenic sites throughout each of these two protein
molecules. Sera raised against the portion of the *erb-B* polypeptide
expressed in bacteria possess the converse advantage of defined
specificity, and the availability of large quantities of the

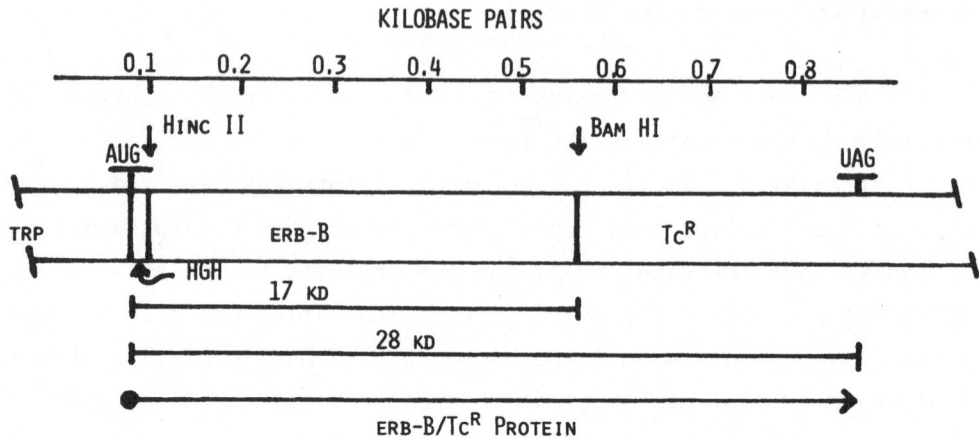

Fig. 2. *Vector for expression in bacteria of an* erb-B *related polypeptide*. A DNA fragment from the *erb-B* region of the molecularly cloned AEV genome was inserted in a bacterial expression vector, between a tryptophan promoter (trp) and a portion of a tetracycline resistance gene (TcR). The expression vector was derived from a plasmid containing human growth hormone (HGH) sequences; 51 base pairs of HGH sequence remain in the *erb-B* plasmid as an artifact of the construction. Expression of the *erb-B* sequences is under the control of the thyptophan promoter. *Escherichia coli* clones possessing this vector can be induced to synthesize high levels of a 28 Kd polypeptide, approximately 17 Kd of this polypeptide is *erb-B* related, as indicated.

bacterial polypeptide allows affinity purification and immuno-competition experiments to be performed.

Serum obtained by either technique is capable of immuno-precipitating the *erb-B* protein from AEV-transformed cells. The results obtained with these sera demonstrated that the *erb-B* protein in AEV-transformed cells is heterogeneous in apparent molecule weight, with major species at approximately 61, 65 and 68 Kd (M. Hayman, personal communication, and M.L.P., J.M.B., J.P. McGrath, and A.D. Levinson, manuscript in preparation). Pulse-chase experiments indicated a precursor-product relationship between p61 and the higher molecular weight forms. We discovered that the conversion of p61 into the higher molecular weight species was completely blocked by the antibiotic tunicamycin, a specific inhibitor of glycosylation. The *erb-B* protein may therefore be glycosylated in transformed cells; consistent with this hypothesis, the viral p65 and p68 viral *erb-B* species appear to be membrane associated (unpublished data).

5. MUTAGENESIS OF AVIAN ERYTHROBLATOSIS VIRUS *IN VITRO*

Results demonstrating the existence of two separate domains within the AEV *erb* region (*A* and *B*), and the separate expression of these two domains as two mRNA's and as two proteins, raise a number of interesting questions. Is the dual target cell specificity of AEV a reflection of these two distinct domains, or are both *erb* domains required to transform both types of target cells? Are the pleiotropic manifestations of cellular transformation due to expression of these different *oncogene* domains? A genetic dissection of the problem appeared valuable.

Both conditional and non-conditional mutants of AEV have been isolated by traditional means (21, 22). The limited number of mutants isolated, the difficulty of definitively mapping these genetic lesions to either the *erb-A* or *-B* domain, and the difficulty of insuring that such lesions represent a single mutational event somewhat limits their usefulness. For these reasons we chose to utilize the molecularly cloned AEV genome to create defined mutations *in vitro*.

We created two forms of mutations in AEV: deletions and frame-shifts. Deletions were generated by cleaving the AEV DNA clone with appropriate restriction endo-nucleases so as to excise a fragment of the genome; ligation of the remaining sequences yielded a deletion of defined end points. Frame-shifts were introduced at defined sites by cleaving with a restriction enzyme that generated

a staggered cut with 5' overhangs; subsequent filling in of the
staggered end with a DNA polymerase and ligation of the resulting
blunt ends introduced four new base pairs into the DNA sequence,
inserting one new codon and a plus one frameshift. The mutations
generated are shown schematically in Figure 1. The mutated DNA
molecules were transfected into chicken cells in the presence of
helper viral DNA molecules; the cells and the viruses released by
them were then characterized for transformation parameters.

Mutations introduced into the *erb-B* domain yielded the most
readily interpreted results (Table 1). Deletion of a large portion
of *erb-B*, or introduction of a frame-shift mutation near the
presumptive N-terminus of the *erb-B* coding region resulted in viruses
incapable of transforming chicken fibroblasts. Fibroblasts infected
by either of these *erb-B* mutants did not form colonies in soft agar,
and possessed a normal morphology. In contrast, cells infected by
the parental virus, or by a virus possessing a frame-shift near
the presumptive C-terminus of the *erb-B* protein formed colonies in
soft agar, and were morphologically transformed. The effects of
the *erb-B* mutations on the ability of the virus to transform erythroblasts are presently under study.

Table 1. *The effects of mutating* erb-A *and* erb-B

Type of mutation	Locus	Fibroblast transformation
wild-type virus		+
deletion	*erb-A*	+
frameshift	*erb-A*	+
deletion	*erb-B*	-
frameshift	*erb-B* (5' region)	-
frameshift	*erb-B* (3' region)	+

An AEV mutant possessing a frameshift that prohibited the expression of *erb-A* was fully capable of transforming chicken fibroblasts by all criteria examined. Deletion mutants in the *erb-A* region were more difficult to analyze. Chicken fibroblasts, infected in mass by an AEV mutant possessing a deletion of portions of *gag* and *erb-A*, demonstrated a near normal morphology and a very low (but reproducible) ability to form colonies in soft agar. These phenomena may be due, however, to poor replication by this mutant compared to either wild-type virus or to the other mutants studied. The lower titers of the deletion mutant may result from a removal of important RNA packaging or processing signals by the deletion. High titer virus stocks of the *erb-A* deletion mutant, isolated from the transformed cell colonies that did appear in the mass infection, demonstrated fibroblast transforming abilities similar to those of the wild-type parental virus.

In summary (Table 1), the results obtained thus far suggest that the *erb-B* domain is necessary for at least some parameters of fibroblast transformation (morphology and anchorage-independent growth). The *erb-A* domain does not appear to be necessary for fibroblast transformation. The effects of these mutations on the ability of AEV to transform erythroblasts and to induce neoplasms in animals is under study.

6. CELLULAR *ERB* SEQUENCES

As mentioned above, sequences homologous to the AEV *erb oncogenes* can be detected in the genomes of normal vertebrate cells (19, 20). These cellular *erb (c-erb)* sequences appear to be indigeneous to cells, and do not appear to be cryptic retroviruses: *c-erb* sequences are present in animals devoid of known endogenous retroviruses (20), the *c-erb* loci contain numerous introns (20), and *c-erb* related sequences are found in a variety of vertebrate species (19). The *c-erb A* and *B* domains of chicken cells were found to be separated by at least 12 Kb of DNA, and may not, in fact be physically linked (20).

The functions of these cellular *oncogenes* in the normal animal remains unclear. One theory, that cellular *oncogenes* are involved in cell differentiation, arises from the observation that many transformed cells appear to be blocked in differentiation relative to their normal cell counterparts (summarized in ref. 31). For example, hematopoietic cells transformed by temperature-sensitive mutants of AEV possess the properties of immature erythroblasts (CFU-E cells) at the permissive temperature, and do not synthesize hemoglobin. If the viral *oncogene* is inactivated by shifting to the non-permissive temperature, the AEV-infected cells appear to further differentiate along the erythrocyte lineage and begin to synthesize hemoglobin. Perhaps viral *oncogenes* induce neoplastic transformation by synthesizing in an uncontrolled manner homologues of proteins functional in normal cell differentiation. If cellular *oncogenes* are indeed involved in differentiation, levels of expression of c-*oncogenes* might be expected to vary among cells committed to different lineages. Although levels of expression of *c-erb* RNA in the chicken was found to vary from one cell type to another, no simple pattern has emerged to date correlating *c-erb* expression with any particular form of differentiation lineage (32).

7. CONCLUSIONS

For many years, avian erythroblastosis virus was ignored by molecular biologists. This situation began to change in the past decade. Although our understanding of AEV is not yet equal to that of its more famous relatives, the Rous sarcoma viruses, the outlines of the molecular biology of AEV are beginning to be discernable. What we already have learned of the avian erythroblastosis virus suggests that there are provocative differences between the *v-erb oncogenes* of AEV and the *v-src oncogene* of the Rous sarcoma viruses. Whether these differences reflect a deeper dichotomy in the actual mechanisms by which AEV and the Rous sarcoma viruses transform cells must be determined by further experimentation. Future characterization of the polypeptides synthesized by the *erb oncogene* region, coupled to a genetic analysis, may lead to an eventual understanding of the exact means by which AEV transforms cells, including the mechanisms behind the intriguing dual target cell transforming abilities of this virus.

8. REFERENCES

1) ROUS, P. (1911). A sarcoma of the fowl transmissible by an agent separable from the tumour cells. J. Exp. Med. 13, 397.

2) ENGELBRETH-HOLM, J. and MEYER, A.R. (1935). On the connection between erythroblastosis (hemocytoblastosis) myelosis and sarcoma in chicken. Acta Path. and Microbiol. Scanda., 12, 352.

3) GRAF, T., ROYER-POKORA, B., SCHUBERT, G.E. and BEUG, H. (1976). Evidence for the multiple oncogenic potential of cloned leukemia virus: *in vitro* and *in vivo* studies with Avian erythroblastosis virus. Virology, 71, 423.

4) OWADA, M., KAMAHORA, T. and YOSHIDA, K. (1978). Replication defective transforming virus of strain R avian erythroblastosis virus. Gann, 69, 857.

5) SHEINESS, D., VENNSTROM, B. and BISHOP, J.M. (1981). Virus specific RNAs in cells infected by avian myelocytomatosis virus and avian erythroblastosis virus: modes of *oncogene* expression. Cell, 23, 291.

6) ANDERSON, S.M., HAYWARD, W.S., NEEL, B.G. and HANAFUSA, H. (1980). Avian erythroblastosis virus produces two mRNAs. J. Virol. 36, 676.

7) LAI, M.M.C., NEIL, J.C. and VOGT, P.K. (1980). Cell-free translation of avian erythroblastosis virus RNA yields two specific and distinct proteins with molecular weights of

75,000 and 40,000. Virology, 100, 475.

8) PAWSON, T. and MARTIN, G.S. (1980). Cell-free translation of
    avian erythroblastosis virus RNA. J. Virol. 34, 280.

9) YOSHIDA, M. and TOYOSHIMA, K. (1980). *In vitro* translation of
    avian erythroblastosis virus RNA; identification of two major
    polypeptides. Virology, 100, 484.

10) VENNSTROM, B., FANSHIER, L., MOSCOVICI, C. and BISHOP, J.M.
    (1980). Molecular cloning of the avian erythroblastosis virus
    gene of chicken cells. J. Virol. 36, 575.

11) ISHIZAKI, R. and SHIMIZU, T. (1970). Heterogeneity of strain
    R avian (erythroblastosis) virus. Cancer Res., 30, 2827.

12) BISHOP, J.M. and VARMUS, H. (1982). In: Molecular Biology
    of Tumour Viruses; RNA Tumour viruses (ed. R. Weiss, N. Teich,
    H. Varmus and J. Coffin), p. 999, Cold Spring Harbor, New
    York.

13) OSKARSSON, M., McCLEMENTS, W.L., BLAIR, D.G., MAIZEL, J.V.
    and VANDE WOUDE, G.F. (1980). Properties of a normal mouse
    cell DNA sequence (sarc) homologous to the *src* sequence of
    Maloney sarcoma virus. Science, 207, 1222.

14) HAYWARD, W.S., NEEL, B.G. and ASTRIN, S.M. (1981). Activation
    of a cellular *onc* gene by promoter insertion in ALV-induced
    lymphoid leukosis. Nature, 290, 475.

15) BISTER, K. and DUESBERG, P.H. (1979). Structure and specific
    sequences of avian erythroblastosis virus RNA: evidence for
    multiple classes of transforming genes among avian tumour
    viruses. Proc. Natl. Acad. Sci. USA, 76, 5023.

16) KAMAHORA, T., SUGIYAMA, H., NOMOTO, A., YOSHIDA, M. and
    TOYOSHIMA, K. (1979). RNA specific for the transformation
    component of avian erythroblastosis virus strain R. Virology,
    96, 291.

17) LAI, M.M.C., HU, S.F. and VOGT, P.K. (1979). Avian erythro-
    blastosis virus: transformation specific sequences form a
    continuous segment of 3.2 Kb located in the middle of the

6 Kb genome. Virology, 97, 366.

18) COFFIN, J.M., VARMUS, H.E., BISHOP, J.M., ESSEX, M., HARDY, W.D., Jr., MARTIN, G.S., ROSENBERG, N.E., SCOLNICK, E.M., WEINBERG, R.A. and VOGT, P.K. (1981). Proposal for naming host cell derived inserts in retrovirus genomes. J. Virol., 40, 953.

19) ROUSSEL, M., SAULE, S., LAGROU, C., ROMMENS, C., BEUG, H., GRAF, T. and STEHELIN, D. (1979). Three new types of viral oncogene of cellular origin specific for hematopoietic cell transformation. Nature, 275, 496.

20) VENNSTROM, B. and BISHOP, J.M. (1982). Isolation and characterization of chicken DNA homologous to the two putative oncogenes of avian erythroblastosis virus. Cell, 28, 135.

21) GRAF, T., ADE, N. and BEUG, H. (1978). Temperature sensitive mutant of avian erythroblastosis virus suggests a block of differentiation as mechanisms of leukemogenesis. Nature, 275, 496.

22) ROYER-POKORA, B., GRIESER, S., BEUG, H. and GRAF, T. (1979). Mutant avian erythroblastosis virus with restricted target specificity. Nature, 282, 750.

23) PRIVALSKY, M.L. and BISHOP, J.M. (1982). Proteins specified by avian erythroblastosis virus: coding region localization and identification of a previously undetected erb-B polypeptide. Proc. Natl. Acad. Sci USA, 79, 3958.

24) BEEMON, K. and HUNTER, T. (1977). In vitro translation yields a possible Rous sarcoma virus src gene product. Proc. Natl. Acad. Sci. USA, 74, 3302.

25) KAMINE, J. and BUCHANAN, J.M. (1977). Cell-free synthesis of two proteins unique to RNA of transforming virious of Rous sarcoma virus. Proc. Natl. Acad. Sci. USA, 74, 2011.

26) BEEMAN, K. and HUNTER, T. (1978). Characterization of Rous sarcoma virus src gene products synthesized in vitro. J. Virol., 28, 551.

27) RETTENMIER, C.W., ANDERSON, S.M., RIEMAN, M.W., and HANAFUSA,
     H. (1979). *Gag*-related polypeptides encoded by replication
     defective avian oncoviruses. J. Virol., 32, 749.

28) HAYMAN, M.J., ROYER-POKORA, B. and GRAF, T. (1979). Defective-
     ness of avian erythroblastosis virus: synthesis of a 75 K
     gag-related protein. Virology, 92, 31.

29) KITCHENER, G. and HAYMAN, M.J. (1980). Comparative tryptic
     peptide mapping studies suggest a role in cell transformation
     for the *gag*-related protein of avian erythroblastosis virus
     strains CM II and MC29. Proc. Natl. Acad. Sci. USA, 77, 1637.

30) BEUG, H., GRAF, T. and HAYMAN, M.J. (1981). Production and
     characterization of antisera specific for the *erb*-portion
     of p 75, the presumptive transforming protein of avian
     erythroblastosis virus. Virology, 111, 201.

31) GRAF, T. and BEUG, H. (1978). Avian Leukemia Viruses. Inter-
     action with their target cells *in vivo* and *in vitro*. Biochim.
     Biophys. Acta, 516, 269.

32) GONDA, T.J., SHEINESS, D.K. and BISHOP, J.M. (1982). Transcripts
     from the cellular homologues of retroviral *oncogenes*: distri-
     bution among chicken tissues. Mol. Cell Biol., 2, 617.

PROTEIN KINASES SPECIFIC FOR TYROSINE RESIDUES AND THE ROLE OF TYROSINE PHOSPHORYLATION OF PROTEINS IN CELL TRANSFORMATION

J. Ghysdael

Laboratoire de Chimie Biologique, Département de Biologie Moléculaire, Université libre de Bruxelles 1640 Rhode St-Genèse, Belgium

1. THE RETROVIRUSES *ONC* GENES

Retroviruses are the etiological agents of various types of neoplasms (sarcomas, carcinomas and leukemias) in many avian and mammalian species (reviewed in ref. 1). On the basis of their pathogenicity in laboratory animals, oncogenic retroviruses have been broadly divided into two classes. Acute viruses induce tumours 15 to 30 days after injection in a susceptible host and are able to transform adequate target cells *in vitro*. Chronic viruses do not induce morphological transformation *in vitro* and require long latency periods (months to years depending on the virus-host considered) to induce tumours. This difference in pathogenicity is the result of the presence in the RNA genome of acute retroviruses of specific nucleotide sequences encoding protein unnecessary for viral replication but required for induction and maintenance of *in vitro* transformation and *in vivo* oncogenic potential. These transformation-specific sequences are referred to as v-*onc* genes. Retroviruses v-*onc* genes show strong sequence homology to specific cellular loci (c-*onc*). The c-*onc* genes are evolutionary conserved and are expressed in normal cells, sometimes in a tissue-specific manner.

209

It is generally believed that v-*onc* genes were derived by
retroviruses from these c-*onc* loci. Since v-*onc* genes are expressed
at much higher levels than their c-*onc* homologues, oncogenesis by
acute retroviruses might be the result of gene dosage. Alternatively
v-*onc* proteins might be functionally similar but not identical to
their c-*onc* homologues. The aberrant behaviour of v-*onc* gene
products might then be the basis for the oncogenic potential of
acute retroviruses. At least 15 different v-*onc* genes have been
identified and, for most of them, their translation product
characterized (reviewed in ref. 2).

2. THE ROUS SARCOMA VIRUS *src* GENE PRODUCT IS A PROTEIN KINASE
    SPECIFIC FOR TYROSINE RESIDUES

The Rous sarcoma virus (RSV) *src* gene encodes a 60 Kd protein
(p60^{src}), that was identified by immunoprecipitation from RSV-
transformed cells with antiserum from rabbits bearing an RSV-induced
tumour (TBR serum; 3) and by *in vitro* translation of subgenomic
portions of the RSV viral RNA (4, 5). P60^{src} is a phosphoprotein
and, present in immune complexes, it is able to rapidly transfer
the γ-phosphate of ATP to the heavy chains of the immunoglobins.
It thus functions *in vitro* as a protein kinase (6, 7). This kinase
has an unusual amino-acid specificity since it was found to
exclusively phosphorylate tyrosine residues (8).

Several lines of evidence indicate that the kinase activity
is intrinsic to p60^{src}. First, the molecule to which the Ig-H
chain phosphorylating activity is associated has the hydrodynamic
properties of a monomeric 60 Kd polypeptide (7, 9, 10). Second,
several schemes have been used to extensively purify the enzyme
responsible for the Ig-H chain phosphorylating activity and, in
each case, the activity was found to copurify with p60^{src} (9-11).
When tested in solution, in addition to Ig-H chain of TBR serum,
the kinase is able to phosphorylate several protein substrates at

tyrosine residues in a time-dependent reaction. Finally, the RSV
src gene together with the *E. coli* lac UV 5 promoter-operator
at its 5' end have been introduced into pBR325. After *E. coli*
transformation, colonies expressing p60^{src} were identified. When
purified from these expressor *E. coli* colonies, p60^{src} was found
to function as a protein kinase specific for tyrosine residues
(12). Since *E. coli* cells are unable to carry our protein phos-
phorylations (13), these experiments unambiguously demonstrate
that the tyrosine-specific transphosphorylating activity detected
in vitro is intrinsic to p60^{src}.

That this activity is essential for cell transformation and
tumour formation by RSV is supported by several observations.
First, mutants of RSV that are temperature-sensitive in their
ability to transform chicken embryo cells have p60^{src} with labile
protein kinase activities (6, 7, 10, 11). Also non conditional
RSV mutants in transformation encode p60^{src} devoid of protein kinase activity (14). Finally, the proteins of cells (avian or
mammalian) transformed by Rous sarcoma virus contain levels of
phosphotyrosine 5-10 times higher than those found in their uninfected homologues and these levels are temperature-dependent
in cells transformed by RSV mutants temperature sensitive for
cell transformation (15).

3. TYROSINE PHOSPHORYLATION OF CELLULAR PROTEINS IS ALSO INVOLVED
    IN THE TRANSFORMATION MECHANISM OF RETROVIRUSES OTHER THAN RSV

The involvement of protein phosporylation at tyrosine residues
in the transformation mechanism of acute retroviruses is not
unique to RSV. The characterization of additional isolates of acute
retroviruses led to the definition of other cell-derived, transformation-specific inserts unrelated to *src* (reviewed in ref. 2).
Expression of at least five of these inserts (Table 1) also results
in a significant increase in the phosphorylation at tyrosine

residues of the transformed cell proteins (32, 33). These phosphorylation levels are temperature-dependent in cells infected by
mutant viruses temperature sensitive for transformation (34).

The transformation-specific inserts of these viruses are
expressed as the carboxyterminal portion of polyproteins that
obtain their aminoterminal domain from a partial viral *gag* gene.
All these polyproteins are phosphoproteins with phosphoserine -
and phosphotyrosine - containing sites located in the transformation-specific domain of the molecules (23, 30, 35). Incubation
of immunoprecipitates containing these transformation-specific
proteins in the presence of γ [^{32}P]ATP and a divalent cation results
in the phosphorylation at tyrosine residues of both the polyprotein
and exogenously added protein substrates (18, 21, 23, 25-27, 29-
31). Except for Ab-MuLV p120 (33), the same major site(s) on the
polyproteins were found to be phosphorylated *in vitro* as those
detected *in vivo*. In every case, it is not known whether the phosphorylation of the transformation-specific protein and that of
the exogenous substrates are carried out by the same protein kinase. With the exception of RSV p60^{src} (see section 2) and Ab-
MuLV p120$^{gag-abl}$, no direct experimental evidence of an activity
intrinsic to the polyproteins themselves has been provided.

Relevant to this question, it should be noted that clear
similarities exist in both the location (36, 37) and structure
of the tyrosine-containing sites of the gene products of the
src, *yes* and *fps* sequences (38, 39). These similarities were found
to be limited to the carboxyterminal domain of the transformation-
specific proteins and are consistent with the divergence of the
corresponding domain of c-*src*, c-*fps* and c-*yes* from a common
ancestor. Also, avian sarcoma viruses *fps* gene has been shown to
be related to the *fes* gene of strains Gardner-Arnstein and Snyder-
Theilen of feline sarcoma viruses (40).

Table 1. *Names of the transformation-specific inserts and translation products whose expression is involved in increased levels of phosphotyrosine in the proteins of transformed cells*

v-onc gene	Viral insert	v-onc protein product	P Tyr in v-onc protein product	Intrinsic protein kinase activity	Increased P Tyr in cellular proteins	References
Avian viruses						
src	RSV-src	p60^{src}	yes	yes	yes	
	B77-src	p60^{src}	yes	yes	yes	6-12
	rASV src	p60^{src}	yes	yes	yes	
fps	FSV-fps	p140$^{gag-fps}$	yes	ND	yes	16-18
	PRCII-fps	p105$^{gag-fps}$	yes	ND	yes	19-21
	PRCIV-fps	p170$^{gag-fps}$	yes	ND	yes	22, 23
	UR-I-fps	p150$^{gag-fps}$	yes	ND	yes	24
yes	Y73-yes	p90$^{gag-yes}$	yes	ND	yes	25
	ESV-yes	p80$^{gag-yes}$	yes	ND	yes	26
ros	UR-2-ros	p68$^{gag-ros}$	yes	ND	yes	27
Mammalian viruses						
abl	Ab-MuLV-abl	p120$^{gag-abl}$	yes	yes	yes	28, 29
fes	ST-FeSV-fes	p85$^{gag-fes}$	yes	ND	yes	30
	GA-FeSV-fes	p110$^{gag-fes}$	yes	ND	yes	31

An approach aiming at the identification of classes of tyrosine-specific protein kinases has been to compare the amino-acid sequences around the target phosphotyrosine of several *in vivo* substrates. The rationale for such experiments comes from the study of the better characterized serine-specific protein kinases where it is known that (1) selection of the target serine residue on the substrate is determined, at least in part, by the amino acid sequence around this site, (2) different kinases can be clearly distinguished on that basis (41-44). Comparison of the sequences around the phosphotyrosine of several transformation-specific proteins has led to the recognition of common features such as (1) the presence of one or more lysine/arginine located at 6-7 residues to the NH_2-terminal side of the target phosphotyrosine and, (2) the presence of one or more acidic residues between the target phosphotyrosine and the lysine/arginine residues (38, 39, 45). The major tyrosine phosphorylation site of the epidermal growth factor receptor (see section 5) also displays these structural features (39), suggesting that enzymes with similar specificities phosphorylate these various substrates. The importance of acidic residues at the amino-terminal side of the target phosphotyrosine is further indicated by the fact that, although a synthetic peptide representing the tyrosine-phosphorylation site of p60[src] is used as a substrate by the EGF-stimulated receptor kinase (see section 5), replacement of the acid residues by basic ones makes the modified peptide a less efficient substrate (46).

Not all tyrosine-specific protein kinases appear to require such an environment to select the target tyrosine residue since the phosphorylation site of the cellular 36 Kd protein (see section 4) has no apparent homology to those just described (39). It should also be mentioned that for the polyprotein phosphorylating activities detected *in vitro*, clear differences exist among the different viruses in parameters such as the ability to use ATP or GTP, the divalent cation requirement, the pH optimas etc. (18, 27, 47).

4. POTENTIALLY IMPORTANT SUBSTRATES OF TYROSINE-SPECIFIC KINASES FOR THE TRANSFORMATION MECHANISM

To determine the pathway(s) by which protein phoshorylation at tyrosine residues leads to cell transformation, it will be necessary (1) to define potentially important substrates in transformed cells, (2) to correlate the phosphorylation of parti- cular proteins with a modulation of their function and, (3) to demonstrate that the modulation of function determines one of the many facets that define the transformed phenotype. Although several potentially important substrates have been defined, no direct experimental evidence exists to correlate phosphorylation of these substrates with a modification of function.

The identification of substrates for the tyrosine-specific kinases has been complicated by the relaxed specificity displayed by these enzymes *in vitro*. For example, *in vitro*, purified p60^{src} will readily phosphorylate proteins such as actin, tubulin or α-casein (10, 48) which are clearly non substrates *in vivo*. Potential substrates were initially search for by comparing the phosphory- lation pattern of proteins from normal and transformed cells using two dimensional gel electrophoresis. This approach led to the discovery by Radke and Martin (49) of a 36 Kd phophoprotein in RSV- transformed cells. The 36 Kd protein is present as a population of non-phosphorylated molecules in the cytoplasm of normal cells, 10% of which becomes rapidly phosphorylated at tyrosine and serine residues upon cell transformation by the viruses listed in Table 1 (50-52). At present, no function has been associated with the 36 Kd protein.

Because tyrosine phosphorylation is a rare modification as compared to serine and threonine phosporylation of cellular proteins even in transformed cells, the identification of additional poten- tial substrates by two-dimensional gel electrophoresis was made

possible only by exploiting the relative stability of the phosphate
ester bond of phosphotyrosine as compared to that of phosphoserine
and phosphothreonine under alkaline conditions. Using this
procedure, Cooper and Hunter (53) have detected, in addition to
the 36 Kd protein, two additional polypeptides which appear to
be phosphorylated *de novo* at tyrosine upon RSV transformation and
four other proteins whose phosphorylation is increased in transformed cells. These proteins appear to be modified in the same
way in cells transformed by *fps*, *yes* and *abl*-containing viruses
(52). Again, no data are available concerning either the localization
or function of any of these proteins. Another approach to define
substrates for tyrosine-specific protein kinases has been to look
for the presence of phosphotyrosine in proteins involved in
cellular functions known to be modified upon transformation. These
modifications are numerous (reviewed in 54, 55) and only some of
these transformation parameters correlate well with tumour-forming
ability (56).

The fact that the organization of microfilaments is disrupted
in RSV-transformed cells (57-59) led Sefton *et al.*, (60) to investigate whether cytoskeletal proteins of RSV-transformed cells
contained high levels of phosphotyrosine. Although several of the
cytoskeletal proteins are phosphoproteins, only filamin, vimentin
and vinculin were found to contain phosphotyrosine in RSV-transformed cells. Low levels of phosphotyrosine (1-2% of the total
phosphoaminiacids of these proteins) were found in filamin and vimentin immunoprecipitated from RSV-transformed cells but not from
normal cells. When isolated from normal cells, vinculin was found
to contain phosphoserine, phosphothreonine and low level of phosphotyrosine (2% of the total phosphoaminoacids of the protein).
However, upon cell transformation by RSV, the level of phosphotyrosine in vinculin increased to approximately 25% of the total
phosphoaminoacids of the protein.

Evidence that vinculin might be phosphorylated directly by p60^{src} comes from the fact that both vinculin (61, 62) and p60^{src} (63) are concentrated in the same region of the cytoplasmic membrane at the ventral surface of cells. In normal cells, vinculin together with α-actinin are concentrated in regions of the cytoplasmic membrane of cell-substratum contact where actin microfilament bundles terminate (focal contacts; 61, 62) and , possibly, as part of a transmembrane complex involving microfilaments and vinculin inside the cells and fibronectin outside cells (64). Upon transformation by RSV, focal contacts are reduced in number and size, actin microfilaments bundles are disrupted and the distribution of α-actinin and vinculin is more diffused and superposed with that of p60^{src} (65, 66). Phosphorylation of vinculin at tyrosine residues might possibly be involved in the disruption of the transmembrane complex with the consequent loss of adherence and flattened morphology of the cell. Consistent with this hypothesis vinculin is also found to be phosphorylated at tyrosine residues in cells transformed by Y73 and Ab-MuLV but not in cells transformed by PRCII (60); cells transformed by Ab-MuLV, Y73 and ESV have indeed a rounded morphology whereas cells transformed by PRCII and FSV are fusiform. Evidence exists that cells exhibiting a fusiform morphology have an adherence intermediate between that of normal cells and rounded transformed cells (reviewed in 54).

Two proteins of M_r 50 and 90 Kd appear to be complexed with p60^{src} (8) and with the transformation specific polyproteins of at least *fps* and *yes* containing viruses (67). In uninfected cells the 50 Kd protein is labelled only at serine residues and becomes further phosphorylated at tyrosine upon avian sarcoma viruses transformation. However, unlike phosphorylation of other potential substrates for p60^{src}, phosphorylation at tyrosine of the 50 Kd protein is not temperature-dependent in cells transformed by RSV mutants that are temperature-sensitive in transformation (67). Recent evidence

indicate that this 50 Kd component is involved in the cytoplasmic
transport of p60^{src} from its synthesis site to its final membrane
location. The 90 Kd protein is only phosphorylated at serine in
both normal and transformed cells and it appears to be identical
to one of the cellular heat-shock proteins (8). The functional
significance of its association with p60^{src} is unknown.

5. INVOLVEMENT OF TYROSINE-SPECIFIC PROTEIN KINASES IN THE METABOLISM OF NORMAL CELLS

The translation products of the cellular equivalents of
the viral transforming genes listed in Table 1 have been identified
for c-*src* (avian and mammalian p60^{c-src}, ref. 69), c-*abl* (mouse
cells NCP 150, ref. 70), c-*fes* (feline cells NCP 92, ref. 31)
and c-*fps* (chicken cells NCP 98, ref. 71). All phosphoproteins
and immunoprecipitates containing these proteins are associated
with tyrosine specific protein kinases. No data is available
as to their role in the metabolism of normal cells.

Binding of several polypeptide growth factors to specific
receptors stimulates the phosphorylation of tyrosine residues of
both membrane and cytoplasmic proteins. Epidermal growth factor
(EGF), a 6 Kd polypeptide purified from the submaxillary gland of
the mouse, interacts with a specific receptor present in many
different cell types to produce a number of phenotypic changes
(reviewed in 72). Rapid changes include, among others, increased
transport of sugars and ions, increased glycolysis and membrane
ruffling. The best characterized long-term effects are initiation
of DNA synthesis and increased cell multiplication. At the molecular level, EGF appears to form a complex with plasma membrane
receptors followed by internalization and degradation of the EGF-
receptor complex. EGF binding to purified plasma membrane stimulates a protein kinase activity, resulting in the phosphorylation
at tyrosine residues both of a 170 Kd membrane protein and exo-

genously added protein substrates such as histones, TBR serum Ig-H chain or a synthetic peptide corresponding to the sequence of the phosphorylation site in p60^{src} (46, 73, 74).

Similarly, binding of platelet-derived growth factor (PDGF; reviewed in ref. 75) to plasma membranes of human glial cell or of cells derived from connective tissues from various sources is accompanied by phosphorylation at tyrosine residues of membrane proteins with M_r of 175 and 130 Kd, unrelated to the 170 Kd component phosphorylated upon EGF binding (76). There is good evidence that the 170 Kd protein is the EGF receptor and that the kinase activity is intrinsic to that polypeptide (77, 78).

Probably as a consequence of receptor phosphorylation (79), EGF appears to stimulate the phosphorylation at tyrosine residues of other cellular proteins. This effect is particularly striking with the A431 human carcinoma cell-line where the addition of EGF results in a rapid 3-4 fold increase in the level of phosphotyrosine associated with cellular proteins (80). At least two proteins of M_r 81 and 39 Kd are phosphorylated *de novo* upon EGF addition (80). The 39 Kd polypeptide is the human homolog of the 34 Kd polypeptide phosphorylated in cnicken cells upon transformation by avian sarcoma viruses (see section 4). In fact, the 39 Kd protein is phosphorylated at the same site in A431 cells infected with Rous sarcoma virus and upon EGF binding (81) indicating that the kinases phosphorylating this polypeptide in these two cases have the same specificity. Similar experiments with human HeLa cells or mouse 3T3 cells show a slight increase in tyrosine phosphorylation of cellular proteins upon EGF addition (80). This difference is possibly the result of the ten fold excess of EGF receptors found on A431 cell membrane as compared to HeLa or 3T3 cells (10^6/cell vs 10^5/cell, respectively) and of the inability of A431 cells to internalize the EGF-receptor complex.

It is unclear at present which of the many effects of EGF are mediated through protein phosphorylation. It should be emphasized that, although highly responsive to EGF-stimulated phosphorylation, A431 cells, in contrast to most other cells, are growth-inhibited upon EGF addition (82). It is possible that the anomalous high level of kinase activation observed in these cells is responsible for this growth inhibition. Consistent with this, addition of EGF to A431 variants selected for their resistance to EGF growth-inhibition results in small or undetectable increase in the level of phosphotyrosine of cellular proteins (79). It remains to be seen whether in properly responsive cells, the low level of EGF-stimulated tyrosine phosphorylation is required for growth stimulation of these cells. As with EGF, the mitogenic effects of $p60^{src}$ also do not appear to require high levels of p60-kinase acitivity (83). Whether tyrosine phosphorylation of appropriate protein-targets is not required for mitogenesis or whether it is required on a small subset of the total target proteins of transformed cells or only transiently is at present unknown.

Recent studies with partial transformation mutants of RSV suggest that high levels of phosphorylation of the 36 Kd protein are necessary for tumour formation (56). Tyrosine phosphorylation of the EGF receptor is also observed upon binding of transforming growth factors (TGF; 84). TGFs are low molecular weight polypeptides produced by several transformed cell-lines including cells transformed by RNA or DNA tumour viruses and cell-lines of human tumour origin (85-87). Although unrelated structurally to EGF, such factors, via the EGF-receptor system, are able to induce normal cells to display phenotypic properties of transformed cells, e.g. the ability to grow in serum-free or in semi-solid media. Their production by transformed cells is believed to regulate the anchorage-independent growth of these cells. Clearly, the study of the molecular basis of the non-transforming and transforming properties of EGF and TGFs, respectively, may help to define the biochemical pathways that distinguish malignant from non-malignant growth.

6. REFERENCES

1) TOOZE, J. The molecular biology of tumour viruses. (1973). Cold Spring Harbor Laboratory.

2) COFFIN, J.M., VARMUS, H.E., BISHOP, J.M., ESSEX, M., HARDY, W.D., MARTIN, G.S., ROSENBERG, N.E., SCOLNICK, E.M., WEINBERG, R.A. and VOGT, P.K. (1981). Proposal for naming host-derived inserts in retrovirus genomes. J. Virology, 40, 953.

3) BRUGGE, J.S. and ERIKSON, R.L. (1977). Identification of a transformation-specific antigen induced by an avian sarcoma virus. Nature (London), 269, 346.

4) BEEMON, K. and HUNTER, T. (1978). Characterization of Rous sarcoma virus *src* gene products synthesized *in vitro*. J. Virology, 28, 551.

5) PURCHIO, A.F., ERIKSON, E. and ERIKSON, R.L. (1977). Translation of 35 S and subgenomic regions of avian sarcoma virus RNA. Proc. Natl. Acad. Sci. USA, 74, 4661.

6) COLLETT, M.S. and ERIKSON, R.L. (1978). Protein kinase activity associated with the avian sarcoma virus *src* gene product. Proc. Natl. Acad. Sci. USA, 75, 2021.

7) LEVINSON, A.D., OPPERMANN, H., LEVINTOW, L., VARMUS, H.E. and BISHOP, J.M. (1978). Evidence that the transforming gene of avian sarcoma virus encodes a protein kinase associated with a phosphoprotein. Cell, 15, 561.

8) HUNTER, T. and SEFTON, B.M. (1980). The transforming gene product of Rous sarcoma virus phosphorylates tyrosine. Proc. Natl. Acad. Sci. USA, 77, 1311.

9) MANESS, P.F., ENGESER, H., GREENBERG, M.E., O'FARRELL, M., GALL, W.E. and EDELMAN, G.M. (1979). Characterization of the protein kinase activity of avian sarcoma virus *src* gene product. Proc. Natl. Acad. Sci. USA, **76**, 5028.

10) LEVINSON, A.D., OPPERMANN, H., VARMUS, H.E. and BISHOP, J.M. (1980). The purified product of the transforming gene of avian sarcoma virus phosphorylates tyrosine. J. Biol. Chem., 11973.

11) ERIKSON, R.L., COLLETT, M.S., ERIKSON, E.L. and PURCHIO, A.F. (1979). Evidence that the avian sarcoma virus transforming gene product is a cyclic AMP-independent protein kinase. Proc. Natl. Acad. Sci. USA, **76**, 6260.

12) GILMER, T.M. and ERIKSON, R.L. (1981). Rous sarcoma virus transforming protein, p60^{src}, expressed in *E. coli*, functions as a protein kinase. Nature (London), **294**, 771.

13) RUBIN, C.S. and ROSEN, O.M. (1975). Protein phosphorylation. Ann Rev. Biochem. **44**, 831.

14) OPPERMANN, H., LEVINSON, A.D. and VARMUS, H.E. (1981). The structure and protein kinase activity of proteins encoded by non conditional mutants and back mutants in the *src* gene of avian sarcoma virus. Virology, **108**, 47.

15) SEFTON, B.M., HUNTER, T., BEEMON, K. and ECKHART, W. (1980). Evidence that the phosphorylation of tyrosine is essential for cellular transformation by Rous sarcoma virus. Cell, **20**, 807.

16) HANAFUSA, T., WANG, L.H., ANDERSON, S.M., KARESS, R.E., HAYWARD, W.S. and HANAFUSA, H. (1980). Characterization of the transforming gene of Fujinami sarcoma virus. Proc. Natl. Acad. Sci. USA, **77**, 3009.

17) LEE, W.H., BISTER, K., PAWSON, A., ROBBINS, T., MOSCOVICI, C. and DUESBERG, P.H. (1980). Fujinami sarcoma virus: an avian RNA tumour virus with a unique transforming gene. Proc. Natl. Acad. Sci. USA, **77**, 2018.

18) FELDMAN, R.A., HANAFUSA, T. and HANAFUSA, H. (1980). Characterization of protein kinase activity associated with the

transforming gene product of Fujinami sarcoma virus. Cell, 22, 757.

19) BREITMAN, M., NEIL, J.C., MOSCOVICI, C. and VOGT, P.K. (1981). The pathogenicity and defectivenes of PRCII. Virology, 108, 1.

20) NEIL, J.C., BREITMAN, M. and VOGT, P.K. (1981). Characterization of a 105 Kd *gag*-related phosphoprotein from cells transformed by the defective avian sarcoma virus PRCII. Virology, 108, 98.

21) NEIL, J.C., GHYSDAEL, J. and VOGT, P.K. (1981). Tyrosine-specific protein kinase activity associated with p105 of avian sarcoma virus PRCIV. Virology, 109, 223.

22) BREITMAN, M., HIRANO, A., WONG, T. and VOGT, P.K. (1981). Characteristics of avian sarcoma virus strain PRC IV and comparison with strain PRCIIp. Virology, 114, 451

23) GHYSDAEL, J., NEIL, J.C. and VOGT, P.K. (1981). Cleavage of four avian sarcoma virus polyproteins with virion protease p15 removes *gag* sequences and yields large fragments that function as tyrosine phosphoacceptors *in vitro*. Proc. Natl. Acad. Sci. USA, 78, 5847.

24) WANG, L.H., FELDMAN, R., SHIBUYA, M., HANAFUSA, H., NOTTER, M. and BALDUZZI, P.C. (1981). Genetic structure, transforming sequence, and gene product of avian sarcoma virus UR 1. J. Virol., 40, 258.

25) KAWAI, S., YOSHIDA, M., SEGAWA, K., SUGIYAMA, H., ISHIZAKI, R. and TOYOSHIMA, K. (1980). Characterization of Y73, a newly isolated avian sarcoma virus. Proc. Natl. Acad. Sci. USA, 77, 6199.

26) GHYSDAEL, J., NEIL, J.C., WALLBANK, A.M. and VOGT, P.K. (1981). Esh avian sarcoma virus codes for a *gag*-linked transformation-specific protein with an associated protein kinase activity. Virology, 111, 386.

27) FELDMAN, R.A., WANG, L.H., HANAFUSA, H. and BALDUZZI, P.C.
    (1982). Avian sarcoma virus UR-2 encodes a transforming protein
    which is associated with a unique protein kinase activity.
    J. Virol., 42, 228.

28) WITTE, O.N., ROSENBERG, N., PASKIND, M., SHIELDS, A. and
    BALTIMORE, D. (1978). Identification of an Abelson murine
    leukemia virus-encoded protein present in transformed fibro-
    blasts and lymphoid cells. Proc. Natl. Acad. Sci. USA, 75,
    2488.

29) WITTE, O.N., DASGUPTA, A. and BALTIMORE, D. (1980). Abelson
    murine leukemia virus protein is phosphorylated *in vitro* to
    form phosphotyrosine. Nature (London), 283, 826.

30) VAN DE VEN, W.J.M., REYNOLDS, F.H. and STEPHENSON, J.R. (1980).
    The non-structural components of polyproteins encoded by
    replication-defective mammalian transforming retroviruses are
    phosphorylated and have associated protein kinase activity.
    Virology, 101, 185.

31) BARBACID, M., BEEMON, K. and DEVARE, S.G. (1980). Origin and
    functional properties of the major gene product of the
    Snyder-Theilen strain of feline sarcoma virus. Proc. Natl.
    Acad. Sci. USA, 77, 5158.

32) BEEMON, K. (1981). Transforming proteins of some feline and
    avian sarcoma virus are related structurally and functionally.
    Cell, 24, 145.

33) SEFTON, B.M., HUNTER, T. and RASCHKE, W.C. (1981). Evidence
    that the Abelson virus protein functions *in vivo* as a protein
    kinase that phosphorylates tyrosine. Proc. Natl. Acad. Sci.
    USA, 78, 1552.

34) PAWSON, T., GUYDEN, J., KUNG, T.H., RADKE, K., GILMORE, T.
    and MARTIN, G.S. (1980). A strain of Fijinami sarcoma virus
    is temperature sensitive in protein phosphorylation and
    cellular transformation. Cell, 22, 767.

35) PAWSON, T., KUNG, T.H. and MARTIN, G.S. (1981). Structure and phosphorylation of the Fujinami sarcoma virus gene product. J. Virology, 40, 665.

36) COLLETT, M.S., ERIKSON, E. and ERIKSON, R.L. (1979). Structural analysis of the avian sarcoma virus transforming protein: sites of phosphorylation. J. Virol., 24, 770.

37) NEIL, J.C., GHYSDAEL, J., VOGT, P.K. and SMART, J.E. (1982). Structural similarities of proteins encoded by three classes of avian sarcoma viruses. Virology, 121, 274.

38) NEIL, J.C., GHYSDAEL, J., VOGT, P.K. and SMART, J.E. (1981). Homologous tyrosine-phosphorylation sites in transformation-specific gene products of distinct avian sarcoma viruses. Nature (London), 291, 675.

39) PATCHINSKY, T., HUNTER, T., ESCH, F.S., COOPER, J.A. and SEFTON, B.M. (1982). Analysis of the sequence of amino acids surroundings sites of tyrosine phosphorylation. Proc. Natl. Acad. Sci. USA, 79, 973.

40) SHIBUYA, M., HANAFUSA, T., HANAFUSA, H. and STEPHENSON, J.R. (1980). Homology exists among the transforming sequences of avian and feline sarcoma viruses. Proc. Natl. Acad. Sci. USA, 77, 6536.

41) KEMP, B.E., BYLUND, D.B., HUANG, T. and KREBS, E.G. (1975). Substrate specificity of the cyclic AMP-dependent protein kinase. Proc. Natl. Acad. Sci. USA, 72, 3448.

42) KEMP, B.E., GRAVES, D.J., BENJAMINI, E. and KREBS, E.G. (1977). Role of the multiple basic residues in determining the substrate specificity of cyclic-AMP-dependent protein kinase. J. Biol. Chem., 252, 4888.

43) TUAZON, P.T., BINGHAM, E.W. and TRAUGH, J.A. (1979). Cyclic nucleotide independent protein kinases from rabbit reticulocytes. Eur. J. Biochem., 94, 497.

44) JESSE-CHAN, K.F., HURST, M.O. and GRAVES, D.J. (1982). Phosphorylase kinase specificity. J. Biol. Chem., 257, 3655.

45) SMART, J.E., OPPERMANN, H., CZERNILOFSKY, A.P., PURCHIO, A.F., ERIKSON, R.L. and BISHOP, J.M. (1981). Characterization of sites for tyrosine phosphorylation in the transforming protein of Rous sarcoma virus (pp60^{v-src}) and its normal cellular homologue (pp60^{c-src}). Proc. Natl. Acad. Sci. USA, 78, 6013.

46) PIKE, L.J., GALLIS, B., CASNELLIE, J.E., BORNSTEIN, P. and KREBS, E.G. (1982). Epidermal growth factor stimulates the phosphorylation of synthetic tyrosine-containing peptides by A431 cell membranes. Proc. Natl. Acad. Sci. USA, 79, 1443.

47) GHYSDAEL, J., NEIL, J.C. and VOGT, P.K. (1981). A third class of avian sarcoma viruses, defined by related transformation-specific proteins of Yamaguchi 73 and Ech sarcoma virus. Proc. Natl. Acad. Sci. USA, 78, 2611.

48) COLLETT, M.S., PURCHIO, A.F. and ERIKSON, R.L. (1980). Avian sarcoma virus-transforming protein, pp60^{src} shows protein kinase activity specific for tyrosine. Nature, (London), 285, 167.

49) RADKE, K. and MARTIN, G.S. (1979). Transformation by Rous sarcoma virus: effects of *src* gene expression on the synthesis and phosphorylation of cellular polypeptides. Proc. Natl. Acad. Sci. USA, 76, 5212.

50) RADKE, K., GILMORE, T. and MARTIN, G.S. (1980). Transformation by Rous sarcoma virus: a cellular substrate for transformation-specific protein phosphorylation contains phosphotyrosine. Cell, 21, 821.

51) ERIKSON, E. and ERIKSON, R.L. (1980). Identification of a cellular protein substrate phosphorylated by the avian sarcoma virus-transforming gene product. Cell, 21, 829.

52) COOPER, J.A. and HUNTER, T. (1981). Four different classes of retroviruses induce phosphorylation of tyrosines present in similar cellular proteins. Mol. Cell Biol. 1, 394.

53) COOPER, J.A. and HUNTER, T. (1981). Changes in protein phosphorylation in Rous sarcoma virus-transformed chicken embryo

cells. Mol. Cell Biol., 1, 165.

54) VOGT, P.K. (1977). Genetics of RNA tumour viruses. In: Comprehen-
     sive Virology (Eds., H. Fraenkel-Conrat and R. Wagner.), 9,
     p. 341, Plenum Press, New York and London.

55) HANAFUSA, H. (1977). Cell transformation by RNA tumour viruses.
     In: Comprehensive Virology (Eds. K. Fraenkel-Conrat and R.
     Wagner), 10, p. 401, Plenum Press, New York and London.

56) KAHN, P., NAKAMURA, K., SHIN, S., SMITH, R.E. and WEBER, M.J.
     (1982). Tumorigenicity of partial transformation mutants of
     Rous sarcoma virus. J. Virol., 42, 602.

57) EDELMAN, G.M. and YAHARA, I. (1976). Temperature-sensitive
     changes in surface modulaing assemblies of fibroblasts
     transformed by mutants of Rous sarcoma virus. Proc. Natl.
     Acad. Sci. USA, 73, 2047.

58) ASH, J.F., VOGT, P.K. and SINGER, S.J. (1976). Reversion from
     transformed to normal phenotype by inhibition of protein
     synthesis in rat kidney cells infected with a temparature
     senstive mutant of Rous sarcoma virus. Proc. Natl. Acad.
     Sci. USA, 73, 3603.

59) WANG, E. and GOLDBERG, A.R. (1976). Changes in microfilament
     organization and surface topography upon transformation of
     chick embryo fibroblasts with Rous sarcoma virus. Proc. Natl.
     Acad. Sci. USA, 73, 4065.

60) SEFTON, B.M., HUNTER, T., BALL, E.H. and SINGER, S.J. (1981).
     Vinculin: a cytoskeletal target of the transforming protein
     of Rous sarcoma virus. Cell, 24, 165.

61) GEIGER, B., TOKUYASU, K.T., DUTTON, A.H. and SINGER, S.J.
     (1980). Vinculin, an intracellular protein localized at
     specialized sites where microfilaments bundles terminate
     at cell membrane. Proc. Natl. Acad. Sci. USA, 77, 4127.

62) BURRIDGE, K. and FERAMISCO, J. (1980). Microinjection and
     localization of a 130 K protein in living fibroblasts: a
     relationship to actin and fibronectin. Cell, 19, 587.

63) ROHRSCHNEIDER, L.R. (1980). Adhesion plaques of Rous sarcoma virus-transformed cells contain the *src* gene product. Proc. Natl. Acad. Sci. USA, 77, 3514.

64) HYNES; R. (1982). Phosphorylation of vinculin by pp60^{src}: what might it mean? Cell, 28, 437.

65) DAVID-PFEUTY, T. and SINGER, S.J. (1980). Altered distributions of cytoskeletal proteins vinculin and α-actinin in cultured fibroblasts transformed by Rous sarcoma virus. Proc. Natl. Acad. Sci. USA, 77, 6687.

66) SHRIVER, K. and ROHRSCHNEIDER, L. (1981). Organization of pp60^{src} and selected cytoskeletal proteins within adhesion plaques and junctions of Rous sarcoma virus-transformed rat cells. J. Cell Biol., 89, 525.

67) BRUGGE, J.S. and DARROW, D. (1982). Rous sarcoma virus-induced phosphorylation of a 50,000 molecular weight cellular protein. Nature (London), 295, 250.

68) OPPERMANN, H., LEVINSON, W. and BISHOP, J.M. (1981). A cellular protein that associates with the transforming protein of Rous sarcoma virus is also a heat-shock protein. Proc. Natl. Acad. Sci. USA, 78, 1067.

69) OPPERMANN, H., LEVINSON, A.D., VARMUS, H.E., LEVINTOW, L. and BISHOP, J.M. (1979). Uninfected vertebrate cells contain a protein that is closely related to the product of the avian sarcoma virus transforming gene (*src*). Proc. Natl. Acad. Sci. USA, 76, 1804.

70) WITTE, O.N., ROSENBERG, N.E. and BALTIMORE, D. (1979). A normal cell protein cross-reactive to the major Abelson murine leukemia virus gene product. Nature (London), 281, 396.

71) MATHEY-PREVOT, B., HANAFUSA, H. and KAWAI, S. (1982). A cellular protein is immunologically cross reactive with and functionally homologous to the Fujinami sarcoma virus transforming protein. Cell, 28, 897.

72) CARPENTER, G. and COHEN, S. (1979). Epidermal Growth Factor.

Ann. Rev. Biochem., 48, 193.

73) USHIRO, H. and COHEN, S. (1980). Identification of phospho-
    tyrosine as a product of Epidermal Growth Factor-activated
    protein kinase in A431-cell membranes. J. Biol. Chem., 255,
    8363.

74) CHINKERS, M. and COHEN, S. (1981). Purified EGF receptor-
    kinase interacts specifically with antibodies to Rous sarcoma
    virus transforming protein. Nature (London), 290, 516.

75) ROSS, R. and VOGEL, A. (1978). The Platelet-derived growth
    factor. Cell, 14, 203.

76) EK, B., WESTERMARK, B., WASTESON, A. and HELDIN, C.H. (1982).
    Stimulation of tyrosine-specific phosphorylation by platelet-
    derived growth factor (PDGF). Nature (London) 295, 419.

77) COHEN, S., CARPENTER, G. and KING, L. (1980). Epidermal growth
    factor-receptor-protein kinase interactions. J. Biol. Chem.,
    255, 4834.

78) BUHROW, S.A., COHEN, S. and STAROS, J.V. (1982). Affinity
    labelling of the protein kinase associated with the EGF
    receptor in membrane vesicles from A431 cells. J. Biol. Chem.,
    257, 4019.

79) BUSS, J.E., KUDLOW, J.E., LAZAR, C.S. and GILL, G.N. (1982).
    Altered epidermal growth factor (EGF)-stimulated protein
    kinase activity in variants A431 cells with altered growth
    responses to EGF. Proc. Natl. Acad. Sci. USA, 79, 2574.

80) HUNTER, T. and COOPER, J.A. (1981). Epidermal Growth Factor
    induces rapid tyrosine phosphorylation of proteins in A431
    human tumour cells. Cell, 24, 741.

81) COOPER, J.A. and HUNTER, T. (1981). Similarities and differen-
    ces between the effects of Epidermal Growth Factor and Rous
    sarcoma virus. J. Cell Biol., 91, 878.

82) GILL, G.N. and LAZAR, C.S. (1981). Increased phosphotyrosine
    content and inhibition of proliferation in EGF-treated A431
    cells. Nature (London), 293, 305.

83) POIRIER, F., GALOTHY, G., KARESS, R.E., ERIKSON, E. and
    HANAFUSA, H. (1982). Role of p60^{src} kinase activity in the
    induction of neuroretinal cell proliferation by Rous sarcoma
    virus. J. Virol. 42, 780.

84) REYNOLDS, F.H., TODARO, G., FRYLING, C. and STEPHENSON, J.R.
    (1981). Human transforming growth factors induce tyrosine
    phosphorylation of EGF receptors. Nature (London), 292, 259.

85) DE LARCO, J.E. and TODARO, G.J. (1978). Growth factors from
    murine sarcoma-virus transformed cells. Proc. Natl. Acad.
    Sci. USA, 75, 4001-4005.

86) TODARO, G.J., FRYLING, C. and DE LARCO, J.E. (1980). Trans-
    forming growth factors produced by certain human tumour cells:
    polypeptides that interact with epidermal growth factor
    receptors. Proc. Natl. Acad. Sci. USA, 77, 5258.

87) OZANNE, B., FULTON, R.J. and KAPLAN, P.L. (1980). Kirsten
    Murine Sarcoma Virus transformed cell lines and a sponta-
    neously transformed rat cell line produce transforming
    factors. J. Cell. Physiol., 105, 163-180.

ENZOOTIC BOVINE LEUKOSIS AND BOVINE LEUKEMIA VIRUS

G. Marbaix*, R. Kettmann*#, J. Deschamps*,
D. Couez*, M. Mammerickx‡, and A. Burny*#

*Department of Molecular Biology, University of
  Brussels (ULB), 1640 Rhode-Saint-Genèse, Belgium

#Faculty of Agronomy, 5800 Gembloux, Belgium

‡National Veterinary Institute, 1180 Uccle, Belgium

1. THE DISEASE AND THE CAUSATIVE AGENT

Enzootic bovine leukosis (EBL) is a lymphoproliferative disease of cattle characterized by persistent lymphocytosis (too high number of lymphocytes) and (or) by the development of lymph nodes tumours (1). EBL is contagious among cattle and experimentally transmissible to cattle, sheep and some other animal species. It affects B lymphocytes and is induced by a virus called Bovine Leukemia Virus (BLV) (2).

BLV is a typical type C (morphological characterization) retrovirus that was originally obtained from short-term cultures of leucocytes from leukemic cattle. It was first detected by electron microscopy and its retroviral nature (presence of reverse transcriptase and of a 70S RNA complex) was clearly proven (3). BLV is now produced by an established fetal lamb kidney cell line (FLK) permanently infected by BLV (4). BLV contains a central spherical core harbouring a complex of two 35-38S genomic RNA

molecules (about 8.3 Kb) associated with a few molecules of reverse
transcriptase and several thousands of "*gag*" (group specific
antigens) proteins (p24, p15, p12, p10). An outer membrane surrounds
the central core, this envelope carries thousands of spikes made
of glycoproteins (*gp*51 and *gp*30 associated together by disulfide
bridges). The genetic information for *gag* antigens, reverse
transcriptase and *gp* proteins is contained in the genome of BLV.
BLV is thus a non-defective virus since its genome contains all
what is needed for its replication (5).

Retroviruses, whose life-cycle includes the integration of
the proviral DNA into the host cell nuclear DNA are known to be
the causative agents of several leukemias, sarcomas and carcinomas
in numerous animal species. A first group of retroviruses is
responsible for the rapid development of sarcomas and of some types
of leukemias (acute leukemias) in various avian and mammalian
species (6). This group also causes cells transformation *in vitro*.
These viruses are defective and depend on a helper retrovirus for
their replication, the only known exception being Rous Sarcoma
Virus which was for this reason the first retrovirus to be discovered about 70 years ago. The other important characteristic of
these viruses is that they all contain an oncogene ("*onc*") gene
(which is generally different from a virus species to another and
different when the kind of caused cancer is different) closely
related to a normal nuclear host gene (7). The overexpression or
the expression at a wrong time of this "*onc*" gene carried by, and
under control of the integrated provirus, is responsible for
tumefaction and cell transformation. So, this first group of
defective retroviruses determines a rapid cancerization due to
the expression of the "*onc*" gene being closely related to a normal
cellular gene.

A second group of retroviruses is responsible for leukemias

which develop after long latent periods (several months) in
several avian and mammalian species. These "slow" viruses are
non defective and are unable to cause cell transformation *in vitro*.
Their genome does not contain any gene related to the host genome
and responsible for tumefaction like the "*onc*" genes for the first
group viruses. A typical example for the retroviruses of this
second group is Avian Leukosis Virus (ALV). How do these retroviruses cause tumefaction? It is proven in many cases that the
integration site in the host genome is important. In some instances
of ALV-induced tumours, the virus causes tumefaction when it
integrates at the 5'-end of a cellular "*onc*" gene ("c *myc*") and
determines by a strong promoter effect, the overexpression of this
gene (8). In other cases, the situation is not as clear but it is
likely that the integration of the virus genome in defined regions
determines the overexpression of cellular "*onc*" genes in the
neighbourhood of the integration site (9). One cannot exclude also,
a priori, the possibility that in some cases, the integration of
the virus could hinder the function of hypothetical genes repressing
the activity of cellular "*onc*" genes.

The question may now be asked: "How does BLV leukemogenize?"
"Does it belong to the first, the second or to another group of
retroviruses?". But let us first briefly describe enzootic bovine
leukosis. It has been possible to follow the development of the
disease in field cases and in experimentally infected animals (10
Soon after infection by BLV, increasing titers of antibodies raised
against the viral envelope glycoprotein and the major core protein
(p24) are observed in the blood plasma of infected cattle. These
animals may then develop (or not) persistent lymphocytosis (PL)
characterized by an increased number of circulating B lymphocytes.
Finally,after long latent periods (from several months to several
years) the infected animals may develop (or not) lymphoid tumours
(after a PL phase or not) which will finally cause the death of the
animals.

More details about the immunological aspects of the BLV
system may be found in references 12-13. General reviews have
also been published (1, 14, 15).

2.   BLV IS COMPLETELY EXOGENOUS TO THE BOVINE SPECIES

One of the first questions that could be answered once it
was possible to prepare a representative radioactive complementary
DNA copy (cDNA) of the BLV RNA genome (using reverse transcriptase)
was: "is BLV completely or partially endogenous to the bovine
species?". Hybridization in liquid medium of [^{3}H] BLV cDNA with
DNA from normal healthy animals or from normal tissues of infected
animals gave a negative answer. On the contrary, BLV sequences
were detected in the DNA from tumours and from lymphocytes of infected
animals (3). However, there was always a slight positive hybridization background with normal bovine DNA due to ribosomal RNA contaminating the BLV RNA preparations used for the synthesis of BLV
[^{3}H] cDNA. The possibility that a small fraction of the BLV genome
is endogenous to the bovine species could thus not be completely
excluded.

On the other hand, it might be that BLV produced by lymphocytes from leukemic animals after short term culture or by FLK
cells chronically infected by BLV contain a genome somewhat
different from the genome of a BLV provirus integrated into the
DNA of a tumour cell. BLV is indeed perhaps only tumorigenic after
recombination with a host cell DNA segment as it is the case for
known acute oncogenic retroviruses (16).

It was thus an absolute necessity to clone a complete BLV
genome after excision from tumour DNA and to use this as a probe
to answer the initial question. Taking advantage of the existence
of a *Sac* I restriction site in the long terminal repeats of the
integrated BLV proviral DNA from a bovine tumour (and of its

absence in the rest of the provirus), the total proviral information could be cloned and amplified in the lambdoid vector λgt Wes
λ B. Use of this labelled cloned probe allowed to definitely rule
out the existence of any normal bovine DNA sequence in the BLV
provirus (17). There is thus no "*onc*" gene, related to a normal
cellular gene, in the proviral BLV DNA. It was also shown in the
same kind of experiments that there is no BLV sequence in the
normal ovine, caprine, murine, feline, chicken and human genomes.

3. TUMOURS INDUCED BY BLV ARE MONOCLONAL; MULTIPLE INTEGRATION
    SITES EXIST

      Since BLV does not cause leukemogenization due to the express-
ion of a viral "*onc*" gene related to a normal cellular gene, it
might be supposed that the integration site of the proviral information into the host cell DNA is important for this process. The
BLV proviral integration sites into the host cell genome of naturally infected animals that were either in the PL or the tumour stage
of the disease were thus studied.

      Eco RI (Figure 1A) or *Bam* HI digestion fragments (Figure 1B)
of tumour cell DNA from different animals were thus separated by
gel electrophoresis, transferred to nitrocellulose sheets and
hybridized with BLV [^{32}P] cDNA. Autoradiograms are showed in
Figure 1. Normal calf thymus DNA was used as a control. It can be
seen that BLV [^{32}P] cDNA hybridizes to two *Eco* RI fragments in the
case of W950 DNA, W82 DNA and to three *Eco* RI fragments in the case
of W15 DNA (a fourth one at 0.6 x 10^6 daltons is too weak to be
detected here). W15 DNA shows, in addition, another hybridization band
at 5.0 x 10^6 daltons also shown by control DNA and corresponding
to ribosomal DNA. Since *Eco* RI makes one single cut into the BLV
proviral DNA, these results show that tumours are monoclonal and
that the BLV integration site is different from one animal to
another (18). One copy of the proviral DNA is present per tumour

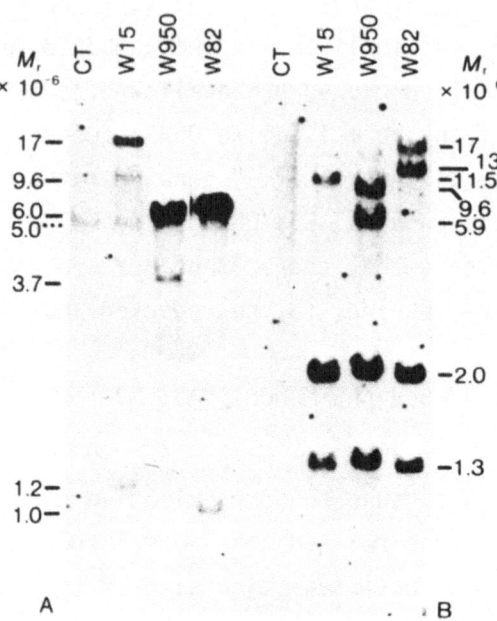

Fig. 1. *Hybridization patterns of BLV [^{32}P] cDNA on DNA restriction fragments from leucocytes of animals with lymph node tumours (DNA restriction fragments from corresponding tumours gave the same hybridization patterns).* Twenty µg each of leukocyte W15, W950, W82 and calf thymus (CT) DNAs were exhaustively digested by either *Eco* RI (A) or *Bam* HI (B) and electrophoresed on a 1% agarose gel. After the restriction fragments were transferred to nitrocellulose sheets, they were soaked in a prehybridization mixture and hybridized for 24 hr with 5 x 10^6 cpm of BLV [^{32}P] cDNA per ml. After extensive washing, autoradiographs were taken. From ref. 18, with authorization.

cell except in the case of W15 where two copies are present. These conclusions are reinforced by the results obtained with the *Bam* HI digestion fragments (Figure 1B). This enzyme makes three cuts into the BLV proviral DNA : the internal viral fragments are detected at 2.0 and 1.3 x 10^6 daltons, the external fragments being different from one animal to another.

In the PL stage of the disease, the integration site of the
BLV provirus is not identical in all circulating lymphocytes of
infected animals; this population is polyclonal (18). From these
experiments, it might be concluded that, in tumour cells, the
integration site of BLV is not identical from one case to another.
However, due to the outbred nature of the cattle population considered, it was not possible to completely rule out integration
into allelic variants of the same integration region.

To test this possibility, *Eco* RI tumour DNA junction fragments
containing the 3' BLV proviral DNA fragment and its cellular
flanking sequences were cloned in Charon 21 A (Figure 2). *Eco* RI
digests of DNA from normal leucocytes and from tumour cells were
hybridized to labelled probes obtained by nick-translation of those
cloned sequences. Figure 3 illustrates, for 8 out of 17 tumour
cell DNAs tested, the results obtained with labelled cloned 1351
DNA as a probe. The weak hybridization bands, present only in
tumour cell DNAs, correspond to the viral fragments revealed by the
LTR moiety of the probe. In all DNAs tested, including the control
leukocyte DNA, the cellular moiety of the probe strongly hybridized

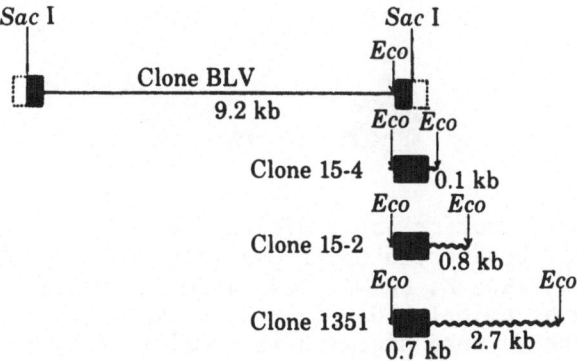

Fig. 2. *Diagrams of the cloned DNAs.* Dark box, LTR DNA; empty box,
LTR DNA absent from cloned BLV DNA; straight line, viral DNA;
wavy line, cellular DNA; *Eco: Eco* RI. From ref. 19, with autoriza-
tion.

to a 2.9 Kb fragment (not revealed by a viral probe). In addition, cloned 1351 DNA hybridized to a 3.4 Kb fragment in the homologous 1351 tumour cell DNA (Figure 3, lane 2), this fragment being also revealed by a viral probe. The same kind of result was obtained with the T15-4 probe. Taken together, these results indicate that the cellular sequences flanking the BLV provirus at its 3'side in tumours T15-4 and 1351 are different and differ from the cellular sequences adjacent to the 3'side of the proviruses in 16 other tumours tested (19).

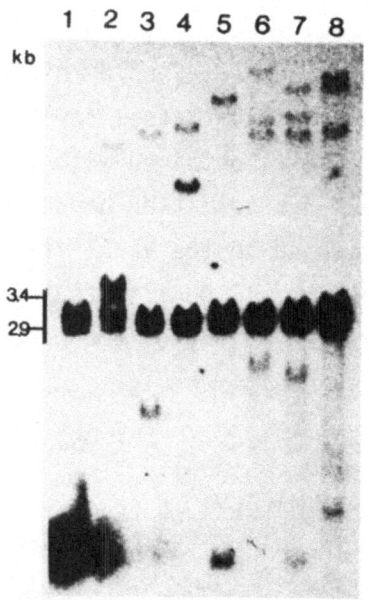

Fig. 3. *DNA hybridization using cloned 1351 DNA as a probe.* Ten μg each of normal bovine leukocyte DNA (lane 1) and bovine tumour DNA (animal 1351, lane 2; animal 104, lane 3; animal 106, lane 4; animal 15, lane 5; animal 3202, lane 6; animal 3261, lane 7 and animal 79-2, lane 8) were digested to completion by *Eco* RI and electrophoresed on a 0.8% agarose gel. The Southern blots were soaked in prehybridization mixture and hybridized for 24 hr with 0.5×10^6 cpm of nick-translated cloned [^{32}P] 1351 BLV DNA (see Figure 2) per ml. After extensive washing, autoradiograph is taken. From ref. 19, with authorization.

The general conclusion of these experiments is that integration
of the BLV provirus in the tumour DNA can take place at many
different sites into only one of the chromosomes of a given pair
(19). This event would be sufficient to switch on the tumorigenic
process. This means either that the activation (or inactivation)
of the expression of many different cellular genes could be responsible for the tumour growth or that the expression of a unique
cellular "*onc*" gene could be affected by a long-distance effect.

4. ABSENCE OF VIRAL RNA IN TUMOURS INDUCED BY BLV

Although BLV does not contain any "*onc*" gene related to a
normal bovine gene, one might ask whether the expression of proviral genome is required for the maintenance of the tumour state.
From studies of tumour BLV proviruses, it was known that in some
cases tumours only harbor defective proviral copies (19), the
deletions being located in the 5' half of the BLV genome. It thus
appeared that the viral information corresponding to the deleted
sequence is not required for maintenance of the neoplastic process
whereas the 3' half of the proviral molecule could be of importance.

Using as a probe the complete cloned BLV proviral DNA [^{32}p]
labelled by nick-translation, we looked for BLV RNA in total RNA
isolated from circulating leukocytes from animals in PL or from
tumour cells. Although the sensitivity of our method (dot blot
hybridization) is such that about one copy of viral BLV RNA per
cell should be seen, we were unable to detect the presence of any
BLV RNA in more than ten cases tested (19, 20). Only one sample
was positive (20).

The absence or the very low level of viral RNA in all tumours
tested strongly suggest that intensive or even moderate expression
of a proviral gene in all cells is not required for maintenance of
the tumour state. These data are however compatible with either

one or both of the following possibilities: all tumour cells
express, at a very low rate, a region of the BLV genome (in the
3' half); or a small proportion of BLV-carrying cells express
the entire or part of the proviral information.

5. LEUKEMOGENESIS BY BLV IS NOT DUE TO IMMEDIATE "DOWNSTREAM
    PROMOTION"

    It is known that the ALV provirus often integrates adjacent
to the 5' end of the cellular "c-*myc*" gene and that transcription,
initiating at the 3' viral LTR promoter, causes enhanced expression
of "c-*myc*" information, thus leading to leukemogenesis (8). Although
the variety of BLV integration sites in bovine tumours does not
favour the existence of a similar mechanism in the case of bovine
leukemia, we looked for the possible occurrence of 3' virus
initiated transcription of cellular DNA sequences. The availability
of cloned cellular sequences flanking the integrated BLV on its
3' side (Figure 2) made possible the search for complementary RNA
transcripts in tumours.

    Nitrocellulose filters containing the same RNA samples as
those which were used for the search of BLV RNA sequences (see
above) were hybridized to a labelled cloned DNA essentially made
of LTR sequences (clone 15-4, Figure 2) and to cloned DNAs containing the 3' viral LTR plus cellular flanking sequences of two
different lengths (clones 15-2 and 1351, Figure 2). For each of the
three probes used, the hybridization results were completely negative (19) even for the RNAs of the tumours from which the DNA
probes had been isolated.

    We can conclude, therefore, that only very little expression
or no expression of cellular sequences flanking the 3' end of the
BLV provirus occurs in BLV-induced tumours or in leukocytes of
animals in PL. The right-end LTR of the BLV provirus thus does not

promote transcription of proximate downstream cellular sequences. However, one cannot rule out expression of sequences upstream to the left-end LTR or further downstream from the right-end LTR. It is also possible that, as proposed by Payne *et al.* (9) for some ALV-induced tumours, viral activation of cellular oncogenes could be the result of a "regional chromosome activation" by a still unknown mechanism.

6. LACK OF EXPRESSION OF CELLULAR HOMOLOGUES OF RETROVIRAL "*ONC*" GENES IN BOVINE TUMOURS INDUCED BY BLV

The transforming genes of retroviruses are derived from normal cellular genes ("c-*onc*") conserved among vertebrates (7,21). There is good evidence that virus-induced transformation is correlated with enhanced levels of expression of these genes (22-24). Using labelled cloned DNA probes containing viral "*onc*" sequences, expression of cellular homologues of retroviral "*onc*" genes was found in human tumour cells (25, 26). Using the same approach, in collaboration with the group of R.C. Gallo, we examined whether or not one of these known "*onc*" genes is expressed at high level during the maintenance of the tumour state in bovine tumours induced by BLV. Four viral "*onc*" genes of avian origin ("*myc*", "*erb*", "*myb*" and "*src*"), two of murine origin ("*abl*" and "*ras*"), one of feline origin ("*fes*") and one of simian origin ("*sis*") were used in this study.

Labelled DNA probes of each viral "*onc*" gene were first ana-lyzed for their ability to detect homologous sequences in bovine DNA. Normal bovine DNA was cleaved with *Eco* RI and the DNA fragments were submitted to agarose gel electrophoresis and to Southern blotting analysis in relaxed hybridization conditions. All "*onc*" probes used detected one or a few DNA fragments in normal DNA. DNA from several tumours induced by BLV were analyzed in parallel. For each "*onc*" probe tested, the patterns obtained were identical to the one observed with normal bovine DNA. These results indicate that in none of the tumour DNAs was there obvious rearrangement of any "*onc*" gene due to, for example, integration of the BLV provirus.

Bovine tumour cells were also tested for the presence of viral "*onc*"-related RNA transcripts both as total and poly (A)-selected RNAs. No difference between the signals observed for the RNA from normal tissue and the RNAs from tumours was observed.

It thus appears that none of the "*onc*" genes tested were implicated in the maintenance of BLV-induced tumours by a mechanism involving enhanced expression of these genes (27).

7. ACKNOWLEDGEMENTS

We thank C. Olson, M. Onuma, A.L. Parodi and M. Van der Maaten for providing bovine tissues. The work described in the present review was supported by the "Fonds Cancérologique" of the "Caisse Générale d'Epargne et de Retraite", Belgium. G.M. is "Maitre de Recherches", R.K. "Chercheur Qualifié" and J.D. "Chargé de Recherches" of the "Fonds National Belge de la Recherche Scientifique".

8. REFERENCES

1) BURNY, A., BRUCK, C., CHANTRENNE, H., CLEUTER, Y., DEKEGEL, D., GHYSDAEL, J., KETTMANN, R., LECLERCQ, M., LEUNEN, J., MAMMERICKX, M. and PORTETELLE, D. (1980). Bovine Leukemia virus: molecular biology and epidemiology. In: Viral Oncology (ed. G. Klein), p. 231, Raven Press, New York.

2) MUSCOPLAT, C.C., JOHNSON, D.W., POMEROY, K.A., OLSON, J.M., LARSON, V.L., STEVENS, J.B., and SORENSEN, D.K. (1974). Lymphocyte surface immunoglobin: Frequency in normal and lymphocytotic cattle. Am. J. Vet. Res., 35, 593.

3) KETTMANN, R., PORTETELLE, D., MAMMERICKX, M., CLEUTER, Y., DEKEGEL, D., GALOUX, M., GHYSDAEL, J., BURNY, A. and CHANTRENNE, H. (1976). Bovine Leukemia virus: an exogenous RNA oncogenic virus. Proc. Natl. Acad. Sci. USA, 73, 1014.

4) VAN DER MAATEN, M.J. and MILLER, J.M. (1976). Replication of bovine leukemia virus in monolayer cell cultures. Bibl. Haematol., 43, 360.

5) GHYSDAEL, J., KETTMANN, R. and BURNY, A. (1979). Translation of BLV virion RNAs in heterologous protein synthesizing systems. J. Virology, 29, 1087.

6) HAYWARD, W.S. and NEEL, B.C. (1981). Retroviral gene expression. In: Current Topics in Microbiology and Immunology, 91, 217, Springer-Verlag, Berlin.

7) STEHELIN, D., VARMUS, H.E., BISHOP, J.M. and VOGT, P.K. (1976). DNA related to transforming gene(s) of avian sarcoma virus is present in normal avian DNA. Nature, 260, 170.

8) HAYWARD, W.S., NEEL, B.C. and ASTRIN, S.M. (1981). Activation of a cellular onc gene by promoter insertion in ALV-induced lymphoid leukosis. Nature, 290, 475.

9) PAYNE, G.S., COURTNEIDGE, S.A., CRITTENDEN, L.B., FADLY, A.M., BISHOP, J.M. and VARMUS, H.E. (1981). Analysis of Avian Leukosis Virus DNA and RNA in bursal tumours: viral gene expression is not required for maintenance of the tumour state. Cell, 23, 311.

10) BEX, F., BRUCK, C., MAMMERICKX, M., PORTETELLE, D., GHYSDAEL, J., CLEUTER, Y., LECLERCQ, M., DEKEGEL, D. and BURNY, A. (1979). Humoral antibody response to Bovine Leukemia Virus infection in cattle and sheep. Cancer Research, 39, 1118.

11) MAMMERICKX, M., PORTETELLE, D., BURNY, A. and LEUNEN, J. (1980). Detection by immunodifusion- and radioimmunoassay-tests of antibodies to bovine leukemia virus antigens in sera of experimentally infected sheep and cattle. Zentralbl. Veterinaermed. Reihe B, 27, 291.

12) PORTETELLE, D., BRUCK, C., MAMMERICKX, M. and BURNY, A. (1980).
    In animals infected by Bovine Leukemia Virus, antibodies to
    envelope glycoprotein _gp_51 are directed against the carbohydrate
    moiety. Virology, 105, 223.

13) BRUCK, C., PORTETELLE, D., BURNY, A. and ZAVADA, J. (1982).
    Topographical analysis of monoclonal antibodies of BLV-_gp_51
    epitopes involved in viral functions. Virology, in press.

14) BURNY, A., BEX, F., CHANTRENNE, H., CLEUTER, Y., DEKEGEL, D.,
    GHYSDAEL, J., KETTMANN, R., LECLERCQ, M., LEUNEN, J.,
    MAMMERICKX, M. and PORTETELLE, D. (1978). Bovine Leukemia
    Virus involvement in enzootic bovine leukosis. Adv. Cancer
    Res., 28, 251.

15) FERRER, J.G. (1980). Bovine Lymphosarcoma. In: Advances in
    Veterinary Science and Comparative Medicine, 24, 1, Academic
    Press, New York.

16) COFFIN, J.M., VARMUS, H.E., BISHOP, J.M., ESSEX, M., HARDY,
    W.D., MARTIN, G.S., ROSENBERG, N.E., SCOLNIK, E.M., WEINBERG,
    R.A. and VOGT, P.K. (1981). Proposal for naming host cell-
    derived inserts in retrovirus genomes. J. Virol., 40, 953.

17) DESCHAMPS, J., KETTMANN, R. and BURNY, A. (1981). Experiments
    with cloned complete tumour-derived bovine leukemia virus in-
    formation prove that the virus is totally exogenous to its
    target animal species. J. Virol., 40, 605.

18) KETTMANN, R., CLEUTER, Y., MAMMERICKX, M., MEUNIER-ROTIVAL, M.,
    BERNARDI, G., BURNY, A. and CHANTRENNE, H. (1980). Genomic
    integration of bovine leukemia provirus: comparison of persis-
    tent lymphocytosis with lymph node tumour form of enzootic
    bovine leukosis. Proc. Natl. Acad. Sci. USA, 77, 2577.

19) KETTMANN, R., DESCHAMPS, J., CLEUTER, Y., COUEZ, D., BURNY, A.
    and MARBAIX, G. (1982). Leukemogenesis by bovine leukemia
    virus: proviral DNA integration and lack of RNA expression of
    viral long terminal repeat and 3' proximate cellular sequences.
    Proc. Natl. Acad. Sci. USA, 79, 2465.

20) KETTMANN, R., MARBAIX, G., CLEUTER, Y., PERTETELLE, D.,

MAMMERICKX, M. and BURNY, A. (1980). Genomic integration
of bovine leukemia provirus and lack of viral RNA expression
in the target cells of cattle with different responses to
BLV infection. Leukemia Research, 4, 509.

21) DUESBERG, P.H. (1979). Transforming genes of retroviruses.
Cold Spring Harbor Symp. Quant. Biol., 44, 13.

22) COLLETT, M.S., BRUGGE, J.S. and ERIKSON, R.L. (1978).
Characterization of a normal avian cell protein related to the
avian sarcoma virus transforming gene product. Cell, 16, 1363.

23) OPPERMANN, H., LEVINSON, A.D., VARMUS, H.E., LEVINTOW, L. and
BISHOP, J.M. (1979). Uninfected vertebrate cells contain a
protein that is closely related to the product of the avian
sarcoma virus transforming gene (src). Proc. Natl. Acad. Sci.
USA, 76, 1804.

24) NEEL, B.G., HAYWARD, W.S., ROBINSON, H.L., FANG, J. and
ASTRIN, S.M. (1981). Avian leukosis virus-induced tumours
have common proviral integration sites and synthesize discrete
new RNAs: oncogenesis by promoter insertion. Cell, 23, 323.

25) EVA, A., ROBBINS, K.C., ANDERSEN, P.R., SRINIVASAN, A., TRONICK,
S.R., REDDY, E.P., ELLMORE, N.W., GALEN, A.T., LAUTENBERGER,
J.A., PAPAS, T.S., WESTIN, E.H., WONG-STAAL, F., GALLO, R.C.,
and AARONSON, S.A. (1982). Cellular genes analogous to retroviral onc genes are transcriped in human tumour cells. Nature,
295, 116.

26) WESTIN, E.H., WONG-STAAL, F., GELMANN, E., DALLA FAVERA, R.,
PAPAS, T.S., LAUTENBERGER, J.A., EVA, A., REDDY, E.P.,
TRONICK, S.R., AARONSON, S.A. and GALLO, R.C. (1982). Expression
of cellular homologues of retroviral onc genes in human hematopoietic cells. Proc. Natl. Acad. Sci. USA, 79, 2490.

27) KETTMANN, R., WESTIN, E.H., MARBAIX, G., DESCHAMPS, J., WONG-
STAAL, F., GALLO, R.C. and BURNY, A. (1983). Lack of expression
of cellular homologues of retroviral onc genes in bovine tumours.
In: Modern Trends in Human Leukemia V (ed. R. Neth), Haematolo-
by and Blood Transfusion, Springer-Verlag, Berlin, in press

GENE TRANSFER INTO CULTURE CELLS AND ITS APPLICATION TO STUDY CELL TRANSFORMATION

A. Graessmann and M. Graessmann

*Institut für Molekularbiologie und Biochemie
der Freien Universität Berlin
Arnimallee 22, 1 Berlin 33
FRG*

1. INTRODUCTION

Many aspects of the intracellular behaviour of macromolecules cannot be studied satisfactorily in cell free systems. During the last decades different techniques were developed to analyze the biological activity of isolated molecules directly in their native environment, the living cell (1).

To learn more about eukaryotic gene regulation, we studied early and late SV40 gene expression in microinjected tissue culture cells. Whereas the underlying stimulus for this effort stems from the oncogenic potential of the virus, the process of cell transformation turned out to be too complex to be understood without complete knowledge of the viral genome organization and regulation of viral gene expression. Since the relatively simple virus depends on cellular enzymes for its replication and protein synthesis, its thorough study has added enormously to and still improves our understanding of the molecular biology of eukaryotic cells. This article presents a brief comparison of the different gene transfer techniques and demonstrates that the biological function of SV40

247

Fig. 1. *Reisolation and detection of SV40 DNA from injected
monkey cells (TC7)*. TC7 cells
grown on small glass-slides (2
x 2 mm) were microinjected with
SV40 DNA I (concentration 1 mg/
ml). For DNA extraction, glass
slides with the recipient cells
were directly transferred into
Eppendorf test tubes with 100 µl
of lysis buffer (1 mM Tris, pH
7.6, 1 mM EDTA, 0.6% NaDodSO$_4$)
and DNA was extracted by the
Hirt method (3). Samples were
electrophoresed through 1.2%
agarose gel, blotted to nitrocellulose filter by the Southern
technique (4), and hybridized
with nick translated SV40 DNA
(specific activity 1-2 x 10^8
cpm/µg). Track 1 marker SV40
DNA, I, II, III; track 2 SV40
DNA extracted from 100 injected
cells; track 3 SV40 DNA extracted from 25 injected cells;
track 4 SV40 DNA extracted from
25 injected cells.

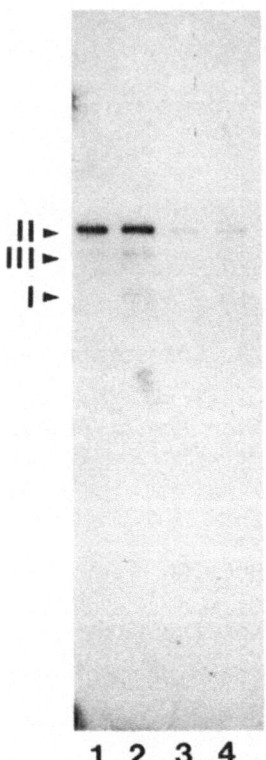

1 2 3 4

DNA fragments and RNA can be efficiently analyzed with a few
microinjected cells.

2. GENERAL ASPECT OF MICROINJECTION

The microinjection technique as developed in our laboratory
(2), is based on the use of small glass capillaries having a tip
diameter of approximately 0.5 µm. A micromanipulator (Leitz)
allows coordinate vertical and horizontal fine movements. Suction
or pressure from a glass syringe (50 ml) connected to the capillary
enables the investigator to fill it through the tip and to deliver
material into the cell. Volumes of 10^{-11} ml can be placed either

into the nucleus or into the cytoplasm of a single cell and up
to 1,000 cells can be injected per hour. If the injected material
is not toxic for the cells up to 100% of the recipient cells allow
expression of the transferred material. Biochemical studies can
be performed on 50 injected cells and transferred material can
be reisolated and further analyzed from about 25-100 recipient
cells (Figure 1).

In order to assess the possibilities of the microinjection
system and to allow a comparison with related transfer methods
(Table 1) general features of the microinjection technique are
summarized:

1. Primary cultured cells or cells from permanent lines
derived from a broad variety of species have been successfully
employed as recipients. Cells usually growing in suspension, e.g.
lymphocytes, are made accessible to microinjection by binding
them to the substrate via a suitable linker (5). This allows us
to assay virtually every biomolecule in its appropriate environment.

2. So far, no limitations have been encountered regarding
the material to be injected: cell organelles (2); virus particles
(6); DNAs with molecular weights (M_r) up to 10^8 (5, 6); RNAs of
different sizes and functions (6, 7, 8); proteins (6, 9, 10, 11)
and molecules unrelated to cellular metabolism have been injected
into cells without afflicting their viability.

3. The site of injection: nucleus or cytoplasm, can be
chosen.

4. Minute amounts of material are required: a sample volume
of about 2 µl is sufficient, and most of this material is pre-
served for other experiments.

5. The cells which have been injected are known.

Table 1. *Properties of the transfer techniques*

Transfer technique	Material transferred	Efficiency in %	Type of recipient cells	Transferred into
I. *DNA mediated gene transfer* (e.g. Ca-phosphate method)	DNA, RNA from plus strand chromosomes	10^{-6}–10^{-4}	Restricted to special cell types (e.g. L-cells, 3T3)	Cytoplasm
II. *Vesicle mediated transfer*				
a) erythrocyte ghosts	RNA, protein	1–20	No cell type restriction	Cytoplasm
b) liposomes	DNA, RNA, protein	10^{-6}–10^{-4}	No cell type restriction	Cytoplasm
c) Reconstituted Sendai virus envelopes	DNA, RNA protein receptors, etc.	20–30	To cells with Sendai virus receptors	Cytoplasm
III. *Microinjection*	Cell organelles, virus particles, DNA, RNA protein	100	No cell type restriction	Nucleus cytoplasm

    6. Injected cells respond efficiently to a variety of macro-
molecules.

3. THE SIMIAN VIRUS 40 (SV40) SYSTEM

    In recent years the small oncogenic DNA virus SV40 has been
the central object of our studies regarding (a) the relationship
between the early viral genome region, its products and their
functions, (b) the regulation of viral gene expression and viral
DNA replication, and (c) stable integration of viral DNA and cell
transformation (for a review, see ref. 6). Some of these results
and preliminary results from investigations currently underway
will serve to illustrate possible applications of the microinjection
technique. SV40 productively infects monkey cells, whereas cells
from other species may be either resistant to viral infection or,
as in the case of rodent cells, will be abortively infected or
eventually become transformed. The covalently closed, superhelical
DNA (DNA I) is made up of about 5,200 base pairs and is divided
into an early and a late viral coding region. The late part contains
the information for the viral capsid proteins, whereas the early
genome region encodes two non-structural viral proteins, SV40
tumour antigens. As shown in Figure 2, the large T-antigen (M_r =
94 Kd), a nuclear DNA binding protein, is coded for by two
discontinuous segments of the viral DNA in contrast to the small
t-antigen (M_r = 17 Kd). The SV40 tumour antigens are multifunctional
proteins, and the oncogenic potency of the virus is linked to
their expression (12).

4. EARLY SV40 GENE EXPRESSION

    Among the many functions attributed to the early viral genome
region (12), the following prominent features observed in virus-
infected cells were of interest to us: expression of two SV40-
associated nuclear antigens, namely T and U antigen; stimulation

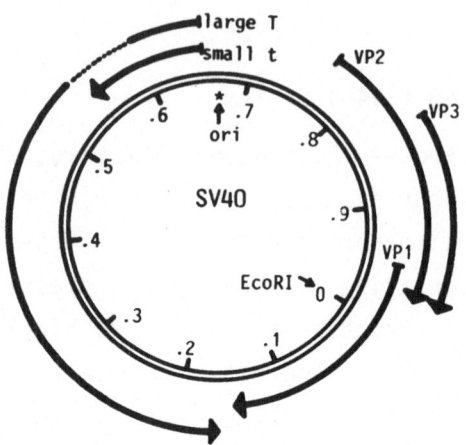

Fig. 2. *SV40 gene map.* Map coordinates refer to the single *Eco* Rı
sites. Arcs with arrowheads cover DNA sequences coding for viral
proteins, with the arrowhead pointing to the carboxy terminals.
Dotted arcs cover large T-antigen intron region,

of cell DNA synthesis; induction of viral DNA replication mediating
late SV40 gene expression; alteration of the cytoskeleton, observed
as the removal of actin cables; a "helper" function for adenovirus
growth in monkey cells; cell transformation. Table 2 summarizes
that all these functions are efficiently exerted in cells microinjected with intact SV40 DNA I. First direct evidence for the now
firmly established fact that SV40 tumour antigens are indeed virus-
coded proteins came from injection experiments employing early
SV40 complementary RNA (cRNA). *In vitro* transcription of SV40
DNA I by *E. coli* DNA dependent RNA polymerase proceeds asymmetri-
cally, yielding mainly transcripts of the early viral DNA strand
with about 50% of the RNA molecules of genome size. After careful
removal of the DNA template molecules, these preparations were
injected into monkey or primary mouse cells where they induced
T-antigen synthesis in up to 75% of the injected cells independent

Table 2. *Early SV40 functions in cells microinjected with various viral nucleic acids or proteins*

Material injected	% of antigen-positive injected cells		% of T-antigen positive cells stimulated for cell DNA synthesis[a]	% of injected cells positive for			Cell transformation[e]
	T	U		Induction of viral DNA replication[b]	Reduction of actin cables[c]	Helper function for Ad2[d]	
SV40 DNA I	100	91	94	nd[f]	80	85	20
SV40 early RNA (cRNA)	72	64	93	45	80	50	nd
T-antigen (D2-protein; SV40-protein)	95	92	80	0	10	80	nd

[a] Induction of cell DNA synthesis was detected by autoradiography in confluent cultures of primary mouse kidney cells

[b] SV40 capsid protein synthesis at the non permissive temperature of 41.5°C in TC7 cells preinfected with SV40 tsA7 or tsA58 was used as a marker for the onset of viral DNA replication

[c] Reduction of actin cables was analyzed in rat 1 and rat embryo fibroblast cells using FITC-conjugated anti-T and rhodamine-conjugated actin antibodies. The percentage of T-antigen positive cells which have lost actin cable structure is given

[d] Ad2 fibre formation in TC7 monkey cells preinfected with adenovirus 2 served as a marker for the helper function

[e] Percentage of injected cells generating cell clones containing the viral DNA covalently integrated and growing in soft agar

[f] not determined

of whether the cells were blocked (chemically) for RNA synthesis
or not. Moreover, T-antigen positive cells were as efficiently
stimulated for cell DNA synthesis as were cells infected with
SV40 or cells microinjected with SV40 DNA I (8). It was shown
later that cRNA injected cells also performed the other early
viral functions cited in Table 2. The direct involvement of the
large T-antigen in the induction of cell DNA synthesis was then
proven by direct microinjection of the purified protein itself
(9).

A more precise mapping of early viral gene functions within
the SV40 genome was achieved by microinjection experiments using
defined DNA fragments covering the early coding region to different
extents. Figure 3 gives a survey of restriction endonucleases used
to generate the desired fragments which, before microinjection,
were extensively purified by preparative agarose gel electrophoresis. The Figure also summarizes the assignment of functions gained
in these investigations, showing above all that stimulation of cell
DNA synthesis is far less dependent on the intact structure of
large T-antigen than is the induction of viral DNA replication
(13). Currently we are analyzing the processing (capping, splicing,
polyadenylation) of early SV40 cRNA in injected cells. However,
experiments in this direction only make sense if it can be assumed
that an intact wild-type (wt) large T-antigen is synthesized in
cRNA-injected cells. Of the experiments cited above (Table 2),
only the complementation by wt cRNA of thermosensitive virus at
the non permissive temperature is a valid hint. It is remarkable
in this respect that both purified large T-antigen proteins, the
SV40 protein isolated from a human SV40 transformed cell line or
the D2 protein extracted from HeLa cells infected with the adeno-
virus-SV40 hybrid Ad2$^+$D2 did not complement early thermosensitive
mutant virus at 41.5^{o}C, most certainly because both proteins
carry deletions. Additional proof that the wt SV40 cRNA brings

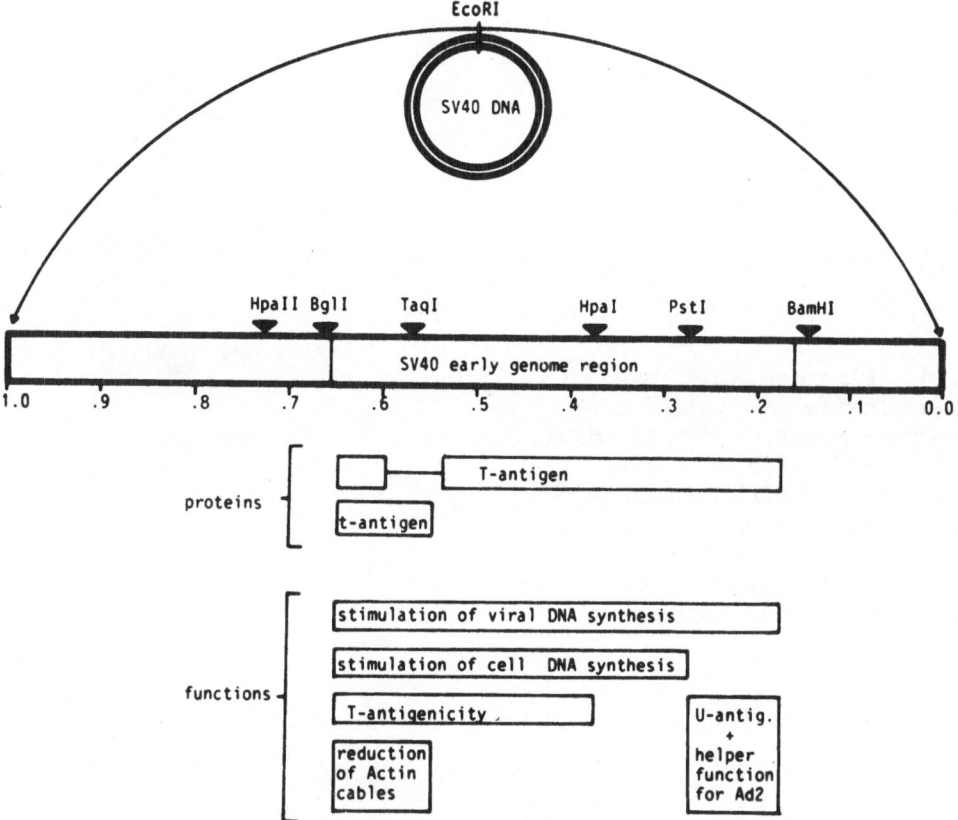

Fig. 3.   *Functional map of the early SV40 coding region (6).*

about the synthesis of intact large T-antigen comes from experiments
involving extraction of proteins from injected monkey cells which
had been labelled with [^{35}S]-methionine. T-antigen reactive proteins
were immunoprecipitated from the crude extract and analyzed by
SDS-polyacrylamide gel electrophoresis. In Figure 4, the mobility
of T-antigens synthesized in monkey cells microinjected with SV40
cRNA is compared with that of normally infected monkey cells and
shown to be undistinquishable.

5.   TRANSCRIPTION CONTROL AND CELL TRANSFORMATION

    SV40 gene expression proceeds strictly biphasic in monkey

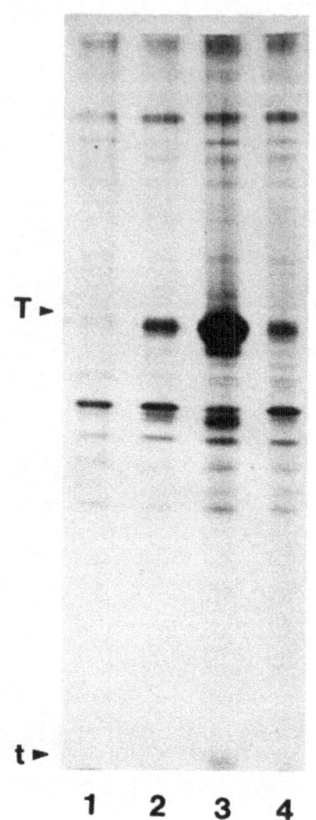

Fig. 4. *Fluorogram of a SDS-
polyacrylamide gel loaded with
immunoprecipitates from (1) mock
injected TC7 cells, (2) TC7
cells injected with SV40 cRNA
into the cytoplasm, (3) TC7
cells infected with SV40, (4)
TC7 cells microinjected with
SV40 cRNA into the nucleus.*

cells conventionally infected or microinjected with SV40 DNA I.
During the early phase of infection transcription of the early
viral genes predominates and viral tumour antigens are accumulated
continously whereas transcripts from the late strand are barely
detectable and capsid protein synthesis is never observed during
this period. By a mechanism still not fully understood T-antigen
then triggers the onset of viral DNA replication which starts
the late phase of infection where both late and early viral genes
are expressed (12). Since these findings argue for a strong signal
on the viral DNA directing this sequence of transcription, SV40
seems to be well suited for the search of a transcription control

unit which, for convenience, will be termed "early promoter".
The analysis and isolation of a strong eukaryotic promoter may
be also important for future experiments aiming at the stable
integration and continuous expression of foreign genes in mammalian
cells.

Some evidence for the assumption that early viral transcription
is possibly regulated by DNA sequences upstream the coding region
and therefore may be located near the origin of replication (0.66
- 0.67 map units) comes from microinjection experiments with SV40
DNA fragments. A fragment covering the entire early coding region
was prepared using the restriction enzymes *Bgl* I and *Bam* HI
(Figures 3 and 5). Cells microinjected with this fragment did not
synthesize SV40 T-antigen in contrast to cells supplied with its
in vitro transcript (13). We then decided to isolate a putative
"promoter" fragment and two coding fragments which upon microinjection should not be expressed on their own. Figure 5 gives
an overview of the fragments prepared and the combinations used
in these experiments. Before injection the fragments were cloned
in *E. coli* using pBR 322 as vector. As anticipated, both the
early *Taq* I - *Bam* HI or the late *Kpn* I - *Bcl* I SV40 DNA fragments
did not induce the synthesis of tumour antigen or capsid protein
related material, respectively, in microinjected monkey or rat
cells. However, cells produced tumour antigen or capsid-reactive
material when microinjected with the appropriate early and late
fragments supplemented with the "promoter" piece, regardless of
whether these mixtures were preincubated with T4 DNA ligase or
not. Control experiments showed that fragments from other regions
of the SV40 genome could not enter this "promoter" function.
These findings not only indicate that the *Hpa* II - *Bgl* I DNA
fragment contains promoter activity, they also suggest the
possibility to link this promoter to other non related DNA sequences which will then be actively transcribed. It is of interest

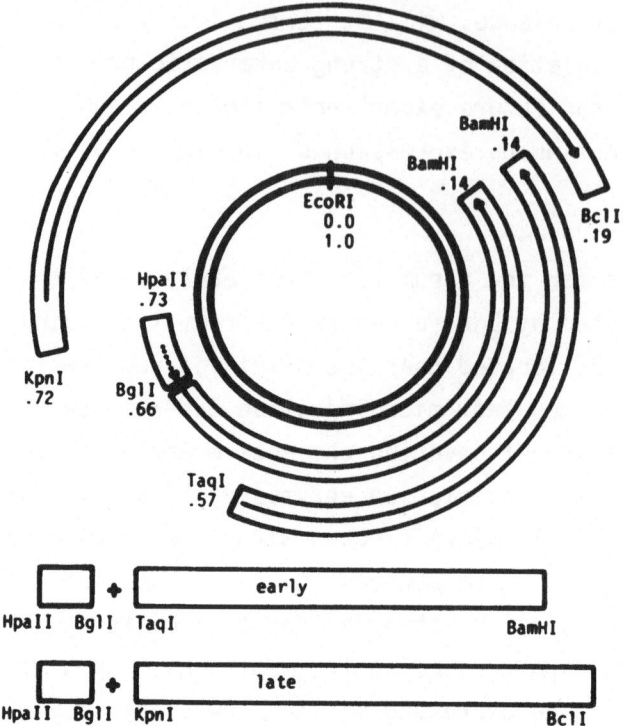

Fig. 5. *Assignment of SV40 DNA fragments used for promoter studies.*
Arrows indicate the direction of transcription, the dotted arrow
shows the orientation of the early promoter.

in this respect that the microinjection technique is well suited
for experiments directed to stably transform mammalian cells.
Up to 20% of rat cells coinjected with the promoter and the
Taq I - *Bam* HI fragment grew out to transformed clones permanently
expressing T-antigen related polypeptides. Properties of these
cell lines are discussed elsewhere (14).

   The high ratio of stable transformants isolated from DNA injected cells is especially remarkable. This efficiency is not restricted to viral oncogenes, since comparable yields have been
obtained upon microinjection of the Herpes simplex Type 1 TK gene
into TK⁻ mouse L cells (15). It remains unclear whether this

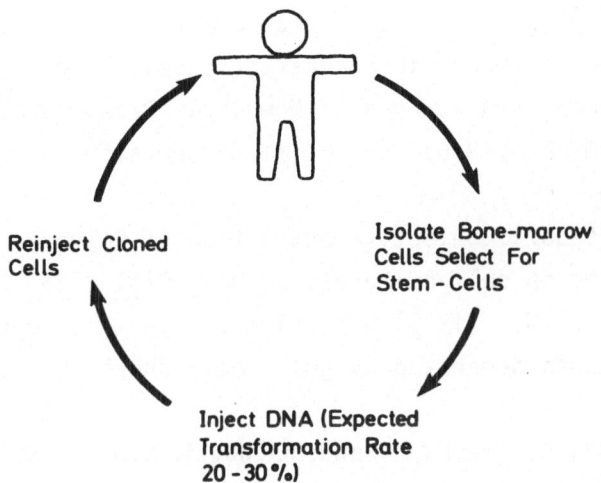

Fig. 6. *Possible application of the microinjection technique.*

efficiency is correlated to the fact that physical microinjection
allows the direct application of DNA into the nuclei of the
recipient cells, while related techniques (Table 1, (1)) only
bypass the cell membrane. However, the microinjection technique
is restricted to the use of cultured cells as recipients, it is
hardly applicable to intact organisms. Yet it seems conceivable to
isolate cells from individuals (e.g. bone marrow cells), to passage
them in culture and to clone out lines from injected cells. The
donor individual could then be the recipient for these cells.
(Figure 6).

6. REFERENCES

1) CELIS, J.E., KALTOFT, K. and BRAVO, R. (1980). Microinjection
    into somatic cells: Direct microinjection with micropipettes
    and PEG erythrocyte ghost mediated microinjection. In: Transfer
    of cell constituents into eukaryotic cell. (eds., J.E. Celis,
    A. Graessmann and A. Loyter) New York, Plenum Press, p. 1.

2) GRAESSMANN, A. (1968). Doctorial dissertation, Free University
    of Berlin

3) HIRT, B. (1967). Selective extraction of Polyoma DNA from
    infected mouse cell cultures. J. Mol. Biol., 26, 365.

4) SOUTHERN, E. M., (1975) Detection of specific sequences among
    DNA fragments separated by gel electrophoresis. J. Mol. Biol.,
    98, 503.

5) GRAESSMANN, A., WOLF, H. and BORNKAMM, G.W. (1980). Expression
    of Epstein-Barr virus genes in different cell types after
    microinjection of viral DNA. Proc. Natl. Acad. Sci. USA, 77
    433.

6) GRAESSMANN, A., GRAESSMANN, M. and MUELLER, C. (1980).
    Microinjection of early SV40 DNA fragments and T-antigens.
    Methods in Enzymology (Acad. Press, N.Y.). 65, 816.

7) GRAESSMANN, A. and GRAESSMANN, M. (1971). Über die Bildung
    von Melanin in Muskelzellen nach der direkten Übertragung
    von RNA aus Harding-Passay-Melanomzellen. Hoppe-Seyler's
    Z. Physiol. Chem., 352, 527.

8) GRAESSMANN, M. and GRAESSMANN, A. (1976). Early simian-virus-
    40-specific RNA contains information for tumour antigen
    formation and chromatin replication. Proc. Natl. Acad. Sci.
    USA, 73, 366.

9) TJIAN, R., FEY, G., GRAESSMANN, A. (1978). Biological activity
    of purified simian virus 40 T-antigen proteins. Proc. Natl.
    Acad. Sci. USA, 75, 1279.

10) KREIS, T.E., WINTERHALTER, K.H., BIRCHMEIER, W. (1979). In
    vivo distribution and turnover of fluorescently labelled

actin microinjected into human fibroblasts. <u>Proc. Natl. Acad. Sci. USA</u>, <u>76</u>, 3814.

11) FERAMISCO, J.R. (1979). Microinjection of fluorescently labelled α-actinin into living fibroblasts. <u>Proc. Natl. Acad. Sci. USA</u>, <u>76</u>, 3967.

12) TOOZE, J. (ed) (1980). DNA Tumour Viruses, Cold Spring Harbor Laboratory, Cold Spring Harbor, New York.

13) MUELLER, C., GRAESSMANN, A. and GRAESSMANN, M. (1978). Mapping of early SV40-specific functions by microinjection of different early viral DNA fragments. <u>Cell</u>, <u>15</u>, 579.

14) GRAESSMANN, A. and GRAESSMANN,M. (1982), Manuscript in preparation.

15) CAPECCHI, M.R. (1980), High efficiency transformation by direct microinjection of DNA into cultured mammalian cells. <u>Cell</u>, <u>22</u>, 479.

EXPRESSION OF CELLULAR PROTEIN IN NORMAL AND TRANSFORMED HUMAN CULTURED CELLS

R. Bravo*, J. Bellatin#, S.J. Fey
P. Mose Larsen and J.E. Celis

Division of Biostructural Chemistry
Department of Chemistry
Aarhus University
DK-8000 Aarhus C

1. INTRODUCTION

Malignant transformation of cultured cells is often characterized by changes in growth properties as well as of cell morphology (1-3). These alterations reflect changes in gene expression that develop through a series of progressive events (4, 5; see also article by L.M. Franks in this volume). Working with the hypothesis that malignant transformation may be due to the abnormal expression of normal genes (6-10) we have carried out a detailed and systematic study of the polypeptides synthesized by normal and transformed cells under a variety of physiological conditions in an effort to reveal cellular proteins that are involved in regulating cell proliferation and that could be used as general markers for malignant transformation (6-11). In this article we will review our studies of normal and transformed human cultured cells. These studies have so far revealed 58 transformation sensitive polypeptides that

Present address: European Molecular Biology Laboratory, Postfach 10.2209, Meyerhofstr. 1, 6900 Heidelberg, FRG.
#Present address: Medical Research Council, Lab. of Molecular Biology, Hills Road, Cambridge, UK.

are present both in normal and transformed cells and that are common
to many cell types analyzed (9). Some of these polypeptides may
correspond to "*oncogenes*" (12-14; see also other articles in this
volume), and the elucidation of their function may lead a better
understanding of the processes that control cell proliferation.

2.   IDENTIFICATION OF TRANSFORMATION SENSITIVE POLYPEPTIDES IN CULTURED EPITHELIAL AND FIBROBLAST CELLS

Given the large number of polypeptides thought to be present
in a somatic cell, we turned our attention to the two dimensional
gel electrophoresis technique described by O'Farrell and co-workers
(15, 16) as this was known to resolve nearly 1000 [^{35}S]-methionine
labelled cellular polypeptides based on their charge and molecular
weight properties. By improving the separation as well as the
labelling techniques, we have been able so far to resolve nearly
1300 [^{35}S]-methionine labelled polypeptides from as few as 50-100
human cultured cells (6, 17, 18). As different cultured human cells
show similar qualitative two dimensional protein patterns, we have
prepared a catalogue of human HeLa cell polypeptides that has been
used to standardize a numbering system for human proteins (i9, 20).

The normal/transformed human cell pairs used in our studies
correspond to epithelial amnion cells and their spontaneously
transformed counterpart (AMA, 9), and to lung fibroblasts (WI-38)
and SV40 transformed WI-38 fibroblasts (WI-38 VA13). Routinely,
between 1000-5000 cells were labelled with [^{35}S]-methionine and
the product of about 1000 cells was analyzed by two dimensional
gel electrophoresis (IEF (15); NEPHGE (16)). For quantitation
the gel spots were cut out from the gels and their radioactivity
determined by scintillation counting.

Figures 1 and 2 show two dimensional gels of basic(NEPHGE)
and acidic (IEF) [^{35}S]-methionine labelled polypeptides from

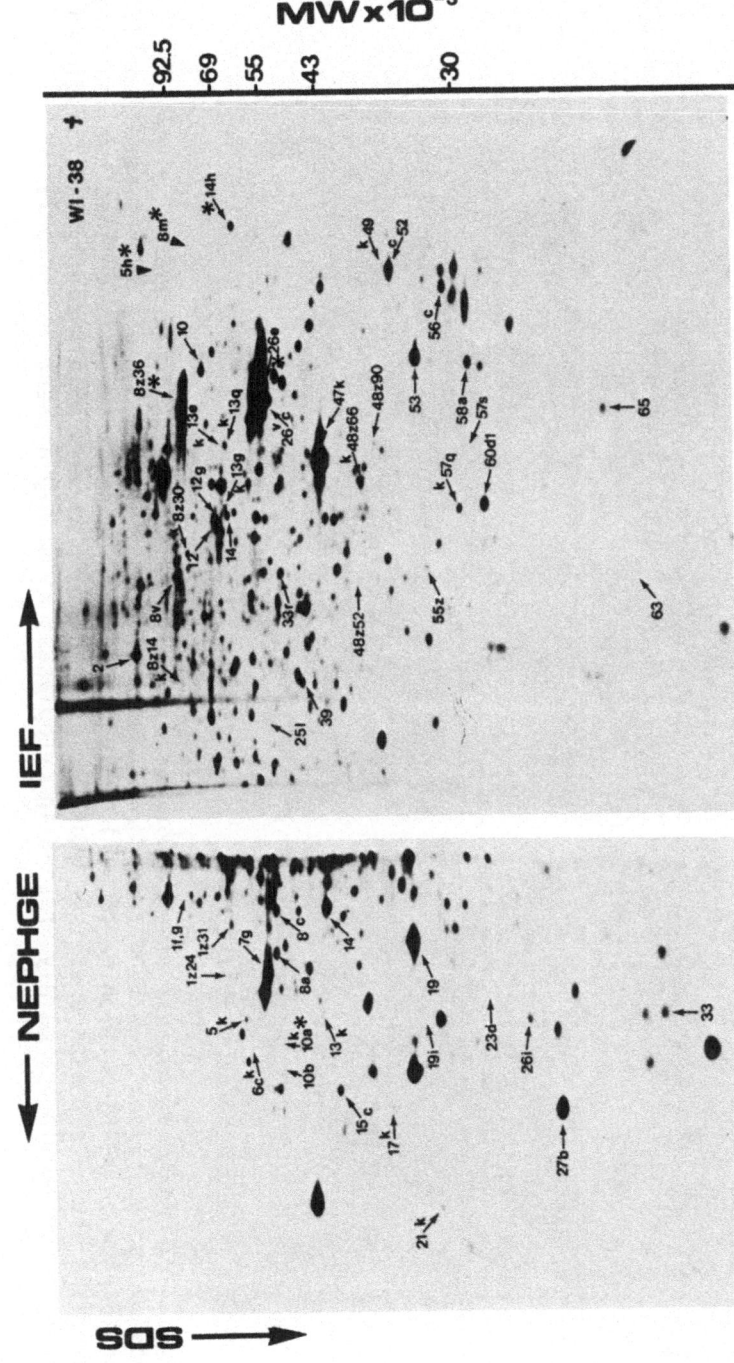

Fig. 1. *Two dimensional gel electroporesis of acidic (IEF) and basic (NEPHGE) [^{35}S]-methionine labelled polypeptides from normal human lung fibroblasts. (9). Sparse cultures were labelled for 16 hr with [35S]-methionine (6, 17). Only transformation sensitive polypeptides are indicated. Asterisks indicate phosphoproteins whose relative proportion vary significantly in transformed cells. v = vimentin; c: mainly present in cytoplasts; k: mainly present in karyoplasts.*

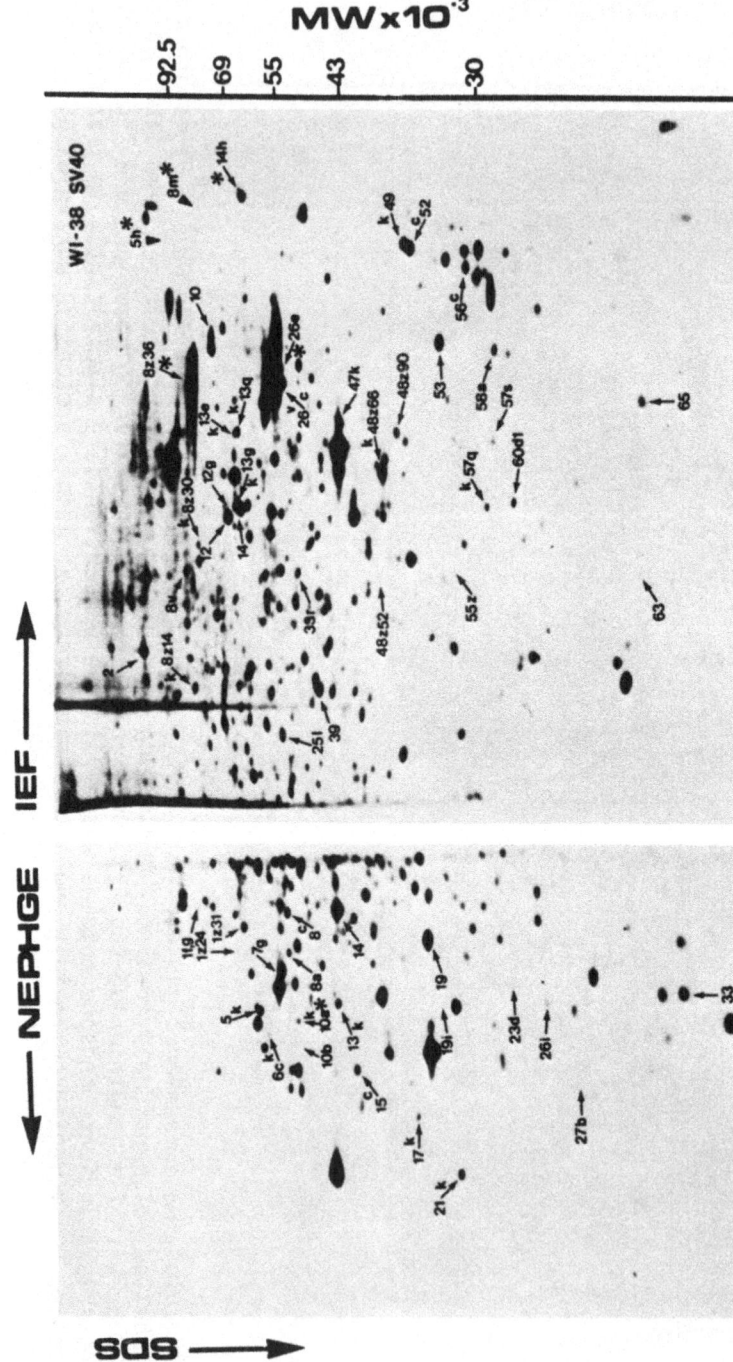

Fig. 2. *Two dimensional gel electrophoresis of acidic (IEF) and basic (NEPHGE) [³⁵S]-methionine labelled polypeptides from SV40 transformed human fibroblasts.* (9). Spase cultures were labelled for 16 hr with [35S]-methionine (6, 17). Only transformation sensitive polypeptides are indicated. Asterisks indicate phosphoproteins whose relative proportion vary significantly in transformed cells. v = vimentin; c̲: mainly present in cytoplasts; k̲: mainly present in karyoplasts.

normal human diploid lung fibroblasts (WI-38, p26) and their SV40
transformed counterparts (WI-38 VA 13). For a given cell type a
total of about 1300 acidic and basic polypeptides could be resolved.
Visual inspection of many films exposed for different times did
not reveal new polypeptides in the transformed cells (9). We
observed, however, significant changes in the abundance of polypeptides that were present both in normal and transformed cells
(Figures 1 and 2). Even though in some cases proteins were synthesized at very low levels in either normal or transformed cells,
they could be detected after prolonged exposure of the gels. A
total of 400 polypeptides were selected for further quantitative
studies (9). Given the reproducibility of the gels and based on
at least 3 sets of quantitations for each cell type, we defined as
variable only those polypeptides whose relative proportion changed
by 40% or more. This arbitrary value was chosen because of the
constancy of the synthesis of many proteins throughout the cell
cycle of HeLa cells (17). Fifty-three polypeptides (22 basic and
31 acidic) common to both cell pairs were found to vary, and these
are indicated with arrows and numbers (HeLa numbering system (19,
20)) in Figures 1 and 2 and are listed in Tables 1 and 2 together
with their coordinates from the HeLa protein catalogue (19, 20). A
few of these polypeptides have been identified and these correspond
to vimentin (IEF 26), cyclin (IEF 49) and a tropomyosin related
polypeptide (IEF 52) (Table 2).

Similar analysis of the amnion/transformed amnion (AMA) pair
(only acidic polypeptides are shown; Figure 3) revealed similar
changes in the 53 variable polypeptides observed in the fibroblast
pair (Tables 1 and 2). These changes have also been observed in
gels stained with silver nitrate (Figure 4, only basic proteins
from amnion and AMA cells are shown in this Figure). We have also
observed many quantitative changes in the relative proportion of
polypeptides that vary only in one of the cell pairs studied and
we will come back to this point further on.

Table 1. *Basic human proteins sensitive to neoplastic transformation*[a]

Polypeptide	Coordinates[•]	cpm-ratio[b] WI-38 VAT3 / WI-38	AMA / AMNION	Relative proportion in transformed cells
NEPHGE				
1f	92/0.22	14.00	11.90	increases
1g	92/0.18	3.66	2.30	increases
1z24	71/0.51	•0.59	0.50	decreases
1z31	71/0.30	4.46	1.79	increases
5[d]	64/0.70	3.04	2.53	increases
6c[d]	61/0.82	2.10	3.83	increases
7g	59.5/0.49	0.52	0.37	decreases
8[c]	57/0.24	0.54	0.22	decreases
8a	57/0.43	0.48	0.14	decreases
10a[d]	54.5/0.76	7.30	7.00	increases
10b	54.5/0.86	7.00	4.00	increases
13[d]	49/0.67	1.82	2.68	increases
14	48.5/0.24	2.18	1.64	increases
15[c]	47/1.00	0.60	0.59	decreases
17[d]	41/1.22	2.59	1.98	increases
19	39.5/0.40	0.28	0.36	decreases
19i[c]	38/0.60	0.42	0.54	decreases
21[d]	35/1.45	2.50	2.70	increases
23d	32.5/0.59	1.90	2.77	increases
26i	28.2/0.71	0.55	0.51	decreases
27b	27/1.11	0.12	0.02	decreases
33	18/0.68	1.84	1.50	increases

[•] From reference 9.

[b] The counts were first standardized against total counts applied to the gels. The standard error of these determinations is in the order of 12%

[c] Present mainly in cytoplasts.

[d] Present mainly in karyoplasts.

Table 2. *Acidic human proteins sensitive to neoplastic transformation*[a]

Polypeptide	Coordinates[a]	cpm-ratio[b] $\frac{WI-38\ VA13}{WI-38}$	$\frac{AMA}{AMNION}$	Relative proportion in transformed cells
IEF				
2	110/0.44	0.44	0.35	decreases
8v	90/0.64	0.34	0.54	decreases
8z14[d]	84/0.32	1.66	2.59	increases
8z30[c]	76/0.72	2.18	1.82	increases
8z36	76/1.27	1.38	1.96	increases
10	72/1.28	2.10	1.60	increases
12	68/0.85	1.57	2.41	increases
12g	68/0.89	<0.05	<0.05	decreases
13e[d]	67/1.03	2.46	2.41	increases
13g[d]	66/0.76	1.87	1.47	increases
13q[d]	66.5/1.03	1.64	3.40	increases
14	66/0.86	3.08	1.66	increases
251	55/0.25	3.35	2.79	increases
26 (vimentin)[c]	54/1.19	0.58	0.35	decreases
33r	50/0.70	0.60	0.32	decreases
39	45.5/0.91	1.65	1.65	increases
47k	43/1.08	0.30	0.09	decreases
48z52	35/0.68	0.50	0.63	decreases
48z66[d]	38.5/0.92	1.76	3.08	increases
48z90	38/1.06	2.32	3.20	increases
49 (cyclin)[d]	36/1.48	9.31	1.72	increases
52 (tropomyosin related)[c]	35/1.43	0.40	0.17	decreases
53	32/1.28	0.21	0.61	decreases
55z	31.5/0.65	0.46	0.14	decreases
56[c]	31/1.41	0.61	0.60	decreases
57s	29.5/1.03	2.05	1.87	increases
57q[d]	30/0.86	0.60	0.33	decreases
58a	29/1.27	0.43	0.20	decreases
60d1	26/0.84	0.21	0.07	decreases
63	20/0.64	1.89	4.00	increases
65	16/1.15	1.86	2.72	increases

[a] From reference 9
[b] The counts were first standardized against total counts applied to the gels. The standard error of these determinations is in the order of 12%
[c] Present mainly in cytoplasts
[d] Present mainly in karyoplasts

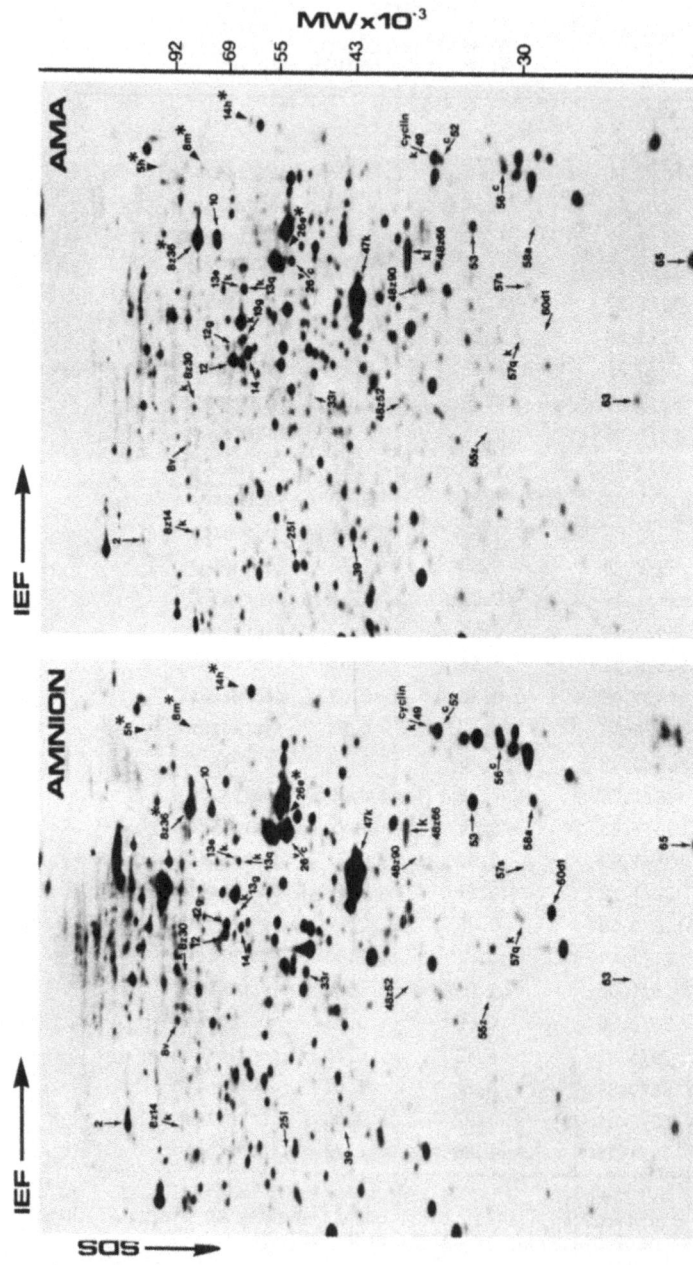

Fig. 3. *Two dimensional gel electrophoresis of acidic (IEF) [^{35}S]-methionine labelled polypeptides from normal and spontaneously transformed human amnion cells (AMA). Sparse cultures were labelled for 16 hr with [35S]-methionine. v = vimentin. Total actin corresponds to the large spot located to the left of 47k. c: mainly present in cytoplasts; k: mainly present in karyoplasts. An asterisk indicates phosphoproteins whose relative proportion varies significantly in transformed cells. From reference 9.*

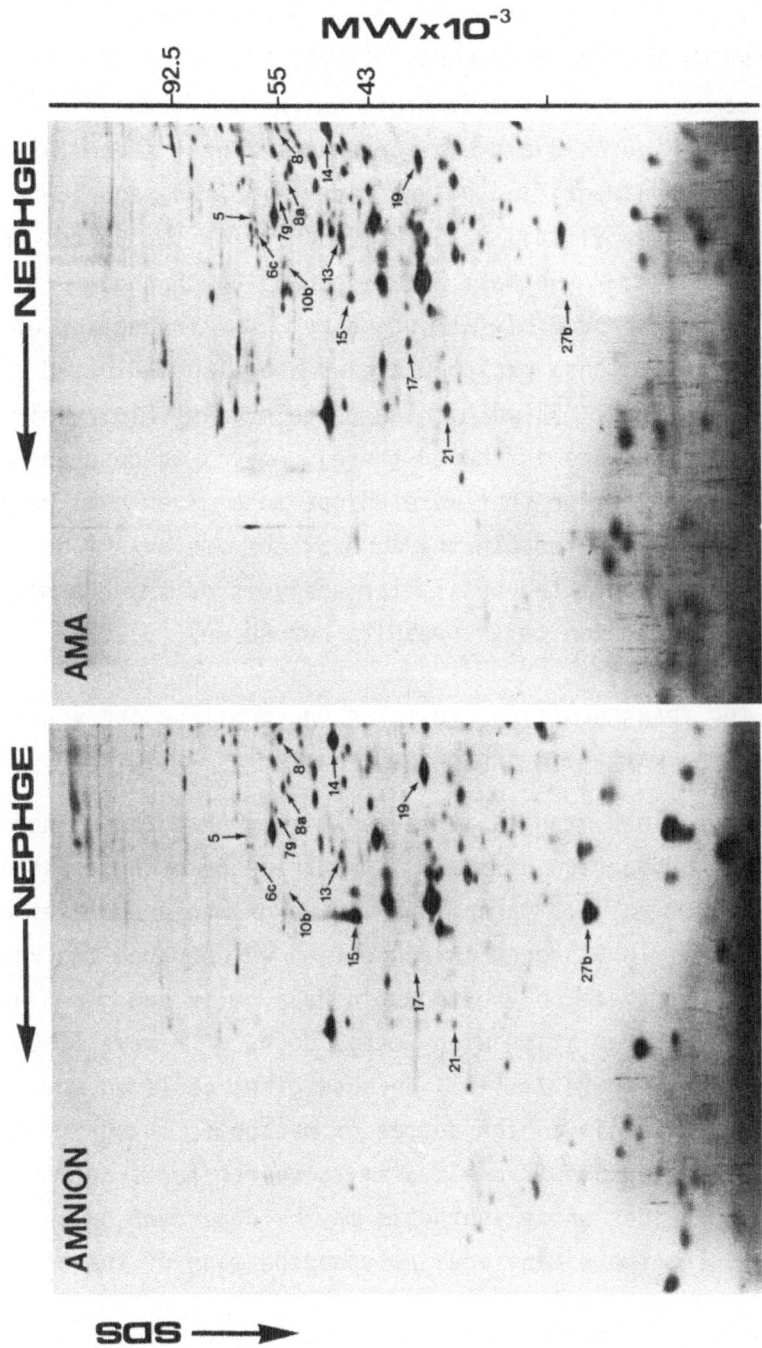

Fig. 4. *Two dimensional gel electrophoresis of basic (NEPHGE) proteins from amnion and AMA cells stained with silver nitrate (21). Only transformation sensitive polypeptides are indicated.*

3. TRANSFORMATION SENSITIVE PHOSPHOPROTEINS

    The analysis of total polypeptides obtained from both cell
pairs labelled with [^{32}P]-orthophosphate for 2 or 8 hr revealed
that of 250 common phosphorylated proteins (WI-38/WI-38 SV40
pair is shown in Figures 5 and 6 as an example) only seven (NEPHGE
10a (M_r 54.5 Kd), unidentified polypeptide (M_r 57 Kd) and IEF's
5h (M_r 96 Kd), 8m (M_r 91 Kd), 8z36 (M_r 76 Kd), 14h (M_r 65 Kd) and
26e (phosphovimentin; M_r 54 Kd); Figures 5 and 6) changed
substantially and reproducibly as judged by visual inspection of
the fluorograms (9). These phosphoproteins are also indicated
with asterisks in the [^{35}S]-methionine protein gels (Figures 1-3;
only six polypeptides are indicated there). Again, we observed
changes in phosphorylation that were unique to a given cell pair,
but these are not indicated. In the case of the amnion/AMA pair,
the data have been confirmed by similar analysis of transformed
Fl-amnion and WISH-amnion cells (results not shown) (9).

4. MOST OF THE TRANSFORMATION SENSITIVE POLYPEPTIDES ARE ALSO
    SENSITIVE TO CHANGES IN THE GROWTH PROPERTIES OF THE CELLS

    From our previous studies it seemed likely that the changes
in the relative proportion of some of the transformation sensitive
polypeptides could reflect variations in the growth properties of
the cells (9, 11). To approach this question we compared the relative proportion of these polypeptides in HeLa cells and giant HeLa
cells produced by irradiation with lethal doses of X-rays (1100
rads; 110 rads/min; Figure 7) (10). Because giant cells do not
proliferate but maintain a high degree of metabolic integrity (23,
24), they represented a suitable system to search for transformation
sensitive polypeptides whose synthesis may be dependent on cell
proliferation. The two dimensional polypeptide maps of the basic

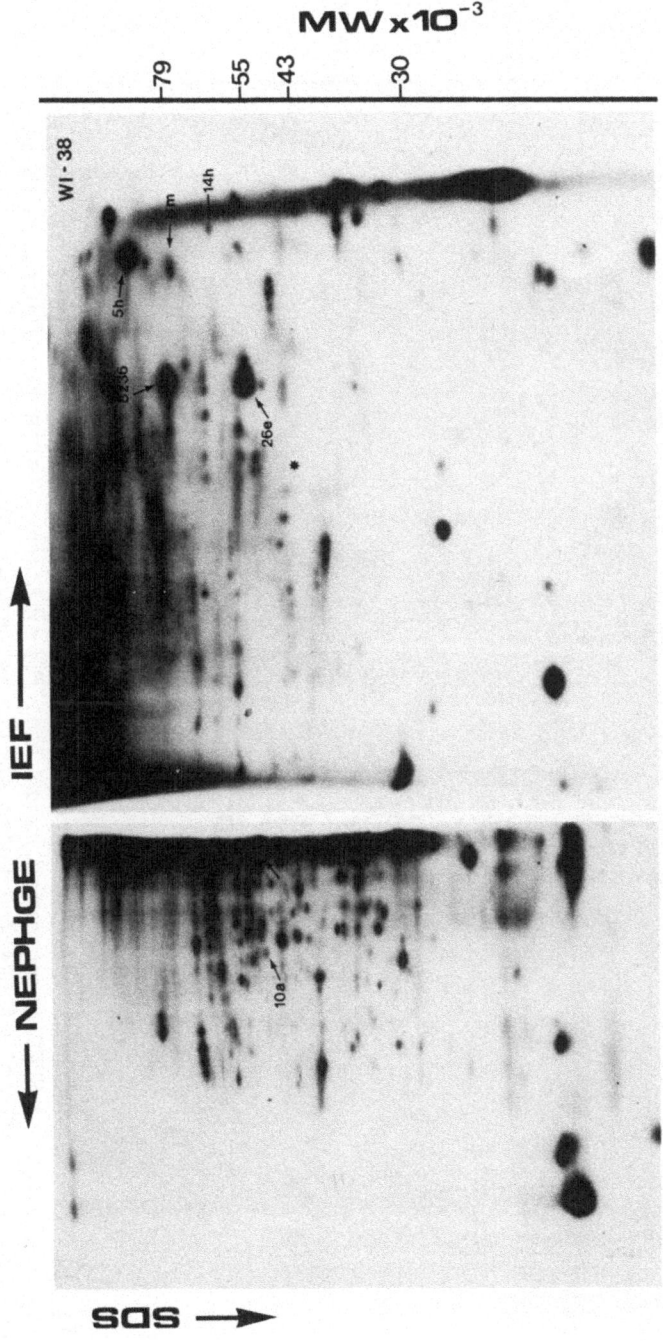

Fig. 5. *Two dimensional gel electrophoresis (NEPHGE, IEF) of total* [³²P]-*orthophosphate labelled po-lypeptides from human lung fibroblasts.* Sparse growing cultures were labelled for 8 hr with 2mCi/ml of [32P]-orthophosphate as previously described (9, 12). The asterisk indicates the position of actin as a landmark (this protein is not phosphorylated). The unnumbered polypeptide has not been identified in the HeLa protein catalogue (19, 20).

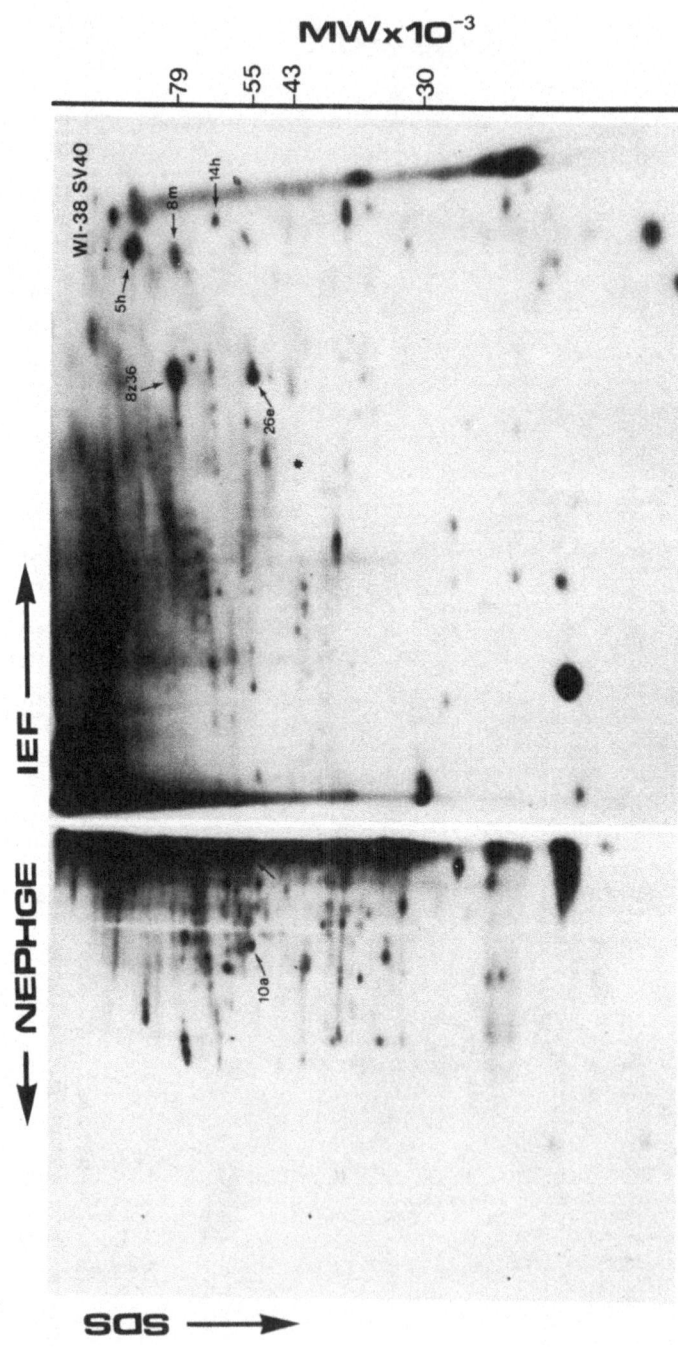

Fig. 6. *Two dimensional gel electrophoresis (NEPHGE, IEF) of total [*^{32}*P]-orthophosphate labelled polypeptides from SV40 transformed human lung fibroblasts.* Sparse growing cultures were labelled for 8 hr with 2mCi/ml of [32P]-orthophosphate as previously described (9, 22). The asterisk indicates the position of actin as a landmark (this protein is not phosphorylated). The unnumbered polypeptide has not been identified in the HeLa protein catalogue (19, 20).

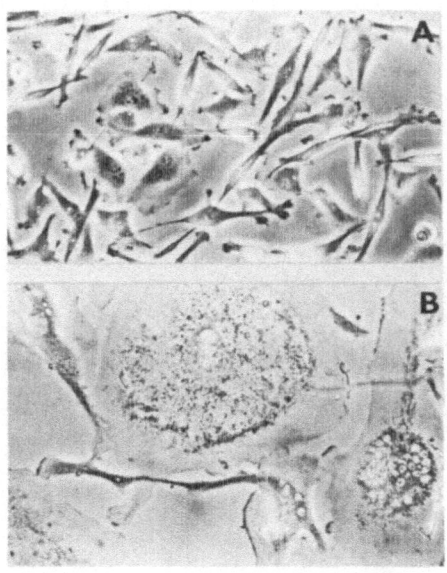

Fig. 7. *Phase contrast photomicrograph of control (A) and irra
diated (B) HeLa cells.* HeLa cells grown on 9 mm^2 coverslips we·
irradiated with 1,100 rads of X-rays. Giant cells were photographed
6 days after irradiation (x 512). Taken from reference 10.

proteins (NEPHGE) of control HeLa cells and giant HeLa cells
labelled for 20 hr with [^{35}S]-methionine are shown in Figure 8.
The respective acidic polypeptides (IEF) are shown in Figure 9.
The giant cells (pure population, Figure 7) were labelled 6 days
after irradiation. Only 47 (27 acidic and 20 basic) of the 58
transformation sensitive polypeptides revealed so far in human
cells were analyzed in this study and are indicated with arrows
and arrowheads in Figures 8 and 9. To compare the relative proportion of the different polypeptides analyzed,the radioactivity
of each spot was standardized against the total radioactivity
applied to the gels. The ratio of the relative proportions for
each spot in control HeLa cells compared to that in giant HeLa
cells is given in Table 3 (10). In many cases, however, the changes
in relative proportion were established by visual inspection of

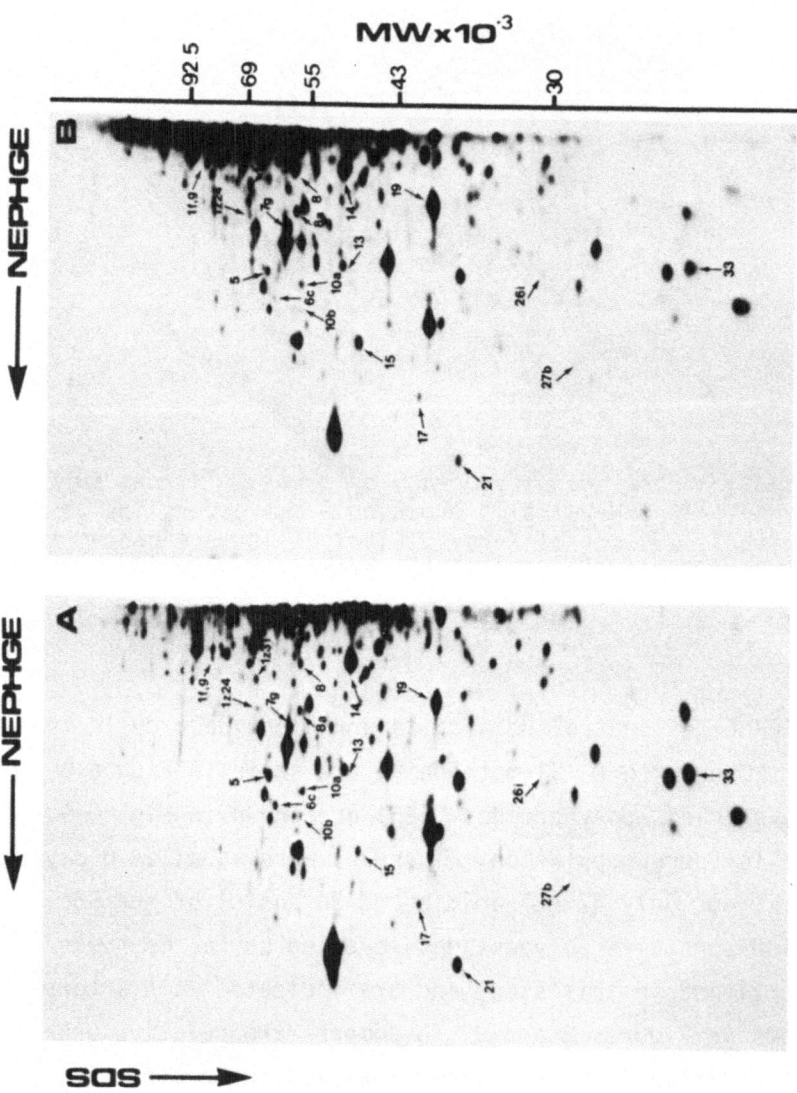

Fig. 8. *High resolution two dimensional gel electrophoresis (NEPHGE) of [35S]-methionine labelled polypeptides from control (A) and giant HeLa cells (B). Control and giant cells were labelled for 20 hr with [35S]-methionine. Only transformation sensitive polypeptides are indicated (9). Taken from reference 10.*

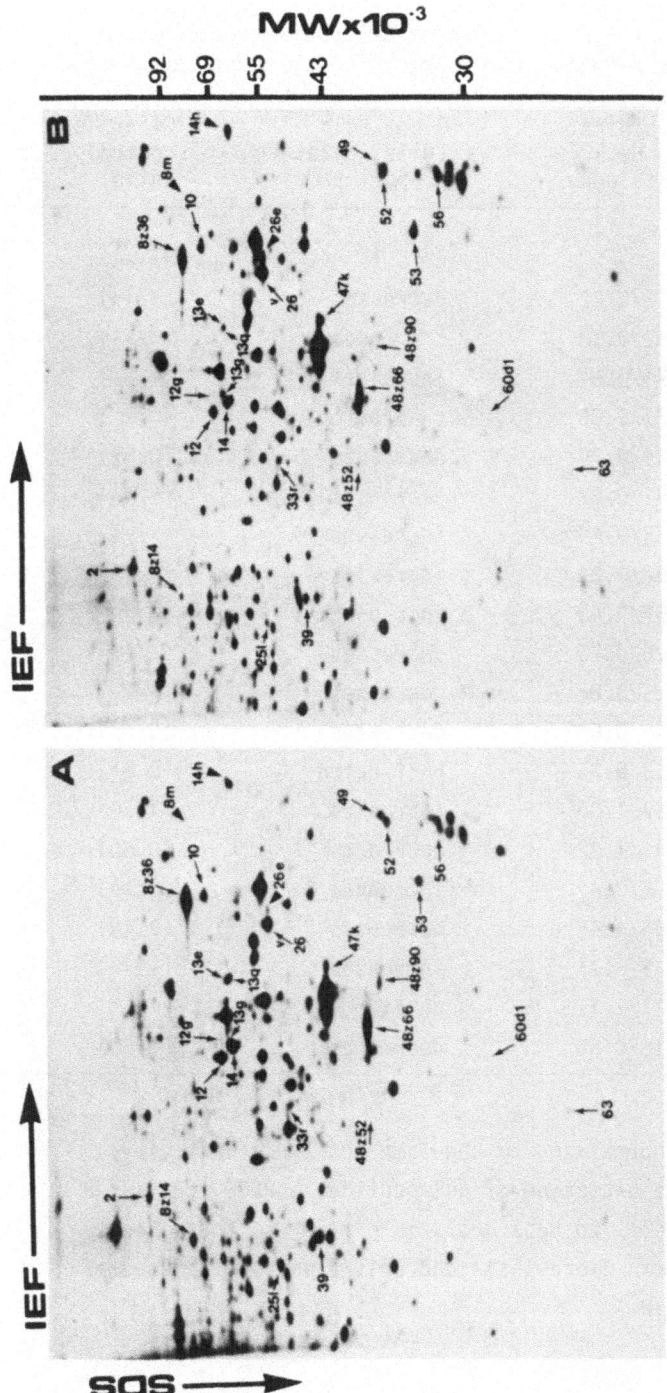

Fig. 9. *High resolution two dimensional gel electrophoresis (IEF) of [^{35}S]-methionine labelled polypeptides from control (A) and giant HeLa cells (B).* Cells were labelled as indicated in Figure 7. Only transformation sensitive polypeptides are indicated (9). Arrowheads indicate phosphorylated proteins (9). v = vimentin. Taken from reference 10.

Table 3. *Basic human proteins sensitive to neoplastic transformations: Relative proportions in giant HeLa cells*[a]

Polypeptide	Coordinates in HeLa catalogue[*]	Giant HeLa cells, relative proportion[b]	Giant cell/ control cell ratio[c]
NEPHGE			
1f	92/0.22	decreases	(+/++)
1g	92/0.18	decreases	(+/+++)
1z24	71/0.51	increases	(++/+)
1z31	71/0.30	unaffected	-
5	64/0.70	decreases	(0.57)
6c	61/0.82	decreases	(+/+++)
7g	59.5/0.49	increases	(+/+++)
8	57/0.24	increases	(1.40)
8a	57/0.43	increases	(++/+)
10a	54.5/0.76	decreases	(+/++)
10b	54.5/0.86	decreases	(+/++)
13	49/0.67	decreases	(0.57)
14	48.5/0.24	unaffected	(0.81)
15	47/1.00	increases	(1.50)
17	41/1.22	decreases	(0.60)
19	39.5/0.40	increases	(1.94)
21	35/1.45	decreases	(0.48)
26i	28.2/0.71	increases	(++/+)
27b	27/1.11	unaffected	-
33	18/0.68	decreases	(0.60)

[*]From refs 10 and 19

[b]All of the gels were normalized to the same total radioactivity

[c]In all cases, with the exception of polypeptides NEPHGE 1z31 and 27b, the ratio of giant HeLa cells to HeLa cells is similar to that observed in normal versus transformed human fibroblasts and epithelial cells (see ref. 10). Taken from reference 10.

Table 4. *Acidic human proteins sensitive to neoplastic transformations: Relative proportions in giant HeLa cells*[a]

Polypeptide	Coordinates in HeLa catalogue[a]	Giant HeLa cells, relative proportion[b]	Giant cell/ control cell ratio[c]
IEF			
2	110/0.44	increases	(++/+)
8z14	84/0.44	decreases	(++/+)
8z36	76/1.27	decreases	(0.68)
10	72/1.28	decreases	(0.67)
12	68/0.85	decreases	(0.59)
12g	68/0.89	unaffected	-
13e	67/1.03	decreases	(0.50)
13g	66/0.76	decreases	(+/++)
13q	66.5/1.03	decreases	(0.50)
14	66/0.86	decreases	(0.67)
251	55/0.25	decreases	(0.41)
26 (vimentin)	54/1.19	increases	(2.28)
33r	50/0.70	increases	(++/+)
39	45.5/0.91	decreases	(0.50)
47k	43/1.08	increases	(++/+)
48z52	35/0.68	increases	(++/+)
48z66	38.5/0.92	decreases	(+/++)
48z90	38/1.06	decreases	(+/++)
49 (cyclin)	36/1.48	decreases	(0.21)
52 (tropomyosin related)	35/1.43	increases	(2.56)
56	31/1.41	increases	(++/+)
60d1	26/0.84	unaffected	-
63	20/0.64	decreases	(0.47)

[a]From refs 10 and 19

[b]All of the gels were normalized to the same total radioactivity

[c]In all cases, with the exception of polypeptides IEF 12g and 60d1, the ratio of giant HeLa cells to HeLa cells is similar to that observed in normal versus transformed human fibroblasts and epithelial cells (see ref. 10). Taken from reference 10.

films from many independent samples and the results are expressed
by using plusses. With the exception of polypeptides IEF 12g and
60d1 and NEPHGE 1z31 and 27b the relative proportion of all the
transformation sensitive polypeptides analyzed changed substantially
and reproducibly in the giant cells. The polypeptides indicated
with arrowheads in Figure 9 (IEF 8m, 14h and 26e (phosphovimentin))
are not listed in Table 3 and correspond to transformation sensitive
phosphoproteins (9) (see also Figures 1-5). The relative proportion
of IEF 8m and 14h decrease in giant HeLa cells while that of
phosphovimentin increases.

Preliminary analysis of the [^{35}S]-methionine labelled polypeptides synthesized by growing (p3) and senescent skin fibroblasts
(p19) showed that the relative proportion of a significant number
of the transformation sensitive polypeptides changed as expected.
Only those polypeptides whose intensity clearly changed are indicated
with arrows in Figure 10 and 11. There are, however, a few transformation sensitive polypeptides whose relative proportion do not
change in senescent cells and these correspond to NEPHGE 15 and
27b (arrowheads in Figure 10), and IEF 10, 56 and 60d1 (arrowheads in Figure 11).

5.   CHANGES IN POLYPEPTIDE SYNTHESIS THAT ARE SPECIFIC FOR A
       GIVEN CELL PAIR

Although our studies have been mainly directed to reveal
polypeptide changes that are common to normal and transformed
cells, we have observed many changes that are not general to all
the cell pairs we have analyzed so far. One such example is the
case of the tropomyosin related polypeptides IEF 52x and 55
(Figure 12). Changes in the synthesis of tropomyosin polypeptides
have been reported in virally transformed non muscle cells (25,
26) as well as in mouse embryonal carcinoma cells induced to
differentiate by hexamethylenebisacetamide (27). While trans-

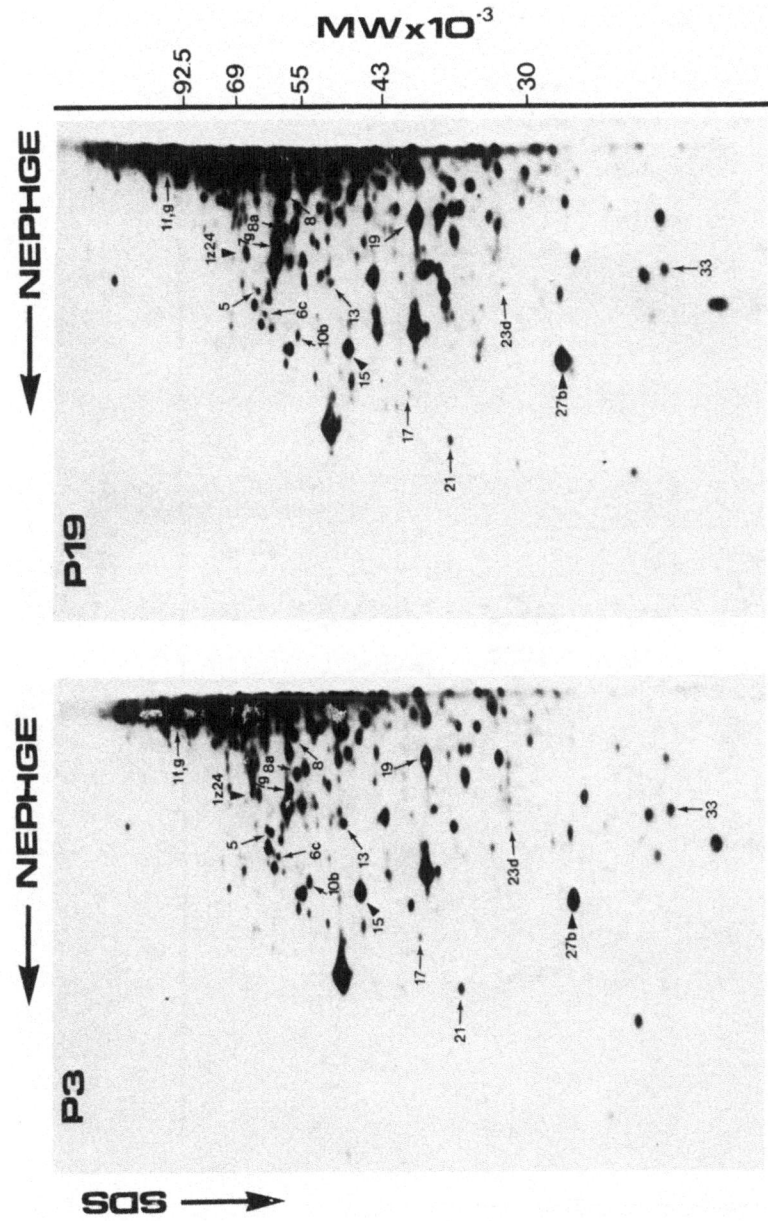

Fig. 10. *Two dimensional gel electrophoresis of basic (NEPHGE) proteins from growing (p3) and senescent (p19) human skin fibroblasts.* Transformation sensitive polypeptides whose intensities clearly change are indicated with arrows. Arrowheads indicate polypeptides that do not change.

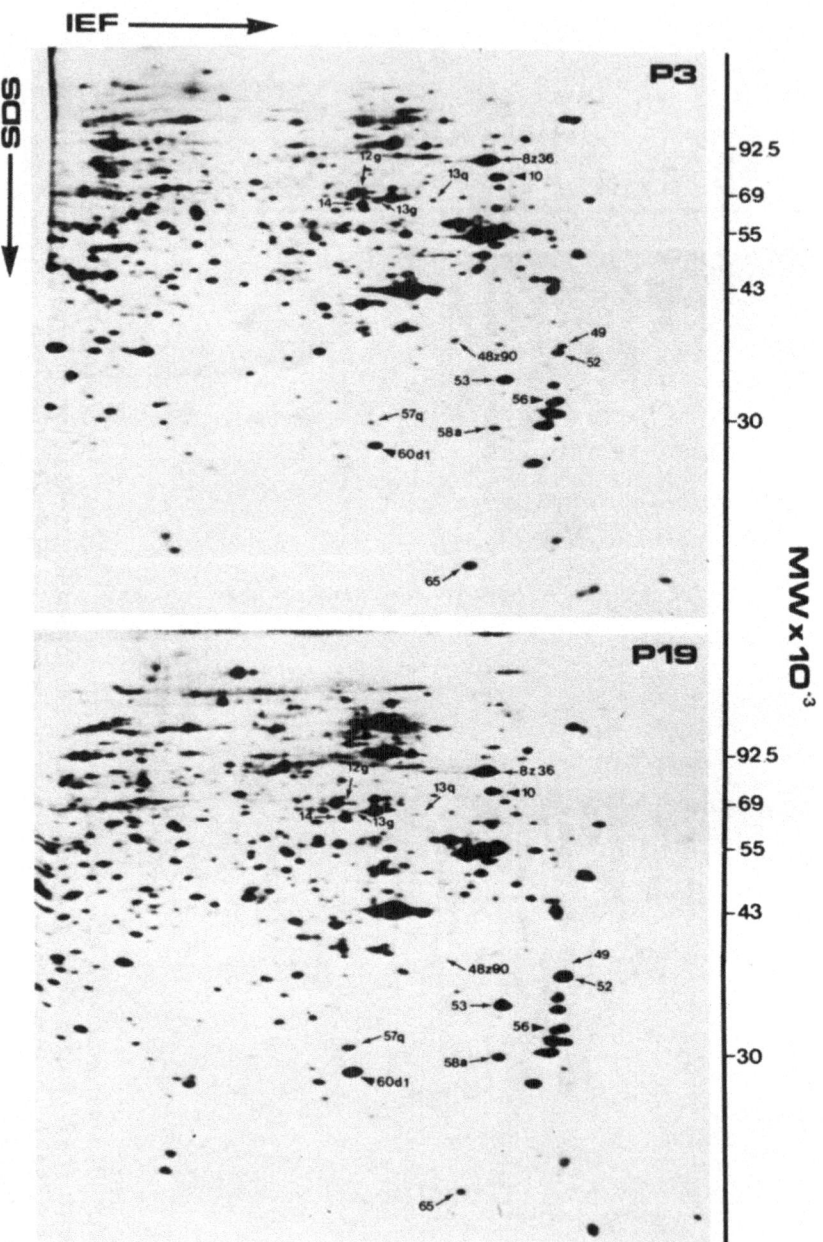

Fig. 11. *Two dimensional gel electrophoresis of acidic proteins (IEF) from growing (p3) and senescent (19) human skin fibroblasts.* Transformation sensitive polypeptides whose intensities clearly change are indicated with arrows. Polypeptides indicated with arrowheads do not change.

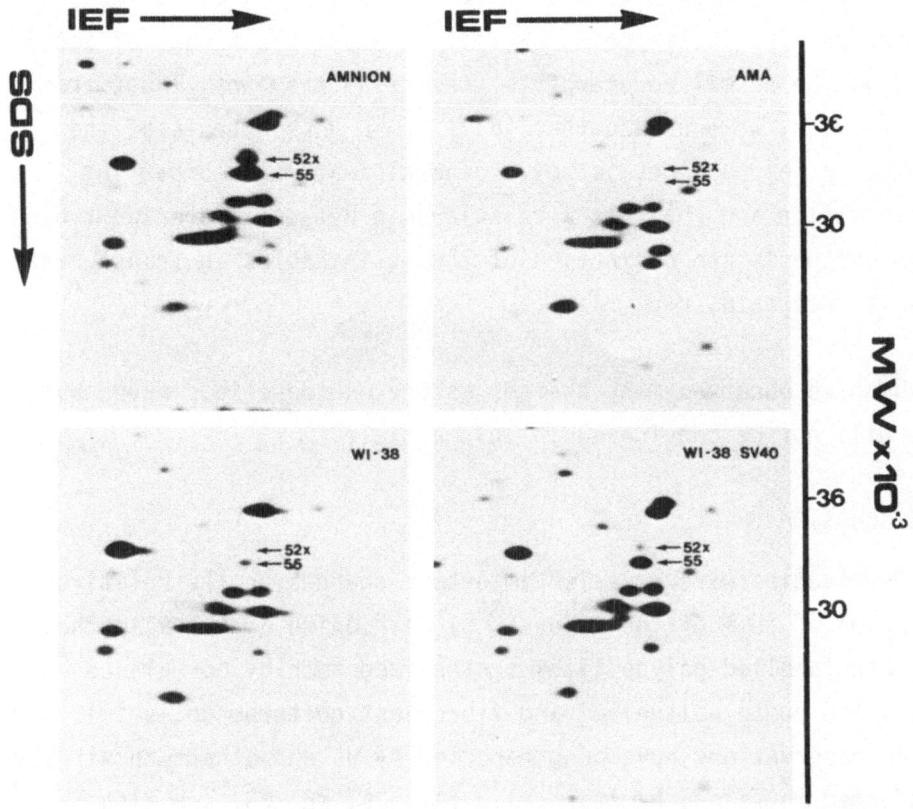

Fig. 12. *Two dimensional gel electrophoresis (IEF) of [^{35}S]-me-thionine labelled polypeptides from normal and transformed cells.* Only a selected area of the fluorograms is shown. Taken from reference 8.

formation generally leads to a decreased synthesis of tropomyosin polypeptides (25), the opposite has been observed during differen-tiation of carcinoma cells (27). Furthermore, from studies of chicken embryo fibroblasts cells transformed with Rous sarcoma virus it has been postulated that the decrease in tropomyosin synthesis may play a role in the disruption of the cellular cyto-skeleton (actin cables) following transformation (25). Our studies, however, have shown that while the synthesis of the tropomyosin related proteins IEF 52x and 55 decreases in many transformed cells

(8, 11) it increases in SV40 transformed human lung fibroblasts
(Figure 12). The latter cells exhibit a comparable number of actin
cables as the normal counterparts (results not shown). Thus, from
these results, we conclude that (a) a decreased synthesis of the
tropomyosin polypeptides is not a general feature accompanying
transformation and (b) that a relationship between decreased tropomyosin synthesis and disruption of the actin cables in transformed
cells is uncertain.

We have observed many changes as the one described above but
these will not be considered in this article.

6. DISCUSSION

Our studies have revealed important changes in the relative
proportion of some of the major [^{35}S]-methionine and [^{32}P]-ortho-
phosphate-labelled polypeptides synthesized both by normal and
transformed human epithelial and fibroblast cultured cells (9).
Similar observations have been reported by us and others in virally
transformed mouse and hamster cells (11, 26, 28, 29; see also
other articles in this volume). In general, these studies support
the notion that transformation results in changes in the relative
proportion of polypeptides that are present both in normal and
transformed cells rather than the appearance of new polypeptides
in transformed cells (11). Our failure to reveal T-antigen in the
SV40 transformed WI38 cells is likely due to either co-migration
with other abundant proteins or to the fact that proteins comprising
less than 0.001% of the total protein label are difficult to
detect (9, 20).

Given the large number of polypeptides resolved and considering
the advantage of identifying transformation sensitive polypeptides
common to many transformed cells, we focused our attention mainly
on those polypeptides changes that were common to both cell types

analyzed. Fifty-three transformation-sensitive polypeptides common
to both cell pairs were detected among 400 [^{35}S]-methionine
labelled polypeptides quantitated (9). Similar proportions of
these polypeptides have been observed in other transformed human
cells such as HeLa, Chang liver, Detroit, FL-amnion, and WISH-
amnion (9). So far,of these polypeptides,only vimentin (IEF 26,
decreases), cyclin (IEF 49) (7. 8. 20) and a tropomyosin variant
(IEF 52, decreases) (7, 8) have been identified. A similar decrease
in the relative proportion of vimentin has been reported in SV40-
and polyoma-transformed mouse 3T3B and hamster BHK21 cells (11, 20).

Seven phosphoproteins sensitive to transformation and common
to both cell types studied have been revealed among the 250
phosphorylated proteins clearly resolved by two dimensional gel
electrophoresis (9). None of these proteins seem to have a molecular mass or isoelectric point similar to that of the phospho-
tyrosine-containing proteins recently reported in chicken embryo
cells transformed by different classes of retroviruses (31). Except
for phosphovimentin (IEF 26e) which decreases in transformed cells,
the relative proportion of the other six transformation sensitive
phosphoproteins increased considerably in transformed cells.
Studies of synchronized cells have shown that phosphorylation of
vimentin may play a role in the process of cell division, and
thus it is possible that the decreased relative proportion of
this protein in transformed cells may reflect subtle changes in
the process of cell division (14, 22, 32, 33).

At present we know little about the cellular function of the
transformation sensitive polypeptides. Analysis of irradiated HeLa
cells showed that the synthesis of many of these polypeptides is
sensitive to changes in the rate of cell proliferation (10). The
synthesis of only three polypeptides (IEF 12g, 60d1 and NEPHGE
27b) was not affected indicating that their reduced amount in

malignant cells could reflect a fundamental change that develops during transformation (10). Preliminary studies of growing and senescent human skin fibroblasts have in general confirmed these observations.

It is likely that some of the transformation sensitive poly-peptides we have revealed correspond to potential cellular "*onco-genes*" or cancer gene products (12-14). The cloning of these genes will be necessary in order to evaluate this possibility. We also envisage that the function of these polypeptides or their genes could be investigated using microinjection techniques (34).

Work is now in progress to prepare antibodies against all the transformation sensitive polypeptides so far identified. These antibodies should be of value for purifying these proteins and will provide a rapid probe to screen large numbers of normal and tumour tissue. Such studies together with parallel analysis of normal and tumorigenic cells using two dimensional gel electropho-resis may ultimately lead to polypeptide markers for neoplastic transformation common to many cell types.

7. ACKNOWLEDGEMENTS

We would like to thank A. Celis and H. Macdonald-Bravo for helpful discussion and O. Jensen for photography. R.B. was a recipient of a fellowship from the Medical and Natural Sciences Research Councils. J.B. was a recipient of a Danida fellowship. S.J.F. is a recipient of a fellowship from the Danish Cancer Foundation. P.M.L. is a recipient of a fellowship from the Aarhus University. This work was supported in part by grants from the Danish Medical and Natural Science Research Council, The Danish Cancer Society, The Carlsberg Foundation, Novo and Euratom.

8. REFERENCES

1) HANAFUSA, H. (1977). Cell transformation by RNA tumour viruses.
    In: Comprehensive virology (eds, H. Fraenkel-Conrat and R.
    Wagner), 10, p. 401, Plenum Press, New York.

2) ERIKSON, R.L., PURCHIO, A.F., ERIKSON, E., COLLET, M.S. and
    BRUGGE, J.S. (1980). Molecular events in cells transformed by
    Rous sarcoma virus. J. Cell Biol. 87, 319.

3) POLLACK, R., OSBORN, M. and WEBER, K. (1975). Patterns of
    organization of actin and myosin in normal and transformed
    cultured cells. Proc. Natl. Acad. Sci. USA, 72, 994.

4) ARMITAGE, P. and DOLL, R. (1957). A two-stage theory of car-
    cinogenesis in relation to the age distribution of human
    cancer. Br. J. Cancer, 11, 161.

5) BARRET, J.C. and TS'O, P.O. (1978). Evidence for the progressive
    nature of neoplastic transformation in vitro. Proc. Natl.Acad.
    Sci. USA, 75, 3761.

6) CELIS, J.E. and BRAVO, R. (1981). Towards cataloguing human
    and mouse proteins. Trends in Biochemical Sciences, 6, 197.

7) BRAVO, R., FEY, S.J., BELLATIN, J., MOSE LARSEN, P., AREVALO,
    J. CELIS, J.E. (1981). Identification of a nuclear and of a
    cytoplasmic polypeptide whose relative proportions are sensitive
    to changes in the rate of cell proliferation. Exp. Cell Res.
    136, 311.

8) BRAVO, R., FEY, S.J., BELLATIN, J., MOSE LARSEN, P. and CELIS,
    J.E. Identification of a nuclear polypeptide "cyclin" whose
    relative proportion is sensitive to changes in the rate of
    cell proliferation and to transformation. In: Proceedings of
    the International Society of Developmental Biology (ed. M.
    Burger), p. 235, Alan R. Liss.

9) BRAVO, R. and CELIS, J.E. (1982). Human proteins sensitive to
    neoplastic transformation in cultured epithelial and fibro-
    blast cells. Clin. Chem., 28, 949.

10) BELLATIN, J., BRAVO, R. and CELIS, J.E. (1982). Changes in
     the relative proportion of transformation sensitive polypep-
     tides in giant HeLa cells produced by irradiation with lethal
     doses of X-rays. Proc. Natl. Acad. Sci. USA, 79, 4367.

11) BRAVO, R. and CELIS, J.E. (1980). Gene expression in normal
     and transformed mouse 3T3B and hamster BHK21 cells. Exp. Cell
     Res., 127, 249.

12) KLEIN, G. (1981). The role of gene dosage and genetic trans-
     positions in carcinogenesis. Nature, 294, 313.

13) BISHOP, J.M. (1982). Oncogenes. Scientific American, March, 69.

14) WEINBERG, R.A. (1982). Oncogenes of human tumour cells. Trends
     in Biochemical Sciences, 7 , 135.

15) O'FARRELL, P.H. (1975). High resolution two dimensional elec-
     trophoresis of proteins. J. Biol. Chem., 250, 4007.

16) O'FARRELL, P.Z., GOODMAN, H.M. and O'FARRELL, P.H. (1977). High
     resolution two dimensional electrophoresis of basic as well
     as acidic proteins. Cell, 12, 1133.

17) BRAVO, R. and CELIS, J.E. (1980). A search for differential
     polypeptide synthesis throughout the cell cycle of HeLa cells.
     J. Cell Biol. 48, 795.

18) BRAVO, R., FEY, S.J., SMALL, J.V., MOSE LARSEN, P. and CELIS,
     J.E. (1981). Coexistence of three major isoactins in a
     single sarcoma 180 cell. Cell, 25, 195.

19) BRAVO, R., BELLATIN, J. and CELIS, J.E. (1981).[^{35}S]-methionine
     labelled polypeptides from HeLa cells. Coordinates and per-
     centage of some major polypeptides. Cell Biol. Int. Rep. 5,
     93.

20) BRAVO, R. and CELIS, J.E. (1982). Up-dated catalogue of HeLa
     cell proteins: Percentages and characteristics of the major
     cell polypeptides labelled with a mixture of 16 [^{14}C]-amino
     acids. Clin. Chem., 28, 766.

21) ANSORGE, W. (1981). Preparation of ultrathin gels (0.1-0.2 mm).
     Applications to protein separation with a new silver staining.
     In: Int. Conf. Electrophoresis. Charleston, S.C., in press.

22) BRAVO, R., SMALL, J.V., FEY, S.J., MOSE LARSEN, P. and CELIS, J.E. (1982). Architecture and polypeptide composition of HeLa cell cytoskeletons. Modification of cytoarchitectural proteins during mitosis. J. Mol. Biol. 121.

23) PUCK, T.T. and MARCUS, P.I. (1956). Action of X-rays on mammalian cells. J. Exp. Med., 103, 653.

24) TOLMACH, L.J. and MARCUS, P.I. (1960). Development of X-ray induced giant HeLa cells. Exp. Cell Res., 20, 350.

25) HENDRICKS, M. and WEINTRAUB, H. (1981). Tropomyosin is decreased in transformed cells. Proc. Natl. Acad. Sci., 78, 5633.

26) FORCHHAMMER, J. (1982). Quantitative changes of some cellular polypeptides in C3H mouse cells following transformation by moloney sarcoma virus. In: Biological markers of neoplastic transformation (ed. C. Nicolini), Plenum, in press.

27) PAULIN, D., PERREU, J., JAKOB, H., JACOB, F. and YANIV, M. (1979). Tropomyosin synthesis accompanies formation of actin filaments in embryonal carcinoma cells induced to differentiate by hexamethylene bisacetamide. Proc. Natl. Acad. Sci. USA, 76, 1891.

28) LEAVITT, J. and MOYZIS, R. (1978). Changes in gene expression accompanying neoplastic transformation of Syrian Hamster cells. J. Biol. Chem. 253, 2497.

29) BRZESKI, J.and EGE, T. (1980). Changes in polypeptide pattern in ASV-transformed rat cells correlates with the degree of morphological transformation. Cell, 22, 513.

30) TUSZYNSKI, G.P., FRANK, E.D., DAMSKY, C.H., BUCK, A.C. and WARREN, L. (1979). The detection of smooth muscle desmin like protein in BHK21/C13 fibroblasts. J. Biol. Chem., 254, 6138.

31) COOPER, J.A. and HUNTER, T. (1981). Four different classes of retroviruses induce phosphorylation of tyrosine present in similar cellular protein. Mol. Cell Biol., 1, 394.

32) BRAVO, R., FEY, S.J., MOSE LARSEN, P. and CELIS, J.E. (1981). Modification of vimentin polypeptides during mitosis. Cold

Spring Harbor Symp. Quant. Biol., vol. 46, 379.

33) ROBINSON, S.I., NELKIN, B., KAUFMAN, S. and VOGELSTEIN, B.
     (1981). Increased phosporylation rate of intermediate fila-
     ments during mitosis. Exp. Cell Res., 133, 445.

34) CELIS, J.E., GRAESSMANN, A. and LOYTER, A. (1980). Transfer
     of cell constituents into eukaryotic cells. Plenum Press.

POLYPEPTIDE SYNTHESIS IN HUMAN SARCOMA AND NORMAL TISSUE

J. Forchhammer*# and H. Macdonald-Bravo*‡

*Department of Chemistry, University of Aarhus
DK-8000 Aarhus C, Denmark and #The Fibiger
Laboratory, Ndr. Frihavnsgade 70, DK-2100
Copenhagen Ø, Denmark

1. INTRODUCTION

Over the past few years the study of quantitative changes in the proteins of transformed cell lines have mainly taken two directions. One line of experimentation has been concerned with the total cellular response to transformation (1-5), while the other has been directed to search for changes in few cellular proteins whose expression is presumed to represent primary lesions in cells transformed with DNA or RNA tumour viruses (6-9). Both experimental approaches have made use mainly of cultured cell lines. At present, the relative proportions of at least 50 poly- peptides have been shown to change in transformed cell lines of human origin (1). Furthermore 13 host genes have been shown to be involved in transmission of different malignancies with different retroviruses (9). (See also other related articles in this volume).

In order to determine whether any of the polypeptide changes observed in transformed human cell lines also occurred in tumour tissue, we have compared by two dimensional gel electrophoresis

‡Present address: European Molecular Biology Laboratory, Postfach
10.2209, Meyerhofstr. 1, 6900 Heidelberg, FRG.

(10, 11) the polypeptide patterns from biopsies of three human sarcoma patients with the patterns from corresponding normal differentiated tissue from the same patients. Samples to be analyzed were
either labelled in a medium containing [^{35}S]-methionine or applied
directly on a gel and developed by silver staining (12, 13). The
labelled samples were prepared from biopsies of less than 1 mm^3
immediately after the operation or after 7-10 days of growth in
cell culture conditions, (which the biopsies seemed to tolerate
well). The results obtained from the silver staining techniques
show several abundant polypeptides which are weakly labelled with
[^{35}S]-methionine indicating either a very low turnover of proteins,
low methionine content or the requirement of factors not supplied
by the medium nor by the cell in the biopsy.

Our results show that at least 1/3 of the polypeptide changes
previously observed in human cell lines (1) could also be found
in two or three of the tumours. These studies also revealed a
corresponding number of additional polypeptide changes that have
not been observed in cultured cells (1).

2.   HISTOPATHOLOGY OF TUMOUR AND CONTROL TISSUE

In comparing the protein patterns of normal and malignant
tissue we have chosen to study mesenchymal tissue for two reasons.
First, it seemed possible to obtain rather homogeneous sarcoma specimens containing one predominant cell type (Figures 1-3), and
secondly, because we expected that it would be easier to grow these
cells in short term tissue cultured as compared to most epithelial
cells. Small pieces of tissue were taken immediately after wide
excision of tumours or removal of the total limb with a tumour.
The control tissue was taken from the same operation specimen and
the histological type of control tissue was selected based on the
type of tumour as diagnosed with a preoperative biopsy. One hour
later the biopsies were further subdivided selecting homogenenous

pieces. These were cut in minute pieces and placed under a small
coverslip in the wells of microtiter plates (Falcon) and short
term cultivated. Other pieces of the same specimen were processed
for histo-pathological examination and representative micrographs
are shown in Figures 1-3. The diagnosis and the general pathological findings are summarized in Table 1.

   The three tumours analyzed in this study contain one predominant type of undifferentiated sarcoma cells with only small
amounts of intercellular substance. Furthermore, there was no in-
flammatory reaction and only few plasma cells and granulocytes
could be observed in the biopsies. In a mesenchymoma which contained
mainly lipoblasts (Figure 1a), minor areas with chondroid (Figure
1c) or osteoid elements (Figure 1d) were present. The tumour was
located below the fascia of the thigh, and subcutanous fat tissue

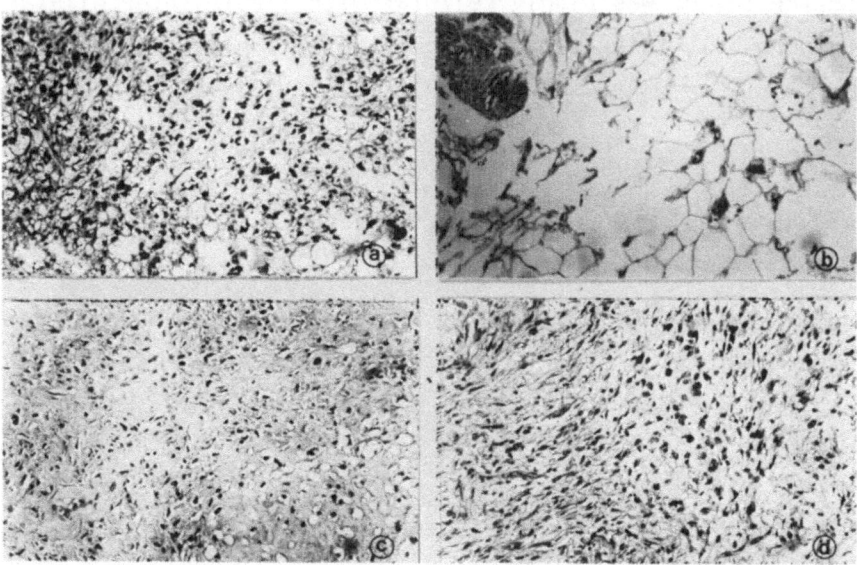

Fig. 1. *Malignant mesenchymoma, grade 3.* Most abundant tumour
element: (a) Liposarcoma; (b) Control tissue: subcutanous fat
tissue. The tumour contains minor areas with (c) chondroid or (c)
osteoid elements. Magnification 100 x.

free of tumour infiltration was taken as control tissue (Figure 1b), since the most abundant component in the tumour was of liposarcoma type

The second tumour studied was a malignant fibrous histiocy-toma (Figure 2a). The tumour composed of a uniform type of cells with only small amounts of intercellular material. This type of tumour has previously been called a fibrosarcoma but was renamed recently (14) reflecting the newer views of the origin of the tumour cells. In this case scar tissue from a previous operation site was used as control. This consisted of fibrous connective tissue without infiltration of any tumour cells Figure 2b).

The third tumour studied here is shown in Figure 3a. The diagnosis was "mesenchymal chondrosarcoma, grade 3", and fibrous connective tissue from the same patient was used as control tissue (Figure 3b). It can be argued that this is not an adequate control, but since the tumour was highly cellular with little intercellular matrix (Figure 3a), this seemed an acceptable sample.

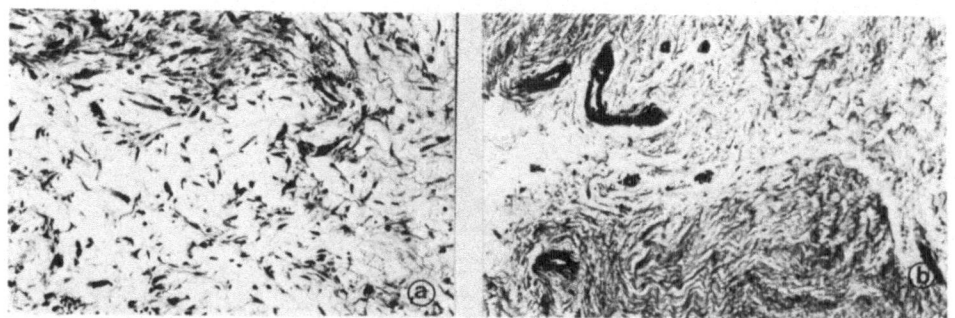

Fig. 2. *Malignant fibrous histiocytoma, myxoid variant, grade 3.*
(a) Tumour tissue; (b) Control tissue: fibrous connective tissue.
Magnification 100 x.

3. POLYPEPTIDES SYNTHESIZED IN TUMOUR AND CONTROL TISSUE

The biopsies described in Table 1 were labelled for 20 hr

with [^{35}S]-methionine both immediately and 7-10 days after the operation (see legends to Figures 4-6). The polypeptides were separated by two dimensional gel electrophoresis (IEF and NEPHGE) as previously described (15).

Table 1. *General aspects of tumour specimens and control biopsies*

Code No.	Tissue	Figure	Diagnosis[a]	Predominant type of tissue	Additional tissue components
OM 2789	Tumour	1a	malignant mesenchymoma (grade 3)	liposarcoma	chondrosarcoma[c] osteosarcoma[d]
	Control	1b	subcutaneous fat tissue	fat tissue	
OM 2818	Tumour	2a	malignant[b] fibrous histocytoma myxoid variant (grade 3)	fibrosarcoma	
	Control	2b	fibrous scar tissue	fibrous connective tissue	
OM 2792	Tumour	3a	mesenchymal chondro-sarcoma (grade 3)	poorly differentiated sarcoma	
	Control	3b	connective tissue	fibrous connective tissue	

[a] The patho-anatomical examinations were kindly performed by O. Myhre Jensen, Aarhus Kommunehospital, DK-8000 Aarhus C.

[b] Formerly referred to as pleomorphic fibrosarcoma (15)

[c] One of the minor areas shown in Figure 1c

[d] One of the minor areas shown in Figure 1d

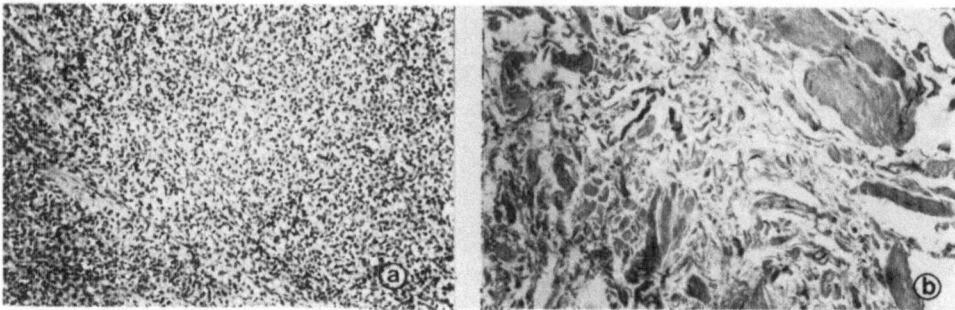

Fig. 3. *Mesenchymal chondrosarcoma, grade 3.* (a) Tumour tissue;
(b) Control tissue: fibrous connective tissue. Magnification 100 x.

    Figures 4a and b show the polypeptide patterns of malignant
mesenchymoma (Figure 1a) and control tissue labelled with [^{35}S]-
methionine immediately after the operation. The polypeptides were
localized by autoradiography and all major spots as well as spots
with apparent differences in intensity between tumour and control
biopsies were cut from the gel and counted in a liquid scintillation counter. The cutting was verified by re-exposing the gel
after all the spots had been cut out. The total gel was then cut
up into 10 x 25 mm pieces and counted. The relative abundance of
cpm in any spot was thereafter expressed as cpm of the spot per
100.000 cpm of the total of the gel (excluding the loading zone,

Fig. 4. *Polypeptides labelled with [^{35}S]-methionine from tissues
from a patient with malignant mesenchymoma.* (a) malignant mesenchymoma, grade 3 (see Figure 1a); (b) control tissue: subcutaneous
fat tissue (see Figure 1b). Pieces of tumour and control tissue
(less than 1 mm^3) were grown in short term cultures in a microtiter plate (Falcon) pressed under a 3 x 3 mm coverslip. The pieces
were labelled as described before (16) with 100 µCi [^{35}S]-methionine
for 20 hr at 36^{0}C beginning 2-3 hr after the operation. The polypeptides were separated by two dimensional electrophoresis and
localized as described in details before (15). Polypeptides are
numbered in accordance to ref. 17. Arrows and arrowheads pointing
upwards/downwards indicate polypeptide increased/decreased more
than 30% in tumour when compared to control. Arrows indicate observation done in all three tumour types. Arrowheads indicate observation only seen in two of the three tumour types. For details
see Table 2. NEPHGE 12: EF-1α and IEF 47: β- and γ-actin is
indicated as marker polypeptides.

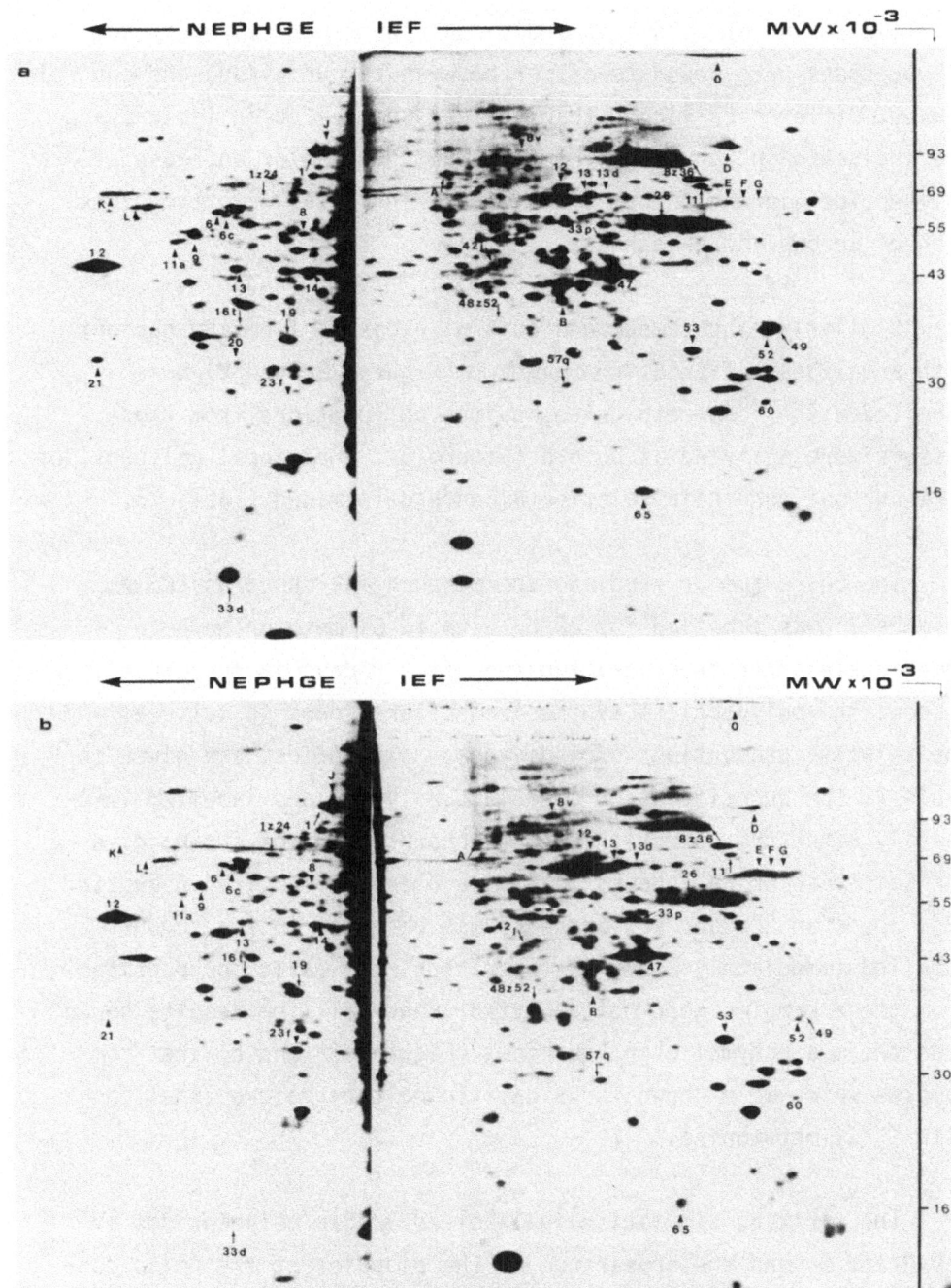

which typically contained $\leq$ 10% of the total counts). When the
relative amount of counts in individual spots was compared, some
polypeptides were found to differ between tumour tissue and control
tissue. Differences larger than 30% are summarized in Table 2 and
are indicated in Figure 4. The relative increase or decrease of
polypeptides in the tumour biopsies are indicated with the direction of arrows and arrowheads.

Similarly, the tumour and control biopsies from the patient
with a malignant fibrous histiocytoma (Figures 2a and b) were
labelled with [^{35}S]-methionine and the polypeptides from these
tissues were analyzed as before (Figure 5). Individual polypeptides
were cut out and their relative amounts determined (Table 2).

The third tumour studied, the mesenchymal chondrosarcoma,
(Figure 3) was labelled for 20 hr with [^{35}S]-methionine both
immediately after the operation and after 9 days in culture. Figure
6 shows the polypeptides synthesized after 9 days in culture, and
the relative proportions of individual polypeptides are given in
Table 2. The quantification of acidic polypeptides labelled immediately after the operation (electrophoretogram not shown) give
the same results as appears in Figure 6 except for IEF 13 and IEF
42j, where an 80% and 40% increase was observed in the tumour
labelled immediately after the operation. The basic polypeptides
from these samples were not analyzed. Hence all the results shown
from the mesenchymal chondrosarcoma (Figures 3a and b) are from
samples which were grown for 9 days in culture before labelling
with [^{35}S]-methionine.

The striking similarity in labelled acidic polypeptides at
day 0 and 9, and the appearance of the biopsies at the phase contrast microscope, indicate that both the control and tumour tissues
grew well in culture when medium was changed every 2-3 days.

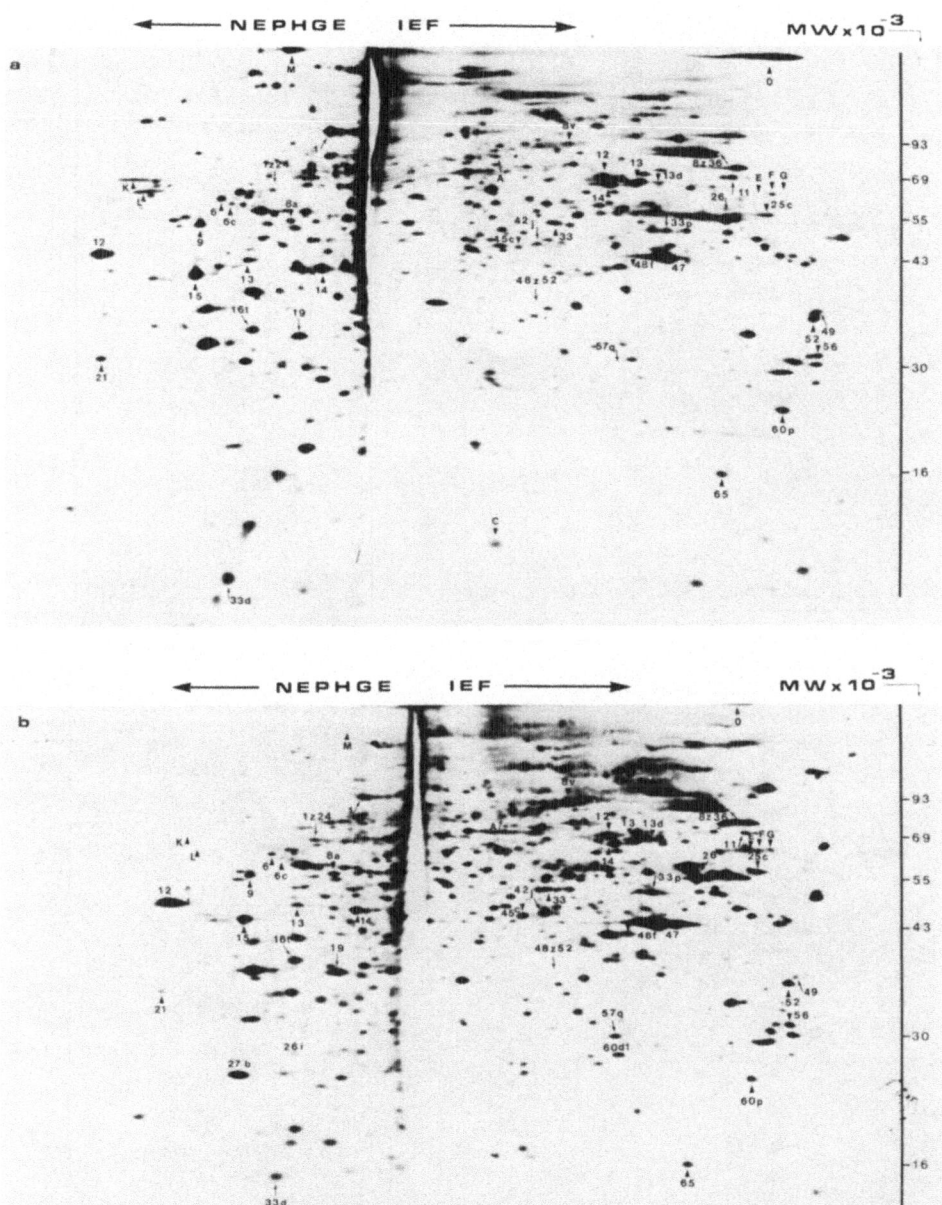

Fig. 5. *Polypeptides labelled with [^{35}S]-methionine*. (a) malignant fibrous histiocytoma, myxoid variant, grade 3 (see Figure 2a). (b) control tissue: fibrous connective tissue (see Figure 2b). Technical details as in Figure 4.

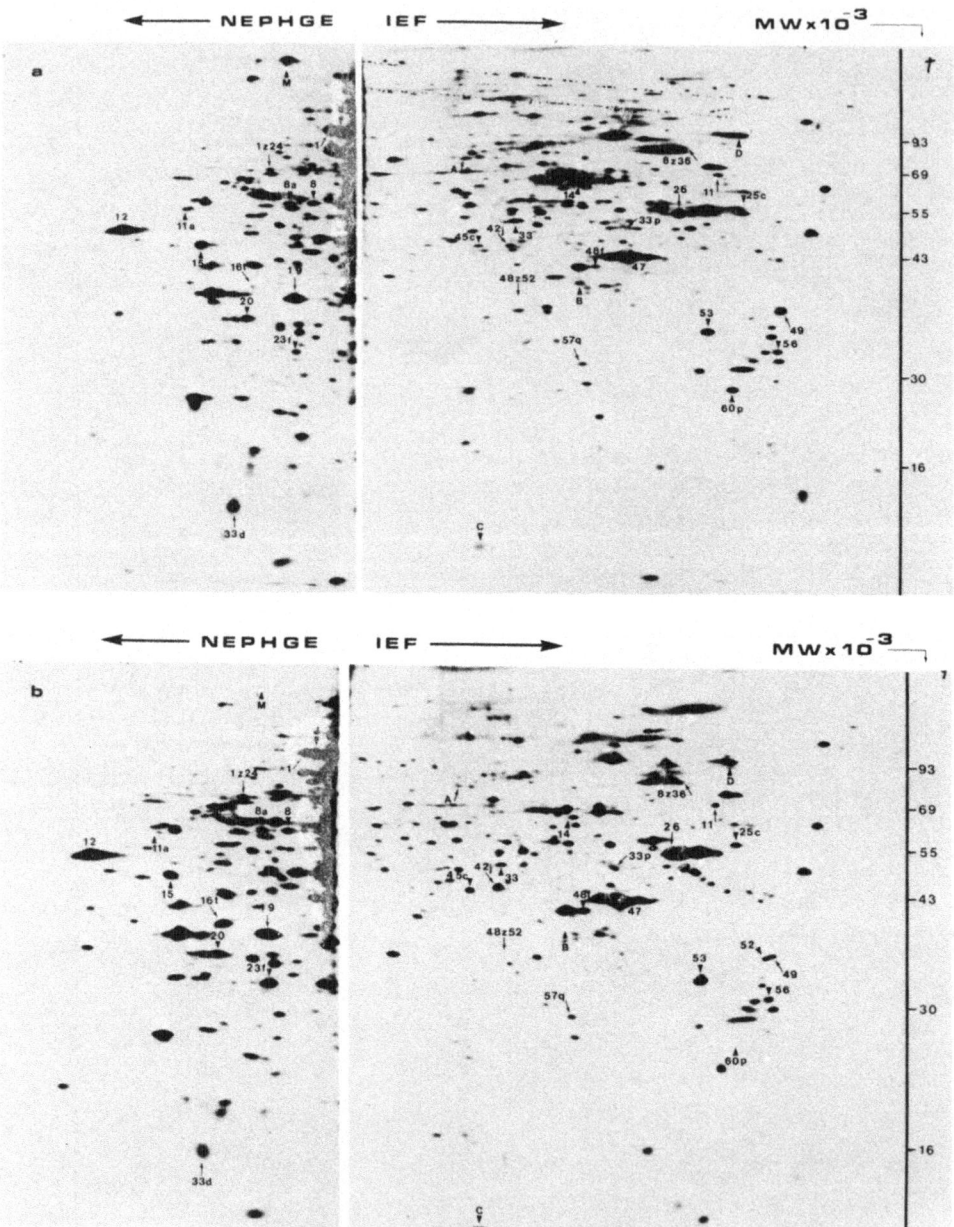

Fig. 6. *Polypeptides labelled with* [^{35}S]-*methionine*. (a) mesenchymal chondrosarcoma, grade 3 (see Figure 3a). (b) control
tissue: fibrous connective tissue (see Figure 3b). Technical details as before except that the biopsy material was labelled
after 9 days in culture, where medium had been changed every 2-3
days.

Table 2. *Polypeptides present in altered amounts in tumours*

Polypeptide[a]	Coordinates[b]	cpm-ratio Fig. 4	cpm-ratio Fig. 5	$\frac{\text{tumour}^e}{\text{control}}$ Fig. 6	Relative proportion in tumours
NEPHGE					
M:	nd/0.43	0.50	19	12	increased
1	$95/0.18^c$	3.9	4.5	6.3	increased
J:	95/0.10	0.23	0.97	0.72	decreased
1z24	$71/0.51^d$	0.29	0.75	0.40	decreased^o
K: 1z17?	72/1.25	10	$++/0^f$	/	increased
L: 3d?	65/1.20	4	$++/0^f$	/	increased
6	$62/0.88^c$	1.6	2.7	0.88	increased
6c	$61/0.82^d$	1.3	1.5	1.1	increased^o
8	$57/0.24^d$	0.77	0.90	0.62	decreased^o
8a	$57/0.43^d$	1.2	0.55	0.62	decreased^o
9	$55/1.00^c$	1.7	1.9	0.84	increased
13	$49/0.67^d$	2.1	1.3	1.2	increased^o
14	$48.5/0.24^d$	2.0	1.8	0.67	increased^o
15	$47/1.00^d$	1.27	1.8*	1.6*	increased*
16t	41.5/0.72	0.16	0.62	0.18	decreased
19	$39.5/0.40^d$	0.67	0.61	0.75	decreased^o
21	$35/1.45^d$	4.0	3.2	0.69	increased^o
23f	32/0.31	0.11	/	0.22	decreased
33d	14/0.80	4.4	1.65	1.56	increased

a: From ref. 17. Individual capital letters were given to polypeptides not homologated to those which appear in ref. 17.
b: Molecular weight/mobility relative to NEPHGE 9
c: Coordinates taken from ref. 17.
d: Coordinates taken from ref. 1
e: Calculated from relative proportion of cpm in the corresponding spots
f: Intensity of spots when absent from one type of extracts. 0: absent; ++ moderate or strong.
o: In agreement with findings in ref. 1.
*: In contrast to findings in ref. 1.
?: Indicates that number assignment is tentative
/: Polypeptide not located.

(continued)

Table 2. *(continued)*

Polypeptide[a]	Coordinates[g]	cpm-ratio		$\dfrac{tumour^e}{control}$	Relative proportion in tumours
		Fig. 4	Fig. 5	Fig. 6	
IEF					
0	nd/1.35[c]	1.3	>2.0	/	increased
D: 8n?	94/1.39	1.8	0/0	1.9	increased
8v	90/0.64[d]	0.32	0.77	/	decreased[o]
A: 8z14?	84/0.32[d]	1.9	1.7	1.5	increased[o]
8z36	76/1.27[d]	1.3	1.8	1.3	increased[o]
11	69/1.33[c]	1.6	2.0	1.4	increased
12	68/0.85[c]	0.62*	0.73*	1.4	decreased*
13	67/0.93[c]	0.50	0.60	1.1	decreased
13d	67/0.98	0.77	0.65	0.98	decreased
14	66/0.86[d]	0.64	1.8	7.2	increased[o]
E: 17j?	64/1.27	0.32	0.5	1.1	decreased
F: 17k?	64/1.32	0.27	0.62	0.45	decreased
G: 17l?	64/1.36	0.37	0.43	1.1	decreased
25c	56/1.34	0.83	0.48	0.62	decreased
26	54/1.19[d]	0.60	0.56	0.63	decreased[o]
33	49.5/0.64	1.05	1.49	1.66	increased
33p	49/0.99	0.67	0.59	0.68	decreased
42j	45/0.56	0.67	0.37	0.38	decreased
45c	44/0.44	0.92	0.61	0.42	decreased
48f·	42/0.88	0.80	0.35	0.48	decreased
B: 48z53	39/0.81	1.65	/	4.1	increased
48z52	37/0.58	0.16	0.58	0.59	decreased
49	36/1.48[d]	5.8	3.4	2.7	increased[o]
52	35/1.43[c]	13*	1.6*	0.1	increased*
53	32/1.28[d]	0.61	0.99	0.77	decreased[o]
56	31/1.41[d]	1.18	0.59	0.73	decreased[o]
57q	30/0.86[d]	0.53	0.20	0.48	decreased[o]
60p	24/1.35	0.90	2.0	10	increased

(continued)

Table 2. *(continued)*

Polypeptide[a]	Coordinates[g]	cpm-ratio		$\dfrac{\text{tumour}^e}{\text{control}}$	Relative proportion in tumours
		Fig. 4	Fig. 5	Fig. 6	
IEF					
65	16/1.15	1.95	1.30	0.74	increased
C: 66h?	nd/0.44	/	0.67	0.40	decreased

g: Molecular weight/mobility relative to β-actin
NEPHGE: non-equilibrium pH gradient electrophoresis
IEF   : isoelectric focussing.

The relative labelling in counts of approximately 250 preselected polypeptides in tumour specimens and the corresponding control biopsies in short term cultures are summarized in Table 2. 49 polypeptides were found to change more than 30% and of these only 10-12 changed more than two-fold (Table 2). An increase in the relative amounts of 24 polypeptides was found in the tumour specimens, whereas a decrease was observed in 25 polypeptides. We did not observe a tendency to a more simple pattern of gene expression in the tumours.

4. TRAUMA RELATED PROTEINS

As shown in Figures 4-6 most of the biopsies gave a polypeptide pattern similar to those previously published using established cell lines (1, 3, 15, 17, 18). All our experiments were done in duplicate and some of the samples were also labelled with a mixture of 16 [^{14}C]-amino acids.

From a total of 60 IEF gels analyzed so far by autoradiography eight showed a remarkably different pattern, an example of which is shown in Figure 7. Two features were common in these 8 samples.

Firstly, the total incorporation of [^{35}S]-methionine was reduced to 3 - 30 fold compared to other biopsies labelled the same day with the same medium. Secondly, the eight gels showed increased amounts of three groups of polypeptides.

Figure 7 shows a gel with only a 3-fold reduction in the relative counts in actin allowing localization of the three strongly labelled groups of polypeptides (tr53/56, tr66/69 and tr69/73), named after their approximate molecular weight. The first group tr66/69, is located to the alkaline side of proteins IEF 12, 13 and 14 and contains 12-43% of the total amount of counts in the IEF gels. The other group, tr53/56 is located to the acidic side of vimentin (IEF 26) and β-tubulin (IEF 25), and contains 1.5-4.5% of the total counts in the IEF gels. The third group, tr69/73, is located in the most acidic part of the gel (pI 4-4.5) and contains 0.2-0.8% of the total counts in the IEF gels.

It should be noted that three of the eight samples came from control tissue without infiltration of tumour cells, whereas the other five were from tumour specimens. We have searched for a common cause to the dramatic change in the protein pattern in these eight biopsies. We suggest that the traumatic effect of cutting the biopsies and pressing them under a coverslip may account for this variation, a manipulation which could have been less gentle for some samples than for others. This is an unproven speculation, but polypeptides with similar location have been observed after heat shock of Drosophila (19) and of HeLa cells (20, R. Bravo, personal communications). This has led us to propose the designation tr for these proteins which might be trauma related and possibly induced by the processing of the biopsy.

With this in mind several gels were reinspected and counted and radioactivity was found in the area of tr66/69 and tr53/56 in

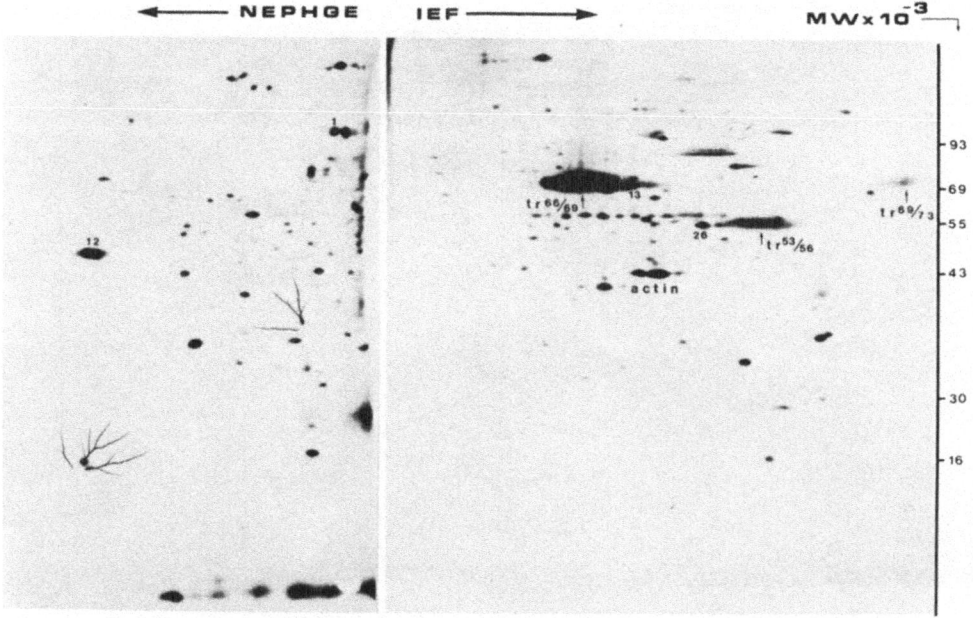

Fig. 7. *Two dimensional gel electrophoresis of total [^{35}S]-
methionine labelled polypeptides.* Cells from a haemorrhagic osteosarcoma in femur. Technical details as in Figure 4, tr is used to
indicate possible trauma related proteins as explained in the text.
Actin, vimentin, (IEF 26) and other spots are given for reference.

several other gels, for instance Figure 4a, 4b and 6a and to a
lesser extent in some of the others.

5. PROTEIN TURNOVER

The preceding experiments using [^{35}S]-methionine for the
labelling of proteins identifies mainly those polypeptides with a
high rate of synthesis. Stable proteins, with a low rate of synthesis may comprise a considerable fraction of the proteins in
tumours as well as in normal tissue. Such polypeptides can be detected after two dimensional gel electrophoresis using different
staining methods.

Hence, from all three tumour specimens as well as control biop-

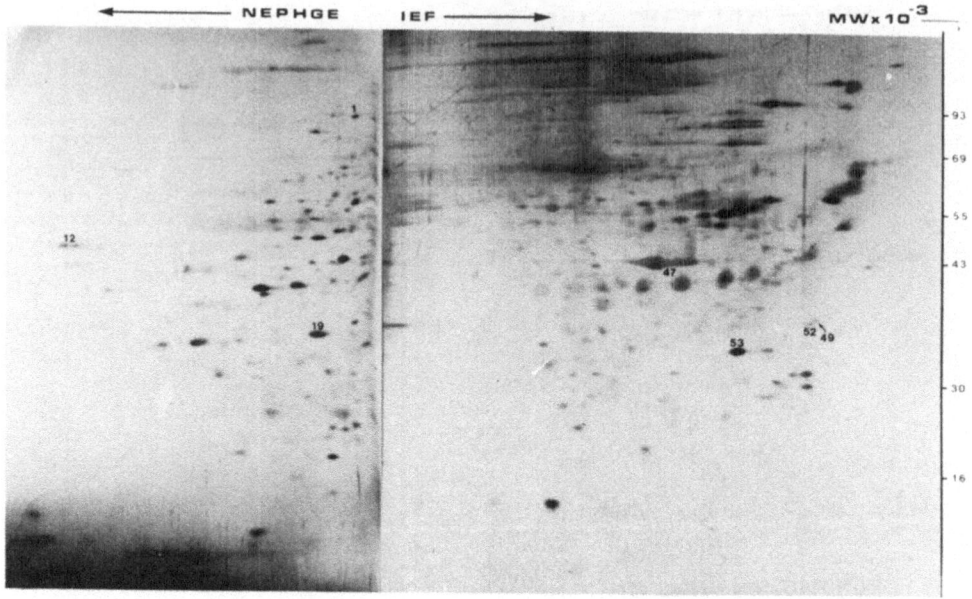

Fig. 8. *Polypeptides from the malignant mesenchymoma, grade 3 stained with silver nitrate (12, 12).* A piece of tumour tissue containing 50-100 µg protein was washed in icecold PBS, then suspended and homogenized on ice in DNase, RNase buffer (50 mM, NaCl, 20 mM Tris, pH 7.3-7.4; DNase 5 x crystallized, 25 µg/ml; preheated pancreatic RNase 5 x crystallized, 25 µg/ml). SDS to a final conc. of 0.3% was then added to the above at room temperature, incubated for 5 min and the total extract was then frozen and lyophilized. The sample was resuspended in lysis buffer (10, 11) and a small aliquot of the radioactive extract from the same tissue was added. This radioactive extract only constituted 1-3% of the protein present in the unlabelled biopsy material and helped to identify the different spots. Electrophoresis was as in Figure 4. The polypeptides were developed with silverstaining (12, 13) as well as autoradiography of the stained gel, and a few polypeptides are indicated on the Figure for reference.

sies, samples containing 25-50 µg protein per gel were stained after electrophoresis with a very sensitive silver staining technique (12, 13). Figure 8 shows the result of the staining of an extract from the malignant mesenchymoma where the biosynthetically labelled polypeptides are shown in Figure 4a. The sample also contained 1-3% of labelled proteins from the same tissue and

autoradiography after the staining helped to identify different
stained polypeptides. Many spots can be identified in both systems
but it is obvious that the intensity varies between homologous
spots detected with the two methods. For instance NEPHGE 6, 9 and
12 as well as IEF 49 and 52 are seen most clearly on the autoradiograms (Figure 4a). The reverse is true for some basic polypeptides of both high and low molecular weight (Figure 8). Also a
series of acidic polypeptides with a molecular weight around 40
Kd can be seen under β- and γ-actin (IEF 47) in Figure 8 but not
in Figure 4a. When stained polypeptides from the three tumours were
compared with the corresponding extracts from control tissue, no
tumour specific polypeptides was localized and no attempts were
made to quantitate the observed individual differences.

6. DISCUSSION

Cell lines transformed chemically or by means of tumour virus
are frequently tumorigenic in *nude* mice or their syngeneic host and
hence regarded malignant. Some of these cell lines are of epithelial origin, whereas other lines are of mesenchymal origin and frequently described as fibroblasts.

Studies of transformed cell lines have revealed changes in
relative abundance of many cellular polypeptides (1-5). Except for
a few virally coded proteins, few if any proteins seem to be synthesized exclusively by transformed cells. The few viral encoded
proteins that have been found belong mainly to the class of fusion-
proteins which appears in many cells transformed by retrovirus (9)
or can be virus specific antigens such as the T-antigen of SV40
transformed cells (21).

In this article we have compared the synthesis of polypeptides
in human sarcomas and in their corresponding normal mesenchymal
tissue. The results have furthermore been related to findings from

cultured human cell lines (1) (Table 2). The histological complexity of tumours raised some questions regarding the interpretation
of the origin of the polypeptides observed in short term labelled
cultures. In all cases the tumour specimens consisted of one dominant cell type which accounted for 80-95% of the total cell
number in the specimen. They contained, in addition, a few small
capilaries, plasma cells and granulocytes. That the tumour tissue
contained few if any epithelial cells is suggested by the total
absence of polypeptide IEF 59 in the gel patterns. This polypeptide which migrates above IEF 60 is found typically in all epithelial cultured cells studied so far (17). We therefore think
that the labelled polypeptides mainly represent mesenchymal
proteins synthesized by the dominant cell type as indicated in
Table 1.

In contrast to this homogeneity, all specimens from a 5 year
old boy with a haemorrhagic osteosarcoma in which a spontaneous
fracture had taken place, were very heterogeneous. The tumour
biopsy consisted of "malignant mesenchymal cells probably osteo-
blasts" plus "hematoma in beginning organization with inflamma-
tory cells". The control biopsy was taken from "normal spongious
bone marrow". The heterogeneity of cell types in these biopsies
precluded interpretation of comparison of polypeptides in these
speciments which are, therefore, not shown here.

Table 2 summarizes the quantitative results from the three
sarcomas with related morphology, namely anaplastic mesenchymal
tumour cells of grade 3 with only small amounts of intercellular
matrix material, capilaries and blood cells.

We found that the relative proportion of 49 out of the 250
preselected polypeptides quantitated, differed more than 30% when
compared to control tissues. 18 of these have been shown previous-

ly to change in transformed human cell lines (1), and are indicated
with o in Table 2. Due to the similarity of our gels with the ones
presented by Bravo *et al.* (1, 15, 17) and to the fact that these
polypeptides have identical relative mobility in most cases, we believe that they correspond to the same polypeptides. Thus it can
be concluded that about one third of the polypeptides that vary in
transformed human cell lines also change in human sarcomas. The 18
polypeptides are, therefore, likely to share functions related to
the changed phenotype of transformed cells as well as the tumorigenicity of sarcoma cells in man.

So far, only a few of these polypeptides have been identified.
IEF 26 corresponds to vimentin, a component of the intermediate
filaments (15, 17). IEF 49 to cyclin, a nuclear protein which increase during the S-phase and cell proliferation (22), and IEF
52 to a tropomyosin related polypeptide (18, 22). Identification
of all of these polypeptides and the study of their function should
receive a high priority in the search for a functional understanding
of malignant growth. It should be noted that the relative proportion of IEF 52 was found to be increased in two of the three sar-
comase in contrast to what has been found in transformed cell
lines (1, 18). The interpretation of this observation, however, is
not clear at the moment.

Three polypeptides which have previously been found to be present in reduced amount in transformed cells (1) also showed reduced
synthesis in the malignant fibrous histiocytoma (Figure 5a) but
were not changed or not observed in the other tumours (Figures 4
and 6) and are hence not listed in table 2. These corresponded to
NEPHGE 26i and 27b and to IEF 60d1 (1).

Our studies revealed 31 polypeptide changes (Table 2) not previously observed in cultured cell lines. This could reflect dif-

ferences in the control mechanisms present in normal tissue which
are lost during the establishment of the cell lines. It could also
reflect small amounts of a normal cell type present in tumour material but absent from the control biopsies. If this is the case,
only major polypeptides in the subpopulation would give rise to the
minor changes in the overall composition of the polypeptide pattern.

An unexpected polypeptide pattern was observed in 8 out of 60
biopsies analysed (see Figure 7). In these the pattern of polypeptides was completely different to that previously described for
many cultured cells of human origin (1, 3, 4, 15, 17, 18, 22). This
new pattern was observed in 5 tumour specimens and 3 control biopsies. These differences are not due to the labelling technique,
because we obtained the typical pattern with parallel samples,
but we are inclined to believe they could have been caused by the
traumatic treatment of cutting and pressing the samples. Of course
the samples were treated practically the same but minor variations
may inevitably occur.

The general observation is the appearance of two strongly
labelled clusters of polypeptides and one less labelled cluster.
One cluster is located to the acidic side of IEF 12, 13 and 14
and has been given the preliminary name tr66/69. This cluster
contains 12-43% of all the cpm present in the IEF gel reflecting
a strong reduction in the synthesis of most other cellular polypeptides. The other strongly labelled cluster is located to the
acidic side of IEF 25 and 26 and has been designated tr53/56. It
is not known whether all cells have been directly affected by the
trauma reducing total labelling by 3-30 fold, or whether only a
limited number of cells have been damaged and then given off a
signal to the remaining cells reducing their normal synthesis
of polypeptides. If the latter is the case, the tr-polypeptides
are likely candidates for mediating these signals to neighbouring
cells.

It is interesting to mention that some of these tr-polypeptides have a similar migration on the IEF gels to some polypeptides
synthesized after heat shock treatment of HeLa cells (20; R. Bravo,
personal communications) and Drosophila (19). Even though the biopsies have not been exposed to temperatures above 36°C, these polypeptides could be the same as trauma induced polypeptides similar
to heat shock proteins, which have been reported in rat tissue
(23, 24).

7. ACKNOWLEDGEMENTS

We would like to thank O. Sneppen for collaboration in col-
lecting the surgical specimens, and O. Myhre Jensen for the patho-
anatomical examinations. Furthermore, we would like to thank R.
Bravo, S. Fey and P. Mose Larsen for discussions of details and
results with regard to the gel electrophoresis. The careful quantitation of cpm in polypeptide spots by Helle Jensen and Birgitte
Rask is also appreciated. Finally, we would like to than A. Celis
and J.E. Celis for hospitality and discussions during our stay
at the Department of Chemistry, Aarhus University.

This study was supported by the Danish Cancer Society, grant
No. 24/81.

8. REFERENCES

1) BRAVO, R. and CELIS, J.E. (1982). Human proteins sensitive to
    neoplastic transformation in cultured epithelial and fibro-
    blast cells. Clin. Chem. 28, 949.
2) STRAND, M. and AUGUST, J.T. (1978). Polypeptide maps of cells
    infected with murine type C leukemia or sarcoma oncovirus.
    Cell, 13, 399.
3) FORCHHAMMER, J. and KLARLUND, J. (1979). Changes in proteins
    from transformed cultures and tumours induced by sarcoma

virus. In: Advances in Medical Oncology, Research and Education (ed. P.G. Margison) Vol. 1, p. 51, Pergamon Press, Oxford.

4) BRAVO, R. and CELIS, J.E. (1980). Gene expression in normal and virally transformed mouse 3T3B and hamster BHK21 cells. Exp. Cell Res. 127, 249.

5) BRZESKI, H. and EGE, T. (1980). Changes in polypeptide pattern in ASV-transformed rat cells are correlated with the degree of morphological transformation. Cell, 22, 513.

6) BRUGGE, J.S., ERIKSON, E. and ERIKSON, R.L. (1981). The specific interaction of the Rous sarcoma virus transforming protein, pp60^{src}, with two cellular proteins. Cell, 25, 363.

7) RADKE, K. and MARTIN, G.S. (1980). Transformation by Rous sarcoma virus: Effects of src-gene expression on the synthesis and phosphorylation of cellular polypeptides. Cold Spring Harbor Symp. Quant. Biol. 44, 975.

8) SHARP, P.A. (1980). Summary: Molecular biology of viral oncogenes. Cold Spring Harbor Symp. Quant. Biol. 44, 1305.

9) COFFIN, J.M., VARMUS, H.E., BISHOP, J.M., ESSEX, M., HARDY, W.D., MARTIN, G.S., ROSENBERG, N.E., SCOLNICK, E.M., WEINBERG, R.A. and VOGT, P.K. (1981). Proposal for naming host cell derived inserts in retrovirus genomes. J. Virol. 40, 953.

10) O'FARRELL, P.H. (1975). High resolution two dimensional electrophoresis of proteins. J. Biol. Chem. 250, 4007.

11) O'FARRELL, P.Z., GOODMAN, H.M. and O'FARRELL, P.H. (1977). High resolution two dimensional electrophoresis of basic as well as acidic proteins. Cell, 12, 1133.

12) ANSORGE, W. (1981). Preparation of ultrathin gels (0.1-0.2 mm). Applications to protein separation with a new silver staining. In: International Conference in Electrophoresis, Charlston, S.C., in press.

13) MOSE LARSEN, P. (1981). An assessment of the potential offered by two dimensional gel electrophoresis and silver

staining for developmental biology. (Thesis: Dept of
Chemistry, Aarhus University, Denmark).

14) HAJDU, S. (1979). Pathology of soft tissue tumours. Lea
     and Fibiger, Philadelphia, ISBN 0-8121-0693-8.

15) BRAVO, R., SMALL, J.V., FEY, S.J., MOSE LARSEN, P. and
     CELIS, J.E. (1982). Architecture and polypeptide composition
     of HeLa cytoskeletons. Modification of cytoarchitectural
     polypeptides during mitosis. J. Mol. Biol. 154, 121.

16) BRAVO, R., FEY, S.J., SMALL, J.V., MOSE LARSEN, P. and
     CELIS, J.E. (1981). Coexistence of three major isoactins
     in a single sarcoma 180 cell. Cell, 25, 195.

17) BRAVO, R. and CELIS, J.E. (1982). Up-dated catalogue of
     HeLa cell proteins: Percentages and characteristics of the
     major cell polypeptides labelled with a mixture of 16 [^{14}C]-
     amino acids. Clin. Chem. 28, 766.

18) FORCHHAMMER, J. (1982). Quantitative changes of some cellular
     polypeptides in C3H mouse following transformation by Moloney
     sarcoma virus. In: "Biological markers of neoplastic trans-
     formation". (ed. P. Chandra), Plenum, in press.

19) BUZIN, C.H. and PETERSEN, N.S. (1982). A comparison of the
     multiple Drosophila heat shock proteins in cell lines and lar-
     val salivary glands by two dimensional gel electrophoresis.
     J. Mol. Biol. 158, 181.

20) HICKEY, E.D. and WEBER, L.A. (1982). Modulation of heat-
     shock polypeptide synthesis in HeLa cells during hyperther-
     mia and recovery. Biochemistry, 21, 1513.

21) TOPP, W.C., LANE, D. and POLLACK, R. (1980). Transformation
     by SV40 and polyoma virus. In: Molecular biology of tumour
     viruses, (ed. J. Tooze) part 2, pp. 205. Cold Spring Harbor,
     New York.

22) BRAVO, R., FEY, S.J., BELLATIN, J., MOSE LARSEN, P. and
     CELIS, J.E. (1982). Identification of a nuclear polypeptide
     (cyclin) whose relative proportion is sensitive to changes in

the rate of cell proliferation and to transformation. <u>In</u>:
Proceedings of the International Society of Developmental
Biology. (ed. M. Burger), Alan R. Liss, New York, In press.

23) CURRIE, R.W. and WHITE, F.P. (1981). Trauma-induced protein
in rat tissues. A physiological role for a "heat shock"
protein?. <u>Science</u>, <u>214</u>, 72.

24) HAMMOND, G.L., LAI, Y.-K. and MARKERT, C.L. (1982). Diverse
forms of stress lead to new patterns of gene expression
through a common and essential metabolic pathway. <u>Proc. Natl.
Acad. Sci. USA</u>, <u>79</u>, 3485.

THE REVERSIBLE MODULATION OF THE SYNTHESIS OF MATRIX COMPONENTS
IN DEFINITIVE CHONDROBLASTS TRANSFORMED BY A *ts*-ROUS SARCOMA
VIRUS MUTANT

M. Pacifici, S.L. Adams, H. Holtzer and D. Boettiger

Departments of Anatomy, Human Genetics and Microbiology
School of Medicine
University of Pennsylvania
Philadelphia, Pennsylvania 19104, USA

1. INTRODUCTION

The differentiation program of several types of terminally-
differentiated embryonic cells can be reversibly altered by
transformation with RNA tumour viruses. For example, definitive
chondroblasts isolated from chick embryo vertebral cartilage and
infected with a temperature sensitive (*ts*)-mutant of Rous sarcoma
virus (*ts*-RSV) lose their characteristic polygonal morphology and
become bipolar in shape when maintained at the permissive temperature for transformation ($36^{0}C$). Concurrently with their morphological alteration, the transformed chondroblasts lose the ability to
synthesize two major intercellular matrix components, the Type IV
sulphated proteoglycan and Type II collagen (1,2). These morphological and biosynthetic alterations induced by viral transformation
are fully reversible; within 1-2 days following a shift to the
non-permissive temperature for transformation ($41^{0}C$), the infected
chondroblast display the characteristic polygonal-epithelioid
appearance and reinitiate the synthesis of Type IV proteoglycan
and Type II collagen.

In addition to vertebral chondroblasts, yolk sac macrophages and retinal pigment cells have been shown to reversibly lose the ability to express their terminal phenotype following retroviral transformation (3,4).

While these aspects of retroviral transformation of definitive embryonic cells are well understood, the mechanisms that induce the reversible modulation of the terminal phenotype are rather obscure. Progress has been recently made in this field by analyzing in molecular terms the effects of retroviral transformation on total chick embryo fibroblasts. Uninfected fibroblasts synthesize, and deposit extracellularly, large amount of Type I collagen and fibronectin (5,6). However, following transformation by wild type RSV, fibroblasts synthesize 80 to 90% less Type I procollagen and fibronectin as compared to companion uninfected fibroblasts (7,8). This reduction of the synthesis of these two major matrix components correlates with proportional decreases in the respective mRNAs translatable *in vitro* by a reticulocyte lysate (9,10). More recently, cDNA hybridization data have confirmed that retroviral transformation produces a reduction in the absolute amounts of mature mRNAs encoding both fibronectin and Type I procollagen subunits (11-14). In addition, Sandmeyer *et al.* (14) have shown that the reduced amount of Type I procollagen mRNAs appears to depend on the reduced rate of transcription of the Type I collagen genes in the transformed fibroblasts.

In the present work, we have analyzed further the reversible effects of *ts*-RSV transformation on the expression of the terminal phenotype of definitive chondroblasts. We will show here that retroviral transformation induces the coordinate decrease of the synthesis of several chondroblast-specific matrix components; this decrease reflects proportional alterations in the levels of the respective mRNAs. Two of these components are the core protein of

Type IV proteoglycan and a newly-identified 60 Kd protein, which
have been identified by means of specific antisera. In addition,
retroviral transformation induces a pronounced increase in the
synthesis of fibronectin, a matrix protein not normally found in
cartilage, which parallels an increase in the fibronectin mRNA
level. Finally, transformation of definitive chondroblasts causes
the accumulation of significant amounts of mRNAs coding for Type
I procollagen subunits, a collagen type not synthesized by normal
definitive chondroblasts but characteristic of several connective
tissue cells including fibroblasts. However, the type I procollagen
mRNAs are not translated by the transformed chondroblasts, thus
suggesting that these mRNAs are under some as yet undefined form
of post-transcriptional control.

2. STRUCTURE OF PROTEOGLYCANS AND COLLAGEN

We will briefly summarize here the structure of proteoglycans
and collagen, two major structural components of the extracellular
matrix of several types of cells including cartilage cells and
connective tissue fibroblasts.

Several types of proteoglycans have been identified in the
last few years. These macromolecules share a basic structure
consisting of a core protein to which various kinds as well as a
variable number of glycosaminglycan chains and oligosaccharides
are covalently attached. The cartilage unique Type IV proteoglycan
(we maintain here the use of our original nomenclature for lack of
a more universally accepted nomenclature) has a core protein of
about 200-300 Kd to which about 80 chains of chondroitin sulphate
and numerous keratan sulphate chains are covalently bound; it has
also been shown that cartilage proteoglycans contain both N- and
O-linked oligosaccharides (15-19). The resulting macromolecule
has an average size of 2.5-4.0 x 10^6 daltons (Figure 1). Single
proteoglycan molecules, termed monomers, bind specifically to

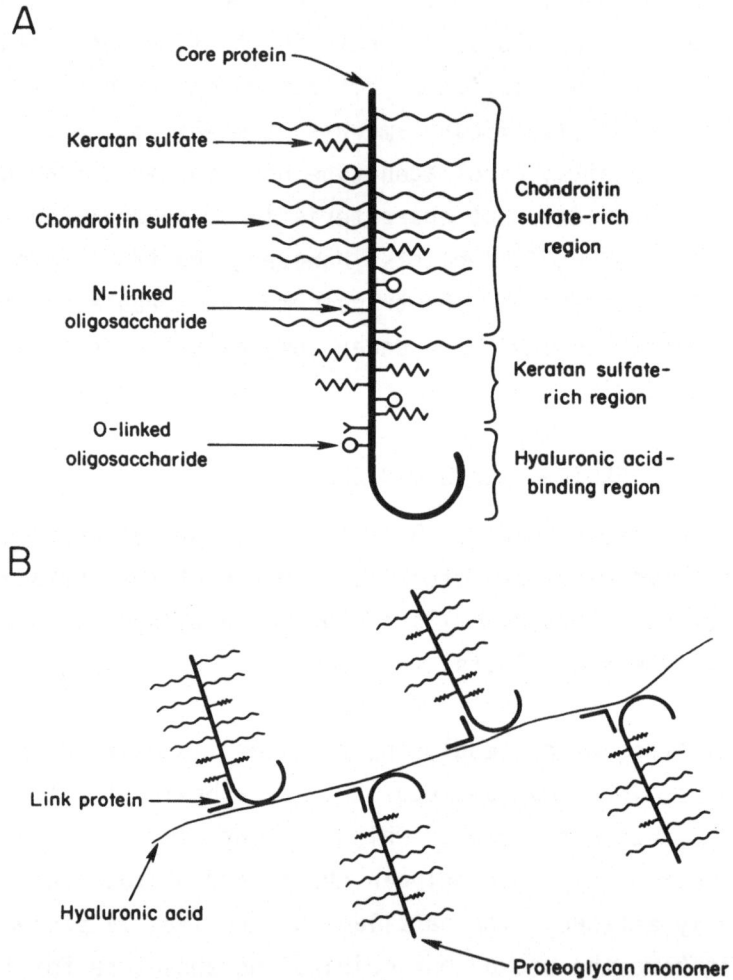

Fig. 1. *Schematical structure of cartilage proteoglycan monomer (A) and aggregate (B)*. Glycosaminglycans are distributed non homogeneously along the central core protein and are clustered in a chondroitin sulphate-rich region and a keratan sulphate rich region. Note that the hyaluronic acid-binding region of core protein has no glycosaminoglycan chains. In B, note how link protein exerts its bridging function between core protein and hyaluronic acid. Oligosaccharides have been omitted for simplicity. (Modeled after Hascall and Hascall (19)).

chains of hyaluronic acid within the cartilage matrix through a specialized region of their core protein, termed hyaluronic acid binding region, which is devoid of glycosaminoglycans but contains complex type oligosaccharides (18, 20, 21). The interaction between proteoglycans and hyaluronic acid is stabilized by link proteins, small proteins of about 44-48 Kd which are believed to interact with both hyaluronic acid and the proteoglycan core protein, thus acting as a stabilizing bridge (22). The resulting multimeric structure is called proteoglycan aggregate and represents the functional proteoglycan unit within the cartilage extracellular matrix (Figure 1).

Skin as well as muscle fibroblasts synthesize and secrete a class of large proteoglycans, Type III, and a class of small proteoglycans, Type I (15, 23-25). The core protein of Type III proteoglycan has a molecule weight of about 400 Kd (our unpublished observations; 26), to which numerous chondroitin sulphate chains are attached (15). Type III proteoglycans appear not to be able to interact with hyaluronic acid (24). It has not yet been determined whether these proteoglycans contain oligosaccharides.

Similar to numerous other kinds of secretory products, proteoglycans are currently believed to follow the classic route of synthesis and secretion, i.e. synthesis of core protein in the rough endoplasmic reticulum; N-linked oligosaccharide addition; transfer to Golgi apparatus where addition of glycosaminoglycans and 0-linked oligosaccharides occurs; packaging into secretory vacuoles; transfer to plasma membrane; secretion into extra cellular milieu (for a review, see ref. 19). It is not clear when the signal peptide (27) is cleaved off. However, direct evidence for this biosynthetic and secretory route is still largely incomplete. We will present direct evidence here for the transit of proteoglycans through the Golgi apparatus in chondroblasts.

Collagen fibres are complex linear macromolecules which are formed by the polymerization of single collagen molecules. In turn, each collagen molecule is constituted of three polypeptides associated with each other in a right-handed triple helix. These three polypeptides are identical in the cartilage characteristic Type II collagen and are known as α 1(II) chains. In contrast, the fibroblast characteristic Type I collagen is formed by two α 1(I) chains and one α 2(I) chain, both of which are structurally different from the cartilage α 1(II) chain (Figure 2). Both in cartilage and in connective tissue fibroblasts, Type II and Type I collagens are synthesized as precursors which are termed pro-collagens. These molecules consist of a central triple-helix region and non-helical extension-peptides (termed pro-peptides) at the NH_2 and the COOH terminals. Following secretion of pro-collagen molecules, most of the two pro-peptide regions are cleaved off and about 20 aminoacids at both terminals are maintained (Figure 2). The resulting collagen molecules polymerize and form cross-striated fibers (see Prockop *et al.* (28)).

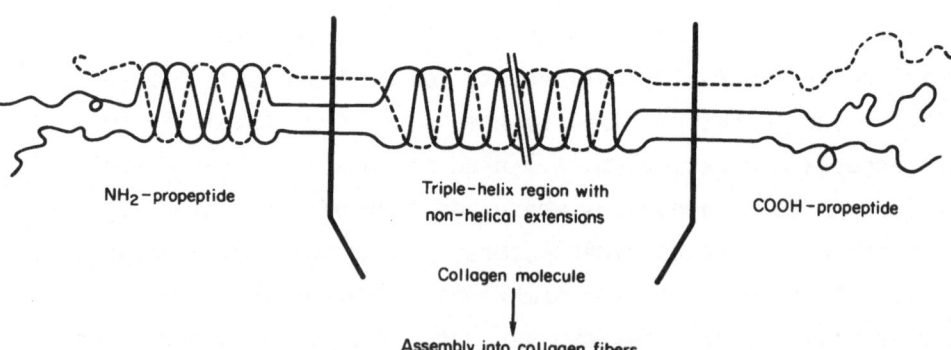

NH2−propeptide   Triple−helix region with   COOH−propeptide
                          non−helical extensions

Collagen molecule

Assembly into collagen fibers

Fig. 2. *Schematical structure of Type I procollagen.* Cleavage of the two propeptides which occurs extracellularly, allows collagen molecules to polymerize into collagen fibres. Note that after cleavage, small portions of the non-helical regions are maintained in the collagen molecules.(Modeled after Prockop *et al.* (28)).

While many authors agree that collagen biosynthesis and
secretion involve the classic route for secretory products, some
argue that collagen secretion may not need the Golgi apparatus
step and collagen may be translocated directly from the rough
endoplasmic reticulum to the plasma membrane for secretion (29).

3. TYPE IV PROTEOGLYCAN ANTISERUM

We will describe here in some detail the preparation and
characterization of this antiserum recently obtained in our laboratory. Type IV proteoglycans were extracted from 6-week-old
chicken sterna by 4M guanidine solutions and purified by successive
associative and dissociative CsCl density gradients (fraction A1D1;
30). Since this fraction usually contains small quantities of
contaminating low molecular weight, cartilage characteristic proteoglycans, the Type IV proteoglycans were further purified
according to size on sucrose gradients (15). The resulting Type
IV proteoglycan preparation, A1D1S1 (S1 stands for bottom sucrose-
gradient fraction) was injected into a rabbit which received a
boost injection one month later. Sera were collected 6, 8 and 10
days following the boost injection; these sera were termed A1,
A2 and A3, respectively. Two weeks later, Type IV proteoglycans
were injected again and sera were collected 6, 8 and 10 days
later; these sera were termed A4, A5 and A6, respectively. Except
when specified, all the experiments reported here were performed
with the A1 serum.

The A1 antiserum was first tested for specificity of staining
the extracellular matrix of cultured vertebral chondroblasts and
skin fibroblasts. As shown in Figure 3, the A1 antiserum stained
highly specifically the extracellular matrix of chondroblasts
but not that of fibroblasts grown under identical conditions. The
proteoglycan matrix appeared to encase individual chondroblasts
and separate them by a few microns. However, the amount of pro-

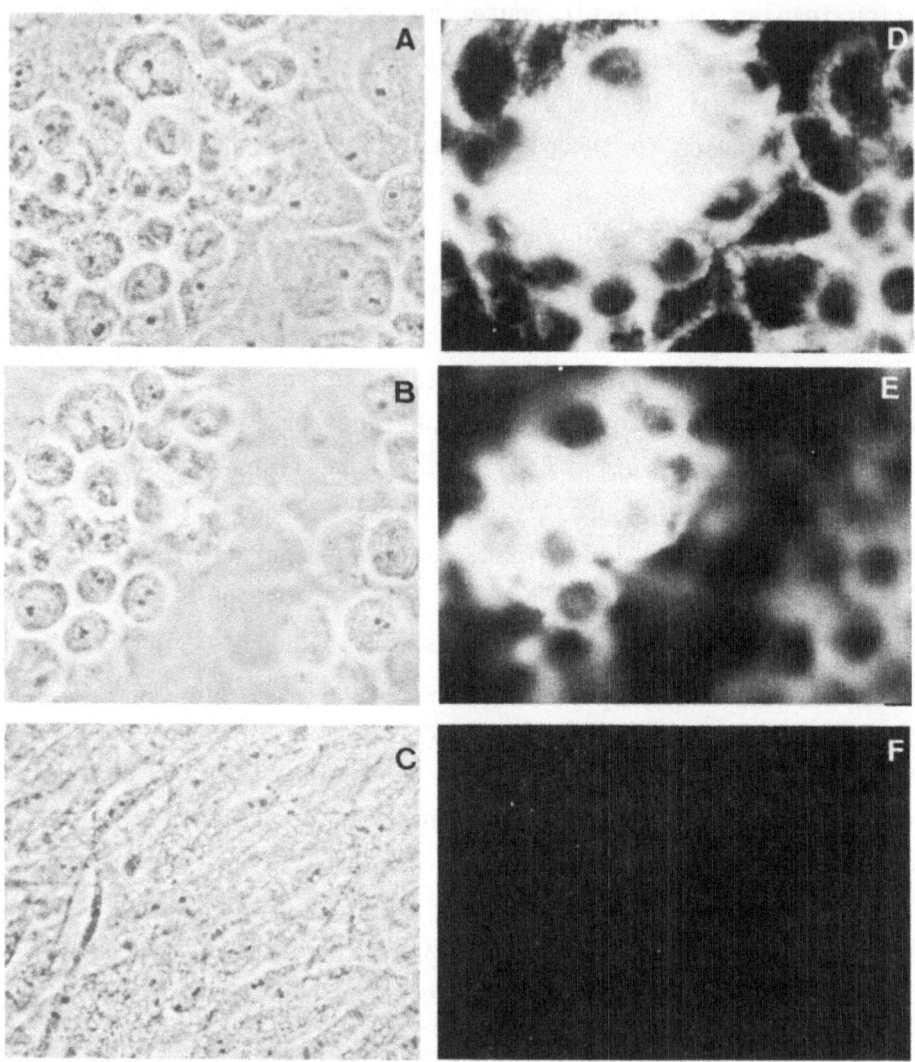

Fig. 3. *Phase (A-C) and immunofluorescent (D-F) micrographs of control chondroblasts and control fibroblasts stained with the A1 rabbit antiserum localized by a rhodamine-conjugated goat-anti-rabbit serum.* A, B, D and E are the same microscopic field of control chondroblasts focused to different levels to demonstrate that rounded condroblasts (B and E) are surrounded by more abundant matrix than flatter condroblasts. Note that control fibroblasts (C and F) are totally negative.

teoglycan matrix varied greatly among different chrondroblasts, as clearly shown in Figure 3. In general, the more rounded chondroblasts had more abundant matrix than the flatter, more epithelioid chondroblasts.

In order to study the intracellular distribution of proteoglycans, chondroblasts were grown for 16-24 hr in low levels of hyaluronidase. As shown in Figure 4, the hyaluronidase-treated chondroblasts exhibited not only the expected reduced amount of extracellular proteoglycan matrix, but also a highly stained structure which was round in shape and in a juxtanuclear position (see also insert in Figure 4B). Evidence will be presented elsewhere deriving from our recent electron immunochemical and electron cytochemical experiments, which demonstrates that the A1 antiserum-positive structure is the Golgi apparatus. The juxtanuclear localization and the large size of the Golgi apparatus in chondroblasts have long been known (31). Thus, the A1 antiserum revealed the presence of Type IV proteoglycans in the Golgi apparatus; to our knowledge, this is the first direct demonstration for the transit of cartilage proteoglycans through the Golgi apparatus.

To further study the specificity of the A1 antiserum, [^{35}S]-methionine labelled Type IV proteoglycans synthesized by chondroblasts and Type III proteoglycans synthesized by skin fibroblasts were purified by CsCl and sucrose gradients. Immunoprecipitation of these macromolecules with the A1 antiserum demonstrated that only the Type IV proteoglycans were recognized by this antiserum (not shown). In a subsequent experiment, aliquots of the [^{35}S]-methionine labelled Type IV and Type III proteoglycans were digested with chondroitinase ABC and/or keratanase. The resulting core proteins were analyzed by gel electrophoresis (see also ref. 32). Following chrondroitinase ABC digestion, Type IV proteoglycan core protein migrated as a sharp band with a molecular weight

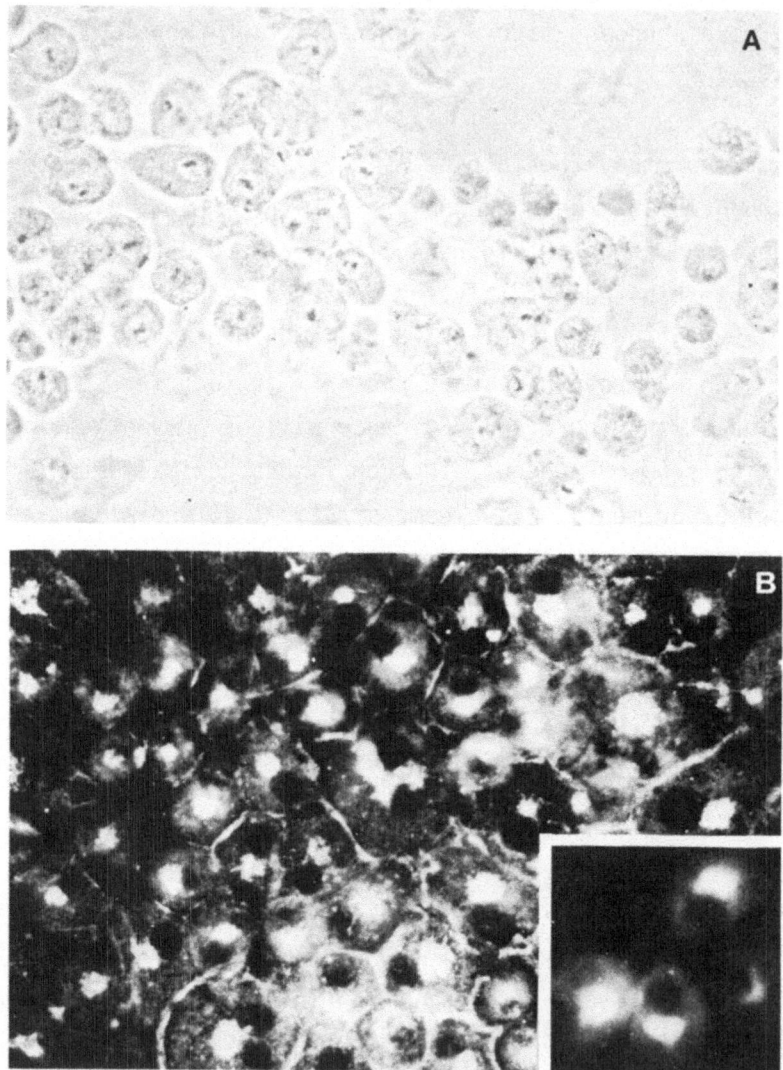

Fig. 4. *Phase (A) and immunofluorescent (B) micrographs of control chondroblasts stained as described in Figure 3 legend following treatment with hyaluronidase.* Note that most of the extracellular matrix proteoglycan matrix has been removed. The enzymatic treatment reveals the highly positive, juxtanuclear structure that we have identified as the Golgi apparatus. In B, insert shows higher magnification of Golgi staining.

higher than 450-500 Kd (Figure 5); as expected, its size decreased
dramatically following digestion of its keratan sulphate chains
by keratanase (not shown). Similarly, Type III core protein migrated
as a sharp band after chondroitinase ABC digestion; however, its
mobility did not change any further after keratanase digestion
because Type III proteoglycans do not contain keratan sulphate
chains. Similar aliquots of Type III and Type IV proteoglycan core
proteins were then immunoprecipitated with the Al antiserum and
the immunoprecipitated material analyzed by gel electrophoresis.
As shown in Figure 5, only the Type IV core protein was immunoprecipitated by the Al antiserum. In conclusion, these data demonstrate that the Al antiserum is directed against the core protein
and/or the oligosaccharides of the Type IV proteoglycan; removal
of both chondroitin sulphate and keratan sulphate chains does not
suppress the ability of this antiserum to immunoprecipitate this
type of cartilage proteoglycan.

4. CULTURE AND INFECTION OF CHONDROBLASTS

Pure populations of chick embryo chondroblasts were selectively harvested from the medium of primary cultures into which they
were induced to float by the accumulation of plasma membrane bound
extracellular matrix (1, 33). Because of the surrounding abundant
matrix, these cells were very difficult to infect, since the matrix
interfered with virus binding and penetration. Treatment with
either trypsin or dispase was followed by infection with 2-4 focus
forming units per cell of ts-LA24A, a temperature-sensitive mutant
of Rous sarcoma virus. More than 90% of the chondroblasts became
transformed following two successive subcultures at the permissive
temperature (1). Transformation caused the loss of the typical
polygonal morphology in chondroblasts and the acquisition of a
typical bipolar shape (compare Figures 6 with Figure 3). As shown
in Figure 6, this morphological alteration was reversed when infected chondroblasts were shifted and grown for a few days at 41°C.

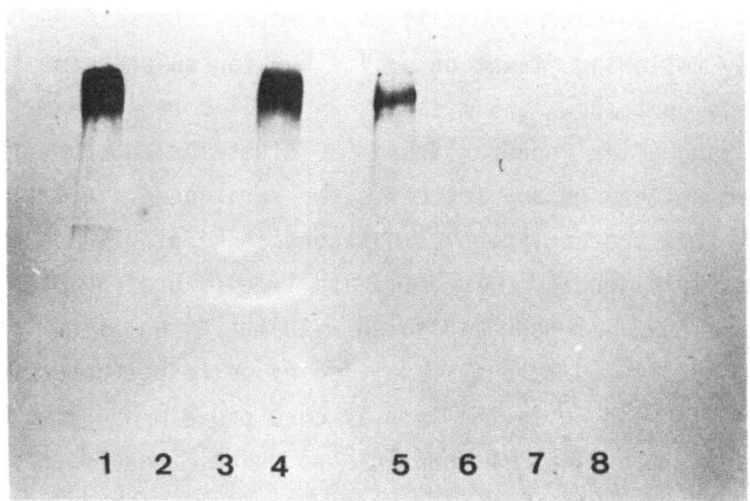

Fig. 5. *Gel electrophoresis analysis of chondroitinase ABC- digested Type IV proteoglycans (lanes 1-4) and Type III proteoglycans (lanes 5-8).* Digested proteoglycans were electrophoresed immediately after digestion (lanes 1 and 5) or following immunoprecipitation with 1) uncoated *Staphylococcus aureus* (Staph A) (lanes 2 and 6); 2) preimmune serum-coated Staph A (lanes 3 and 7); and 3) A1-antiserum coated Staph A (lanes 4 and 8). Note that only the digested Type IV proteoglycan is immunoprecipitated by the A1 antiserum.

When infected chondroblasts were stained with the A1 antiserum the epithelioid, polygonal chondroblasts grown at 41°C exhibited bright staining of their extracellular matrix, thus suggesting that these cells were synthesizing Type IV proteoglycans and depositing them extracellularly. In contrast, the spindle-shaped, transformed chondroblasts grown at 36°C failed to stain the A1 antiserum, thus suggesting that extracellular accumulation of Type IV proteoglycans had ceased. These data fully confirm a previous report from our laboratory in which biosynthesis of Type IV proteoglycans was shown to be turned off in transformed chondroblasts at 36°C and was re-initiated following a shift to 41°C (1).

The biochemical experiments to be described in the following sections were also performed with third-passage infected chondroblasts. In these experiments, cells were maintained for two passages at 36°C, then subcultured and grown in suspension cultures either again at 36°C or shifted at 41°C. Cultures were usually harvested on day three or four. Even in suspension cultures, infected chondroblasts exhibited remarkable morphological differences depending upon the temperature of growth. At 41°C, the normalized cells grew as spherical, single refractile cells, exhibiting some variation in size and closely resembling control cells. At 36°C, the transformed chondroblasts grew as large clusters of numerous, roundish, less refractile cells (2).

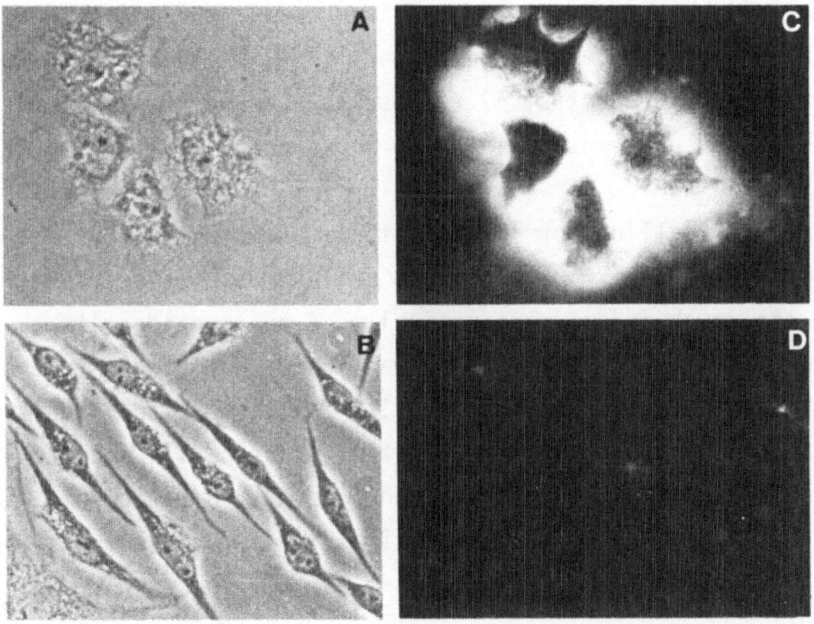

Fig. 6. *Phase (A and B) and immunofluorescent (C and D) micrographs of* ts-RSV *infected chondroblasts grown at 41°C (A and C) or 36°C (B and D) and stained with the A1 antiserum as described in Figure 3 legend.* Note the spindle-spahed morphology of transformed chondroblasts at 36°C (B) and lack of staining (D); in contrast, note the re-acquired polygonal-epithelioid morphology of infected chondroblasts grown at 41°C (A) and the positive staining (C).

5. SYNTHESIS OF MATRIX COMPONENTS BY CONTROL CHONDROBLASTS

Synthesis of various matrix components was established by labelling third-passage floating control chondroblasts with radio-active proline or methionine for 15 min. The one-dimensional electrophoretic analysis of proline-labelled chondroblasts homogenates revealed one major protein band of about 165 Kd following short autoradiographic exposure (Figure 7, lanes 1-2). This band was identified as α 1(II) procollagen subunit, the single component of chondroblast characteristic Type II procollagen, because it was intensely labelled with proline and co-migrated with the major, proline-labelled polypeptide synthesized by organ-cultured intact cartilages. A longer exposure of the gel revealed the presence of other proline-labelled polypeptides, some of which may represent the minor collagen components of hyaline cartilage (34). When control chondroblasts were labelled with [^{35}S]-methionine, the α 1(II) procollagen subunit still represented one of the most intensely labelled polypeptides (Figure 7, lanes 3-4).

By pulse-labelling for 15 min, it is possible to label the core protein of Type IV proteoglycans while it is being synthesized in the rough endoplasmic reticulum. A polypeptide with a molecular weight higher than 400 Kd was indeed detectable in the stacking region of electrophoresis gels. The large size of this protein is likely due to the fact that oligosaccharide addition occurs co-translationally in the rough endoplasmic reticulum. The identity of this polypeptide was confirmed by immunoprecipitation with the A1 antiserum (see below Figure 10).

Among the proteins highly labelled with radioactive methionine was a polypeptide of about 60 Kd. The discovery of this protein was accidental. In fact, of the six antisera collected from the sam rabbit following the two bost injections of Type IV proteglycans, particularly the last one, A6, was capable of specific binding to the 60

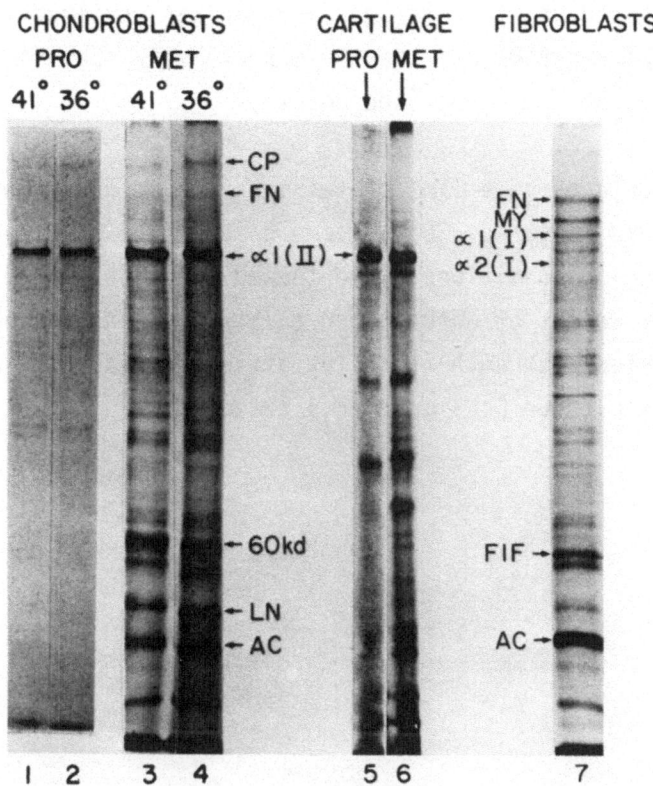

Fig. 7. *Identification of proteins synthesized by chondroblasts, cartilage and fibroblasts*. Cells and tissue were pulsed for 15 min and electrophoresed on 8% SDS polyacrylamide gels. See text for details. CP, Type IV proteoglycan core protein; FN, fibronectin; LN, link protein; AC, actin; FIF, fibroblast intermediate filament subunit; mY, myosin heavy chain; α 1(II), pro α 1(II) collagen; α 1(I), pro 1(I) collagen and α 2(I), pro α 2(1) collagen.

Kd protein. Figure 8 shows that after separation of total condroblast and skin fibroblast homogenates by one dimensional electrophoresis and transfer to nitrocellulose (35), the A6 antiserum specifically recognized the 60 Kd only among chrondroblast proteins and not among fibroblast proteins. When total chondroblast homo-

genates were subjected to two dimensional electrophoresis, the 60
Kd protein exhibited charge heterogeneity and was resolved into
three spots of pI 8.7-8.8. All the three spots were specifically
recognized by the A6 antiserum (see below).

Proteoglycan aggregate link proteins have a molecular weight
ranging from 44 Kd to 48 Kd (36). By one dimensional electrophoresis under reducing conditions, we detected one major protein
with a molecular weight of about 44 Kd (Figure 7, lane 4). This
protein co-migrated with authentic link protein isolated by associative and dissociative CsCl gradients (36, 37).

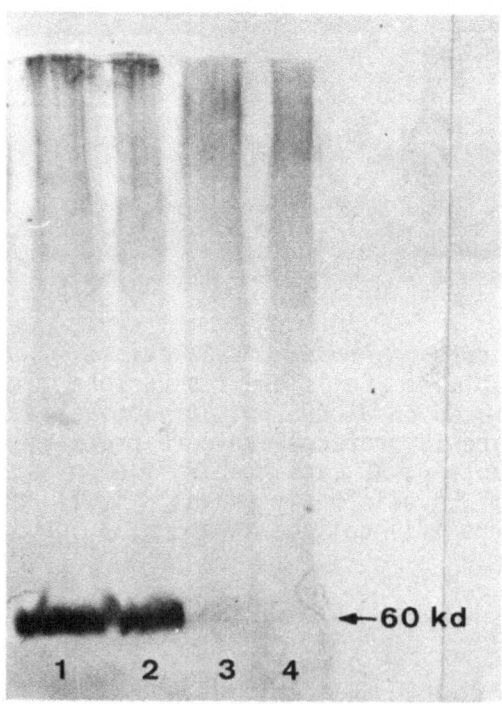

Fig. 8. *Immunological analysis of total chondroblasts (lanes 1
and 2) and fibroblast (lanes 3 and 4) homogenates which were first
electrophoresed on polyacrylamide gels, then transferred to nitrocellulose filters and finally stained with the A6 antiserum
localized by the peroxidase reaction.* Note that the A6 antiserum
reveals the presence of 60 Kd protein only in chondroblasts.

Normal chondroblasts in culture have been shown to synthesize
fibronectin, a matrix component not found and non synthesized by
chondroblasts *in vivo* (38), but characteristic of several other
kinds of connective tissue cells. We have confirmed these data
and have detected low levels of fibronectin synthesis in control
floating chondroblasts (Figure 7). The identity of this protein
was confirmed by immunoprecipitation (Figure 7).

In conclusion, among the proteins newly synthesized by control
floating chondroblasts, we have identified the Type IV proteoglycan core protein, the α 1(II) procollagen subunit, the 60 Kd
protein, the link protein and fibronectin.

6. SYNTHESIS OF MATRIX PROTEINS IN TRANSFORMED CHONDROBLASTS

Infected chondroblasts grown at 41°C for 3-4 days were
labelled with [^{35}S]-methionine for 15 min. One dimensional electrophoretic analysis of the newly synthesized proteins revealed
a pattern of synthesis of matrix proteins virtually identical to
that of control uninfected chondroblasts (Figure 9). When, however,
a similar analysis was performed on transformed chondroblasts
grown at 36°C, a dramatic and coordinate reduction of the synthesis
of all the chondroblast-specific matrix polypeptides was detected.
The reduction of Type IV proteoglycan core protein synthesis was
confirmed by immunoprecipitation with the A1 antiserum (Figure 10).
Thus these data confirmed the immunofluorescence observations reported above obtained with the same antiserum. Densitometric
quantitation indicated that the rate of synthesis of 1(II) procollagen subunit was approximately 20% that of control chondroblasts or infected chondroblasts at 41°C.

Radioactive samples were also analyzed by two dimensional
electrophoresis (Figure 11). By this technique, we were able to
identify only the α 1(II) procollagen subunit, the 60 Kd protein

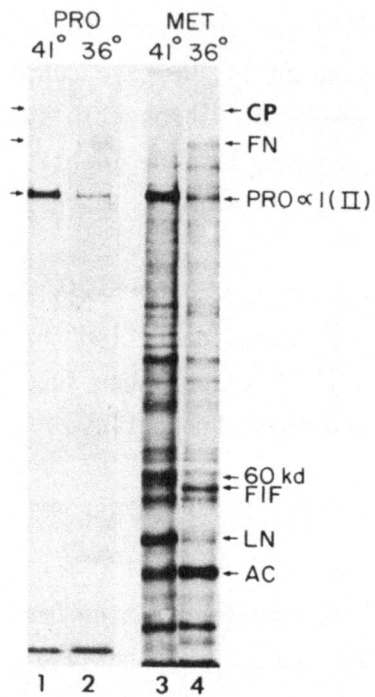

Fig. 9. *Electrophoretic analysis of the effects of transformation on synthesis of chondroblast-specific proteins.* Abbreviations as in Figure 7. Note the reversible reduction of the synthesis of CP, PRO α 1(1I) and 60 Kd, and the reversible increase of the synthesis of FN in transformed chondroblasts grown at 36°C.

and link protein. The 60 Kd protein was identified immunologically by the A6 antiserum (Figure 11). This analysis confirmed the drastic reduction in the synthesis of these three proteins in transformed chondroblasts at 36°C. Interestingly, these three proteins exhibited some charge heterogeneity and pIs of approximately 8.9, 8.7 and 9.2, respectively. Their basic properties

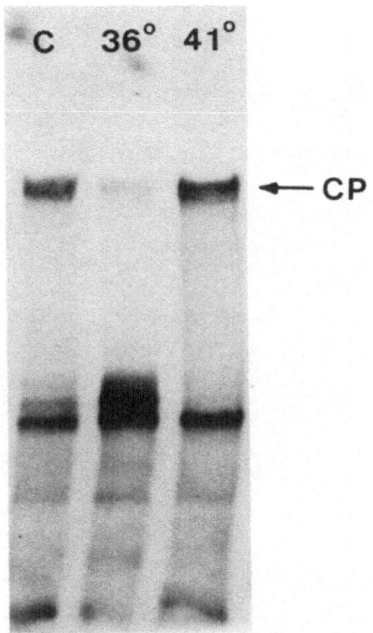

Fig. 10. *Electrophoretic analysis of immunoprecipitated Type IV core protein (CP)*. Control (C) chondroblasts and *ts*-RSV-infected chondroblasts grown at 36°C or 41°C were labelled with radioactive methionine for 15 min; aliquots containing equal amounts of radio-activity were then immunoprecipitated with the A1 antiserum conjugated to Staph A. Note the drastic but reversible reduction of the synthesis of Type IV proteoglycan core protein in trans-formed chondroblasts at 36°C. Other bands precipitated by this procedure were also precipitated by uncoated Staph A and thus represent contaminating material.

may thus favour electrostatic interactions between these proteins and the abundant acidic proteoglycans within the cartilage extra-cellular matrix.

    As pointed out above, control uninfected chondroblasts may initiate the synthesis of fibronectin once they are cultured. While, as just shown above, the rate of synthesis of the various chondroblast-specific matric components decreased in transformed chondroblasts at 36°C, that of fibronectin was at least five

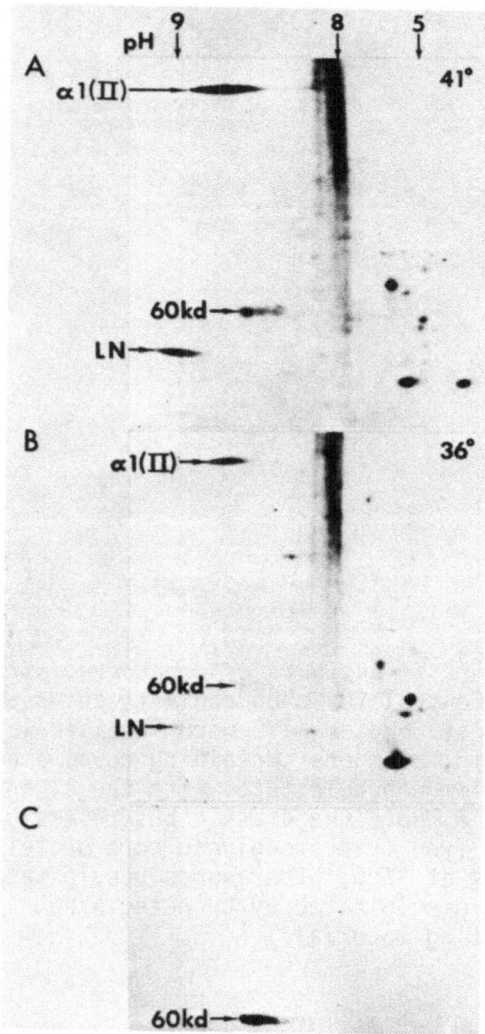

Fig. 11. *Two-dimensional electrophoretic analysis of [^{35}S]-methionine labelled proteins synthesized by ts-RSV infected chondroblasts grown at 41°C (A) or 36°C (B). In C, immunological identification of the 60 Kd protein after transfer to nitrocellulose filter. Abbreviations as in Figure 7.*

times higher (Figure 9). Interestingly, the levels of fibronectin synthesis in transformed chondroblasts matched that of normal uninfected fibroblasts.

7. MESSENGER RNA ANALYSIS

The synthesis of the chondroblast matrix proteins was correlated with the levels of the respective mRNAs by isolation of
total cellular RNA and translation *in vitro* in a reticulocyte
lysate in the presence of radioactive methionine.

mRNAs obtained from infected chondroblasts grown at 41°C for
3-4 days directed the synthesis of several proteins (Figure 12).
A 340 Kd polypeptide was tentatively identified as the Type IV
proteoglycan core protein; its large molecular weight prior to
carbohydrate addition is due to the presence of the signal peptide.
While we have not yet performed any immunological study of this
protein, Upholt *et al.* (39) have recently shown such immunological
identification following *in vitro* translation of cartilage mRNAs.

In addition to the 340 Kd protein, the mRNAs of infected
chondroblasts at 41°C directed the synthesis of the subunit of
Type II procollagen, the prepro α 1(II) collagen (the prefix "pre"
indicates the presence of the signal peptide). The identity of
this protein was based on its susceptibility to collagenase digestion and on co-migration with a polypeptide of 165 Kd present
among the translation products of total cartilage mRNAs. A similar
identification has been recently presented by others (39-42).

Another collagenase-sensitive protein was detected and identified as prepro α 1(I) collagen, one of the two subunits of Type
I procollagen. This polypeptide comigrated with authentic prepro
α 1(I) collagen, whose synthesis was directed *in vitro* by purified
Type I collagen mRNA from chick embryo calvaria (Figure 12, lane
5). However, the other subunit of Type I collagen, prepro α 2(I)
was not detected.

Other collagenase-sensitive polypeptides were detected but

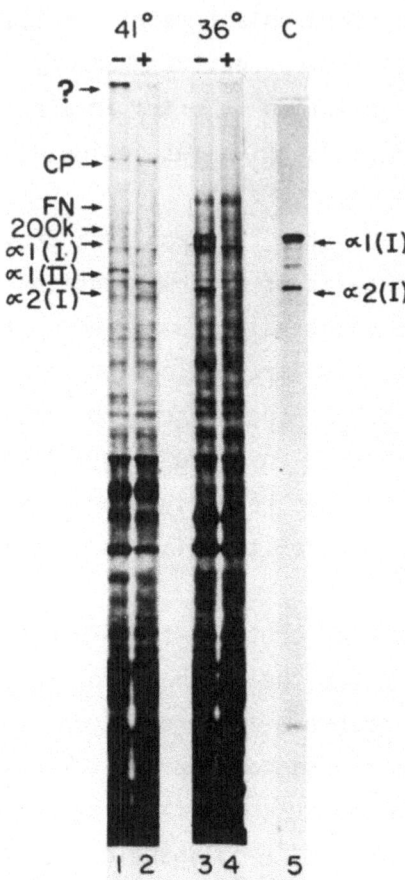

Fig. 12. *Electrophoretic separation of [^{35}S]-methionine labelled translation products.* RNAs were isolated from infected chondroblasts grown at 41°C or 36°C; following *in vitro* translation, aliquots were treated with collagenase (+) or buffer (-). In C, translation products of Type I procollagen mRNAs isolated from chick embryo calvaria. Abbreviations are listed in the legend of Figure 7.

were not conclusively identified; they may represent the subunit of Type V collagen (200 Kd) and the minor collagen components reported by Reese and Mayne (34).

We were unable to detect fibronectin among the translation products of the mRNAs of infected chondroblasts at $41^{o}C$, though the synthesis of this protein was detected at low levels in intact cells.

When mRNAs were isolated from transformed chondroblasts grown at $36^{o}C$ and translated *in vitro*, they revealed marked qualitative and quantitative alterations in the levels of specific mRNAs. The amount of translatable mRNA for Type IV proteoglycan core protein was drastically decreased, as was that of chondroblast characteristic prepro α 1(II) collagen and several other collagenase-sensitive polypeptides. In addition to these decreases in chondroblast-specific mRNAs, increase in fibronectin mRNA was apparent and correlated well with protein labelling data. This analysis also revealed the presence of large amounts of mRNAs coding for both Type I, fibroblast-characteristic procollagen subunits, prepro α 1(I) and prepro α 2(I). The presence of these mRNAs was unexpected based on the lack of synthesis of pro α 1(I) and pro α 2(I) subunits in intact transformed chondroblasts at $36^{o}C$, as shown in Figure 9.

8. CONCLUSIONS

The data presented here clearly show that transformation of definitive chondroblasts by the *ts*-RSV mutant LA24A causes the coordinate inhibition of the synthesis of Type IV proteoglycan core protein, α 1(II) procollagen subunit, the 60 Kd protein and link protein. The *in vitro* translation data confirm that the inhibitory effects of RSV transformation on the synthesis of at least Type IV core protein, α 1(II) procollagen subunit and the minor collagens reflect reductions in the levels of the respective mRNAs. The latter experimental approach has also revealed the unexpected presence of substantial amounts of mRNAs coding for both prepro α 1(I) and α 2(I) subunits of Type I collagen in transformed chondroblasts at $36^{o}C$. Because these two proteins appear not to be

synthesized at appreciable rates by the intact cells, we must conclude that these mRNAs are under an as yet uncharacterized post-
transcriptional control(s) which prevents their utilization in
intact cells. Recently, another group has shown that while chick
chondroblasts transformed by RSV synthesize a highly reduced level
of Type II procollagen, they do not initiate the synthesis of
Type I procollagen (42). The induction of transcription and accumulation of Type I procollagen mRNAs in transformed chondroblasts
is in sharp contrast with the known inhibitory effects of RSV on
Type I procollagen gene transcription and mRNA levels in transformed fibroblasts (9-11, 13, 14, 44, 45).

Both the protein synthesis and the *in vitro* translation data
have clearly shown that fibronectin synthesis is rather low in
control chondroblasts as well as in infected chondroblasts at $41^{o}C$,
but it increases at least five-fold following expression of transformation at $36^{o}C$. This finding contrasts with several reports in
which fibronectin synthesis was consistently found to be depressed
in fibroblasts transformed by RNA tumour viruses (8, 46). It is of
great interest that the rate of fibronectin synthesis in transformed chondroblasts at $36^{o}C$ following a 15 min pulse is practically identical to that found in normal, uninfected fibroblasts
grown under identical conditions.

The above biochemical data correlate well with the immunofluorescent staining pattern we have obtained with the Al antiserum. The extracellular matrix and the Golgi apparatus of only
control chondroblasts and infected chondroblasts grown at $41^{o}C$
were stained by this antiserum; in contrast, the spindle-shaped,
transformed chondroblasts grown at $36^{o}C$ failed to react with this
antiserum. Other authors have reported the preparation of antisera
to Type IV proteoglycans and the immunolocalization of these
molecules extracellularly and intracellularly in cultured chondro-

blasts (47). However, the intracellular staining reported by others
does not reveal the presence in chondroblasts of a highly stained
structure in a juxtanuclear position which we have identified as
the Golgi apparatus. Differences in fixation and in culturing
as well as in the characteristics of the antiserum may well explain
this lack of Golgi staining. It is noteworthy that our immuno-
licalization is the first direct evidence of the transit of Type
IV proteoglycans through the chondroblast Golgi apparatus.

The experiments designed to prove the specificity of the A1
antiserum have revealed that the core protein of Type IV proteoglycan is different from the core protein of the fibroblast
characteristic Type III proteoglycan. Our data confirm similar
previous claims largely based on amino acid analysis of the different core proteins and on differences in hyaluronic acid-binding
capacity (15, 24, 48).

The effects of ts-RSV induced transformation on the biosynthetic program of definitive chondroblasts reported here appear
to be reversible following a shift from permissive to non-permissive
temperature. Thus, these effects must depend upon the direct or
indirect action of the RSV src gene product on the chondroblast
regulatory mechanisms. The current view on the molecular mechanisms
by which retroviruses alter the biosynthetic program of cultured
cells points to an indirect action of the RSV src gene product.
This interpretation has been applied to the suppression by RSV of
the synthesis of fibronectin and Type I procollagen in fibroblasts.
However, this indirect mechanism(s) must also explain how in definitive chondroblasts RSV transformation does indeed inhibit the
synthesis of chondroblast-unique molecules but concurrently induces a five-fold increase in fibronectin and a substantial accumulation of untranslated Type I procollagen mRNAs. Clearly, RSV
transformation cannot induce similar responses in phenotypically

different cells such as vertebral chondroblasts and chick embryo
fibroblasts; rather, the responses to viral transformation strictly
depend upon the phenotypic program of the cell and possibly upon
its position along its differentiation lineage. This latter possibility, i.e. stage-specific transformation, has been proposed for
limb bud chondrogenic cells and macrophages (49); it is also
strongly suggested by recent findings showing that the target cells
for transformation by avian erythroblastosis virus are those stage-
specific progenitor erythroid cells termed burst-forming units-
erythroid (50).

We have reported elsewhere that while RSV transformation blocks
the synthesis of Type IV proteoglycans in transformed chondroblasts,
it concurrently induces the synthesis of a smaller proteoglycan
which is similar in structure to the fibroblast type III proteoglycan (1). Thus, based also on the results reported here, RSV
transformation appears to cause the activation of all the metabolic routes for the synthesis of Type III proteoglycans in transformed chondroblasts but activates only the mechanism for the
transcription but not for the translation of the mRNAs coding for
both Type I procollagen subunits. Further work on this intriguing
finding will be of profound importance for the study of gene expression in differentiating cells.

Besides RSV, several agents can reversibly alter the phenotype of definitive chondroblasts in culture (51-54). For example,
growth in 5-bromodeoxyuridine (BrdUrd) induces cell flattening
and inhibition of the synthesis of Type II collagen and Type IV
proteoglycans; simulaneously, the altered chondroblasts initiate
the synthesis of Type I and Type I trimer $\alpha\ 1(I)_3$ collagens.
Pawlowski *et al*. (42) have recently confirmed the latter results
at the level of mRNA. Clearly, the mechanisms for gene regulation
induced by RSV and BrdUrd in cultured chondroblasts must be

different. BrdUrd-treated chondroblasts synthesize Type I collagen, whereas the RSV-transformed chondroblasts are capable of transcribing but not translating the Type I procollagen mRNAs. It will be most illuminating to search for possible "factors" that may selectively block the translation of these mRNAs. It will be of interest also to determine where these untranslated mRNAs are located and whether their structure is somehow peculiar. Other examples of post-translational regulation are emerging in different eukaryotic cell systems. For example, Type I procollagen mRNA levels are high throughout sheep embryogenesis while Type I collagen synthesis drops off rapidly around 3-4 months of gestation (54). Moreover, the N-terminal propeptides of Type I collagen appear to inhibit specifically *in vitro* translation of collagen mRNAs, thus suggesting that the level of N-terminal propeptides within a given tissue may inhibit collagen biosynthesis by a feedback mechanism (40, 56-58). The latter mechanism might explain the data presented in this paper.

Further work on the effects of RSV and other agents on chondroblasts as well as other types of embryonic cells may not only shed light on the mechanisms of action of the *src* gene product on the expression of the terminal phenotype, but also on the mechanisms of gene regulation during embryogenesis and possibly on the role of endogenous *oncogenes* on normal cell differentiation.

9. ACKNOWLEDGEMENTS

We are grateful to R.J. Focht, D. Shanley, R. Soltesz and G. Thal for assistance with many experiments reported here. This work was supported by grants from the National Institute on Ageing (to M.P.); National Institutes of Health (to S.A. and H.H. - the University of Pennsylvania Genetics center and the Pennsylvania Muscle Institute); from the National Cancer Institute (to H.H. and D.B.); from the W.W. Smith Charitable Trust (to S.A.); and by

Basil O'Connor Grant No. 5-309 from the March of Dimes Birth
Defects Foundation (to S.A.). D.B. is a Leukemia Society Scolar.

9. REFERENCES

1) PACIFICI, M., BOETTIGER, D., ROBY, K. and HOLTZER, H. (1977).
    Transformation of chondroblasts by Rous sarcoma virus and
    synthesis of the sulphated proteoglycan matrix. Cell, 11, 891.

2) ADAMS, S.L., BOETTIGER, D., FOCHT, R.J., HOLTZER, H., and
    PACIFICI, M. (1982). Regulation of the synthesis of extra-
    cellular matrix components in chondroblasts transformed by a
    temperature sensitive mutant of Rous sarcoma virus. Cell,
    30, 373.

3) DURBAN, E.M. and BOETTIGER, D. (1981). Differential effects of
    transforming avian RNA tumour viruses on avian macrophages.
    Proc. Natl. Acad. Sci. USA, 78, 3600.

4) BOETTIGER, D., ROBY, K., BRUMBAUGH, J., BIEHL, J. and HOLTZER,
    H. (1977). Transformation of chicken embryo retinal melano-
    blasts by a temperature sensitive mutant of Rous sarcoma virus.
    Cell, 11, 881.

5) ROUSLAHTI, E., ENGVALL, E. and HAYMAN, E.G. (1981). Fibronec-
    tin: Current concepts of its structure and functions. Collagen
    and Related Res., 1, 95.

6) BORNSTEIN, P. and SAGE, H. (1980). Structurally distinct
    collagen types. Ann. Rev. Biochem., 49, 957.

7) LEVINSON, W., BHATNAGAR, R.S. and LIU, T. (1975). Loss of
    ability to synthesize collagen in fibroblasts transformed by
    Rous sarcoma virus. J. Natl. Cancer Inst., 55, 807.

8) OLDEN, K. and YAMADA, K.M. (1977). Mechanism of the decrease
    in the major cell surface protein of chick embryo fibroblasts
    after transformation. Cell, 11, 957.

9) ADAMS, S.L., SOBEL, M.E., HOWARD, B.H., OLDEN, K., YAMADA, K.M.,
    PASTAN, I. and DE CROMBRUGGHE, B. (1977). Levels of translat-
    able mRNAs for cell surface protein, collagen precursors, and

two membrane proteins are altered in Rous sarcoma virus-transformed chick embryo fibroblasts. Proc. Natl. Acad. Sci. USA, 74, 3399.

10) ROWE, D.W., MOEN, R.C., DAVIDSON, J.M., BYERS, P.H., BORNSTEIN, P. and PALMITER, R.D. (1978). Correlation of procollagen mRNA levels in normal and transformed chick embryo fibroblasts with different rates of procollagen synthesis. Biochemistry, 17, 1581.

11) ADAMS, S.L., ALWINE, J.C., DE CROMBRUGGHE, B. and PASTAN, I. (1979). Use of recombinant plasmids to characterize collagen RNAs in normal and transformed chick embryo fibroblasts. J. Biol. Chem. 254, 4935.

12) SANDMEYER, S. and BORNSTEIN, P. (1979). Declining procollagen mRNA sequences in chick embryo fibroblasts infected with Rous sarcoma virus. J. Biol. Chem.,254, 4950.

13) FAGAN, J., PASTAN, I. and DE CROMBRUGGGE, B. (1980). Sequence rearrangement and duplication of double stranded fibronectin cDNA probably occurring during cDNA synthesis by AMV reverse transcriptase and *E. coli* DNA polymerase I. Nucl. Acids Res., 8, 3055.

14) SANDMEYER, S., GALLIS, B., and BORNSTEIN, P. (1981). Coordinate transcriptional regulation of Type I procollagen genes by Rous sarcoma virus. J. Biol. Chem. 256, 5022.

15) OKAYAMA, M., PACIFICI, M. and HOLTZER, H. (1976). Differences among sulphated proteoglycans synthesized in nonchondrogenic cells, presumptive chondroblasts, and chondroblasts. Proc. Natl. Acad. Sci. USA, 73, 3224.

16) DE LUCA, S., HEINEGARD, D.K., HASCALL, V.C., KIMURA, J.H. and CAPLAN, A.I. (1977). Chemical and physical changes in proteoglycand during development of chick limb bud chondrocytes grown *in vitro*. J. Biol. Chem. 252, 6600.

17) HEINEGARD, D.K. and AXELSSON, I. (1977). Distribution of keratan sulphate in proteoglycans. J. Biol. Chem., 252,1971.

18) DE LUCA, S., LOHMANDER, L.S., NILSSON, B., HASCALL, V.C. and
     CAPLAN, A.I. (1980). Proteoglycans from chick limb bud chondro-
     cyte cultures: Keratan sulphate and oligosaccharides which
     contain mannose and sialic acid. J. Biol. Chem., 255, 6077.

19) HASCALL, V.C. and HASCALL, G.K. (1981). Proteoglycans. In:
     Cell biology of extracellular matrix (ed., E.D. Hay), p. 39,
     Plenum Press, New York.

20) LOHMANDER, L.S., DE LUCA, S., NILSSON, B., HASCALL, V.C.,
     CAPUTO, C.B., KIMURA, J.H. and HEINEGARD, D. (1980). Oligo-
     saccharides on proteoglycans from the Swarm rat chondro-
     sarcoma. J. Biol. Chem., 255, 6084.

21) LOHMANDER, L.S., NILSSON, B., DE LUCA, S. and HASCALL, V.C.
     (1981). Structures of N- and O-linked oligosaccharides from
     chondrosarcoma proteoglycan. Seminars in Arthritis and Rheu-
     matism 11 (Supplement), 12.

22) HEINEGARD, D.K. and HASCAL, V.C. (1979). The effect of dan-
     sylation and acetylation on the interaction between hya-
     luronic acid and the hyaluronic binding region of cartilage
     proteoglycans. J. Biol. Chem., 254, 912.

23) LOWE, M.E., PACIFICI, M. and HOLTZER, H. (1978). Effects of
     phorbol-12-myristate-13-acetate on the phenotypic program
     of cultured chondroblasts and fibroblasts. Cancer Res., 38,
     2350.

24) PACIFICI, M. and HOLTZER, H. (1980). 12-O-Tedradecanoylphor-
     bol-13-acetate-induced changes in sulphated proteoglycan
     synthesis in cultured chondroblasts. Cancer Res. 40, 2461.

25) PACIFICI, M. and MOLINARO, M. (1980). Developmental changes
     in glycosaminoglycans during skeletal muscle cell differenti-
     ation in culture. Exp. Cell Res., 126, 143.

26) CARLSTEDT, I., COSTER, L. and MALMSTROM, A. (1981). Isolation
     and characterization of dermatan sulphate and heparan sulphate
     proteoglycans from fibroblast cultures. Biochem. J., 197, 217.

27) BLOBEL, G. (1978). Mechanisms for the intracellular compart-
     ment of newly synthesized proteins. FEBS (Fed. Eur. Biochem.
     Soc.) Proc. Meet., 43 (A2), 99.

28) PROCKOP, D.J., KIVIRIKKO, K.J., TUDERMAN, L. and GUZMAN, N.A.
     (1979). The biosynthesis of collagen and its disorders. N.
     Eng. J. Med., 301, 13 and 77.

29) NIST, C., VON DER MARK, K., HAY, E.D., OLSEN, B.R., BORNSTEIN,
     P., ROSS, R. and DEHM, P. (1975). Localization of procollagen
     in chick corneal and tendon fibroblasts with ferritin-conju-
     gated antibodies. J. Cell Biol. 65, 75.

30) HEINEGARD, D.K. (1972). Hyaluronidase digestion and alkaline
     treatment of bovine tracheal cartilage proteoglycans. Isolation
     and characterization of different keratan sulphate proteins.
     Biochem. Biophys. Acta, 285, 193.

31) MOSKALEWSKI, S., THYBERG, J., LOHMANDER, S. and FRIBERG, U.
     (1975). Influence of colchicine and vinblastine on the Golgi
     complex and matrix deposition in chondrocyte aggregates. Exp.
     Cell Res., 95, 440.

32) KIMURA, J.H., THONAR, E.J.-M., HASCALL, V.C., REINER, A. and
     POOLE, A.R. (1981). Identification of core protein, and inter-
     mediate in proteoglycan biosynthesis in cultured chondrocytes
     from the swarm rat chondrosarcoma. J. Biol. Chem. 256, 7890.

33) CHACKO, S., ABBOTT, J., HOLTZER, S. and HOLTZER, H. (1969).
     The loss of phenytypic traits by differentiating cells. VI.
     Behaviour of the progeny of a single chondrocyte. J. Exp.
     Med. 130, 417.

34) REESE, C.A. and MAYNE, R. (1981). Minor collagen of chicken
     hyaline cartilage. Biochemistry, 20, 5443.

35) TOWBIN, H., STAEHELIN, T. and GORDON, J. (1979). Electropho-
     retic transfer of proteins from polyacrylamide gels to nitro-
     cellulose sheets: Procedure and some applications. Proc. Natl.
     Acad. Sci. USA, 76, 4350.

36) BAKER, J.R. and CATERSON, B. (1979). The isolation and characterization of the link proteins from proteoglycan aggregates
of bovine nasal cartilage. J. Biol. Chem.,254, 2387.

37) TANG, L. H., ROSENBERG, L., REINER, A., POOLE, A.R. (1979).
Proteoglycans from bovine nasal cartilage. Properties of a
soluble form of link protein. J. Biol. Chem., 254, 10523.

38) DESSAU, W., SASSE, J., TIMPL, R., JILEK, F. and VON DER MARK,
K. (1978). Synthesis and extracellular deposition of fibronectin in chondrocyte cultures. J. Cell Biol., 79, 342.

39) UPHOLT, W.B., VERTEL, B.M. and DORFMAN, A. (1979). Translation
and characterization of mRNAs in differentiation of chicken
cartilage. Proc. Natl Acad. Sci. USA, 76, 4847.

40) CHEAH, K.S.E., GRANT, M.E. and JACKSON, D.S. (1979). Translation of Type II procollagen mRNA and hydroxylation of the
cell-free product. Biochem. Biophys. Res. Commun. 91, 1025.

41) PAGLIA, L.M., WIESTNER, M., DUCHENE, M., QUELLETTE, L.A.,
HORLEIN, D., MARTIN, G.R. and MULLER, P.K. (1981). Effects
of procollagen peptides on the translation of Type II collagen
mRNA and on collagen biosynthesis in chondrocytes. Biochemistry
20, 3523.

42) PAWLOWSKI, P.J., BRIERLEY, G.T. and LUKENS, L.N. (1981).
Changes in the Type II and Type I collagen messenger RNA population during growth of chondrocytes in 5-bromo-2-deoxy-
uridine. J. Biol. Chem. 256, 7695.

43) YOSHIMURA, M., JIMENEZ, S.A. and KAJI, A. (1981). Effects of
viral transformation on synthesis and secretion of collagen
and fibronectin-like molecules by embryonic chick chondrocytes in culture. J. Biol. Chem., 256, 9111.

44) HOWARD, B.H., ADAMS, S.L., SOBEL, M.E., PASTAN, I. and DE
CROMBRUGGHE, B. (1978). Decreased levels of collagen mRNA in
Rous sarcoma virus-transformed chick embryo fibroblasts. J.
Biol. Chem., 253, 5869.

45) FAGAN, J.B., SOBEL, M.E., YAMADA, K.M., DE CROMBRUGGHE, B. and PASTAN, I. (1981). Effects of transformation on fibronectin gene expression using cloned fibronectin cDNA. J. Biol. Chem., 256, 520.

46) HYNES, R.O. (1976). Cell surface proteins and malignant transformation. Biochim. Biophys. Acta, 458, 73.

47) VERTEL, B.M. and DORMAN, A. (1979). Simultaneous localization of Type II collagen and core protein of chondroitin sulphate proteoglycan in individual chondrocytes. Proc. Natl Acad. Sci. USA, 76, 1261.

48) KITAMURA, K. and YAMAGATA, T. (1976). The occurrence of a new type of proteochondroitin sulphate in the developing chick embryo. FEBS Lett., 71, 337.

49) DURBAN, E.M. and BOETTIGER, D. (1980). Progenitor-cell populations can be infected by RNA tumour viruses, but transformation is dependent on the expression of specific differentiated functions. Cold Spring Harbor Symposium on Quant. Biol., 44, 1249.

50) GAZZOLO, L., SAMARUT, J., BOUABDELLI, M. and BLANCKET, J.P. (1980). Early precursors in the eythroid lineage are the specific target cells of avian erythroblastosis virus *in vitro*. Cell, 22, 683.

51) ABBOTT, J. and HOLTZER, H. (1968). The loss of phenotypic traits by differentiated cells. V. The effect of 5-bromodeoxyuridine on cloned chondrocytes. Proc. Natl Acad. Sci. USA, 59, 1144.

52) SCHILTZ, J.R., MAYNE, R. and HOLTZER, H. (1973). The synthesis of collagen and glycosaminoglycans by dedifferentiated chondroblasts in culture. Differentiation, 1, 97.

53) MAYNE, R., VAIL, M.S. and MILLER, E.J. (1975). Analysis of changes in collagen biosynthesis that occur when chick chondrocytes are grown in 5-bromo-2'-deoxyurdine. Proc. Natl. Acad. Sci. USA, 72, 4511.

54) PACIFICI, M. and HOLTZER, H. (1977) Effects of tumour-
     promoting agents on chondrogenesis. Am. J. Anat., 150, 207.

55) TOLSTOSHEV, P., HABER, R., TRAPNELL, B.C. and CRYSTAL, R.G.
     (1981). Procollagen mRNA levels, activity and collagen
     synthesis during the fetal development of sheep lung, tendon,
     and skin. J. Biol. Chem. 256, 9672.

56) PAGLIA, L., WILCZEK, J., DIAZ DE LEON, L., MARTIN, G.R.,
     HORLEIN, D. and MULLER, P. (1979). Inhibition of procollagen
     cell-free synthesis by aminoterminal extension peptides.
     Biochemistry, 18, 5030.

57) WIESTNER, M., KRIEG, T., HORLEIN, D., GLANVILLE, R.W.,
     FIETZEK, P. and MULLER; P.K. (1979). Inhibiting effect of
     procollagen peptides on collagen biosynthesis in fibroblast
     cultures. J. Biol. Chem. 254, 7016.

58) HORLEIN, D., McPHERSON, J., GOH, S.H. and BORNSTEIN, P. (1981).
     Regulation of protein synthesis: translational control by
     procollagen-derived fragments. Proc. Natl. Acad. Sci. USA,
     78, 6163.

PROTEINS AFFECTED BY CHROMOSOME 21 AND AGEING *IN VITRO*

M.L. Van Keuren*#, C.R. Merril* and D. Goldman#

Laboratory of General and Comparative Biochemistry and #Laboratory of Clinical Science, National Institute of Mental Health, Bethesda, Maryland 20205, USA

1. INTRODUCTION

An extra copy of chromosome 21 is the cause of the disorder known as Down syndrome. The mechanisms underlying the pathogenesis of this syndrome are unknown. Some of the features of Down syndrome are mental retardation, a high incidence of cardiac defects and predisposition for development of acute myeloblastic and acute lymphocytic leukemia (1). Individuals with this disorder display features of premature ageing (2, 3); those who survive into their 30's and 40's express dementia, accompanied by neuropathological features identical to those found in Alzheimer's disease (4-6).

Seven genes have been mapped to chromosome 21. They are the genes for 18S and 28S rRNA (located at the short arms of chromosomes 13-15 and 21-22) (7); the soluble, or cytoplasmic form of superoxide dismutase (SOD-1) (8); the interferon receptor (8), two enzymes of the purine *de novo* biosynthetic pathway, phosphoribosyglycineamide synthetase (9), and phosphoribosylaminoimidazole synthetase (10), and the liver form of phosphofructokinase (11).

=*Present address: E. Roosevelt Inst. for Cancer Res. (Box B-129) 4200 East Ninth Av., Denver, Colo. 80262, USA.*

Chromosome 21 is one of the smallest chromosomes; the pair contains only 1.6% of the genomic DNA (12). Complete monosomy for chromosome 21 has been reported in only a few liveborn infants, and is associated with severe growth and mental retardation and failure to thrive (13-15).

What are the biochemical consequences of the presence of an excess of this small amount of normal genetic material? A transcriptional dosage effect has been demonstrated by quantitative DNA hybridization for chromosome 21 in skin fibroblasts (16). Protein and functional dosage effects have been demonstrated for three genes mapped to chromosome 21, including: superoxide dismutase (SOD-1) (17, 18), the interferon receptor (19, 20), and the purine synthetic enzyme, phosphoribosylglycineamide synthetase (21, 22). It is possible that an overabundance of these gene products, or of these yet to be mapped to chromosome 21, plays a role in the development of the features of Down syndrome. Also, it is possible that an excess of chromosome 21 gene products leads to widespread secondary effects, disturbing the expression of genes on other chromosomes. At least eleven enzymes with structural loci assigned to chromosomes other than 21 may have elevated activity in Down syndrome (23).

Studies on gene expression in Down syndrome are limited by the window of gene products and effects which can be examined. With the development of new techniques such as two dimensional electrophoresis, highly sensitive silver staining, and computerized microdensitometry, one can perform surveys of large numbers of proteins. Two dimensional electrophoresis has been used to identify protein alterations associated with diseases such as the Lesch-Nyhan syndrome (24, 25) and Huntington's disease (26) and phenomena such as neoplastic transformation of cells in culture (27).

The highest resolution of large numbers of proteins can be achieved by the two dimensional electrophoresis method as described by O'Farrell (28), in which proteins are separated in the first dimension by isoelectric point (pI) and in the second dimension by molecular weight. Proteins can be visualized by staining, or, if they have been radioactively labelled, by autoradiography, (see also other articles in this volume).

The most commonly used general protein stain has been Coomassie blue. Recently, silver staining methods have been developed which are 200 times more sensitive than Coomassie blue (29-32). Both silver staining and autoradiography of labelled proteins can be performed with the same gel (33). Autoradiography displays net protein synthesis during the labelling period and silver staining shows total detectable protein.

These new sensitive and powerful methods were applied to search for proteins affected by chromosome 21, by direct and secondary effects, and proteins affected by ageing *in vitro*.

2. SUPEROXIDE DISMUTASE: LOCALIZATION AND QUANTITATION

The location of the soluble form of superoxide dismutase (SOD-1) was found with the use of purified enzyme (a gift from John Sykes) and two cell lines (gifts from Frank Ruddle). The cell lines were a human/mouse hybrid containing chromosome 21 as the only human chromosome, and the A9 mouse parental cell line. Cell lines and strains used for experiments are described in Table 1.

Purified SOD-1 contained several protein species as revealed by two dimensional electrophoresis followed by silver staining. The enzyme is a dimer, containing two identical subunits of molecular weight 16 Kd (34). The region of the gel corresponding to this molecular weight is shown in Figure 1. The arrow points to

Table 1. *Description of fibroblast cell lines and strains*

A. Hybrid and mouse fibroblast lines

Line	Description	Source
WAVR4dF9-4a	human/mouse hybrid which contains chromosome 21 as only human chromosome	F.H.R.*
A9	mouse parental cell line	F.H.R

*Gifts from Dr. F.H. Ruddle (New Haven, Connecticut)

B. Human Skin Fibroblasts

Strain	Donor Age	Donor Sex	Chromomsomes	Source	Passage when Obtained	Additional P.D.**
GM10	12 fetal week	male	46,XY	IMR†	4	6 and 36
GM1603	12 fetal week	male	46,XY	IMR	4	5
GM969A	2 years	female	46,XX	IMR	8	6
GM1381	12 fetal week	male	46,XY	IMR	5	6
GM2504	1 month	male	47,XY,+21	IMR	7	5 and 35
GM2767B	14 years	female	47,XX,+21	IMR	8	6
CCL54	2 month	male	47,XY,+21	ATCC††	13	4
CCL84	child	female	47,XX,+21	ATCC	15	3
GM230	14 month	female	45,XX,-21	IMR	10	5

**P.D.= population doublings
†IMR= Institute for Medical Research (Camden, New Jersey)
††ATCC= American Type Culture Collection (Rockville, Maryland)

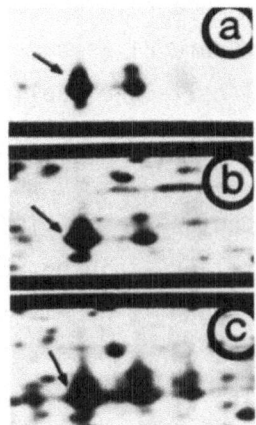

Fig. 1. *Localization of SOD-1 on silver-stained two dimensional gels.* (a) 4 μg purified SOD-1. (b) 4 μg purified SOD-1 co-electrophoresed with proteins extracted from WAVR4dF9-4a hybrid cells.(c) 4 μg purified SOD-1 co-electrophoresed with proteins extracted from human fibroblasts (GM2504) (In all photographs, the acidic is to the right).

the major, most densely stained species in the purified sample. This major species co-electrophoresed with a protein consistently seen in silver stained gels and autoradiograms of proteins from hybrid cells. There was only a very faint spot sometimes seen in this location on silver stained gels and autoradiograms of proteins from the mouse parental line (Figure 2).

The density of SOD-1 was measured in silver stained gels and autoradiograms of human fibroblast strains trisomic, disomic and monosomic for chromosome 21 (GM2504, GM10 and GM230). Densities were normalized for gel to gel comparisons by the slope and y-intercept from linear regression analysis of density versus density graphs of thirty proteins on each gel (35). Density ratios (Table 2) closely approximate the previously reported enzyme activity ratios of 1.5 : 1.0 : 0.5 for fibroblasts trisomic, disomic and monosomic for chromosome 21 (17, 18), This has recently been corroborated by Brown *et al.* in a similar set of experiments (36).

Table 2. *Percent ratios of densities of superoxide dismutase (soluble) in human fibroblasts*

	Silver stained gels	Autoradiograms
Trisomy 21 (GM2504)	161.7	164.7
Normal (GM10)	100.0	100.0
Monosomy 21 (GM230)	68.2	33.0

3. PROTEIN DIFFERENCES BETWEEN THE HYBRID AND MOUSE CELL LINES

Double-label autoradiography was performed to demonstrate proteins expressed in the hybrid but not the mouse cell line. [^{3}H] labelled hybrid cell proteins were electrophoresed together with [^{14}C] labelled mouse cell proteins as described previously (38). Using the digitized images on the computer, positive and

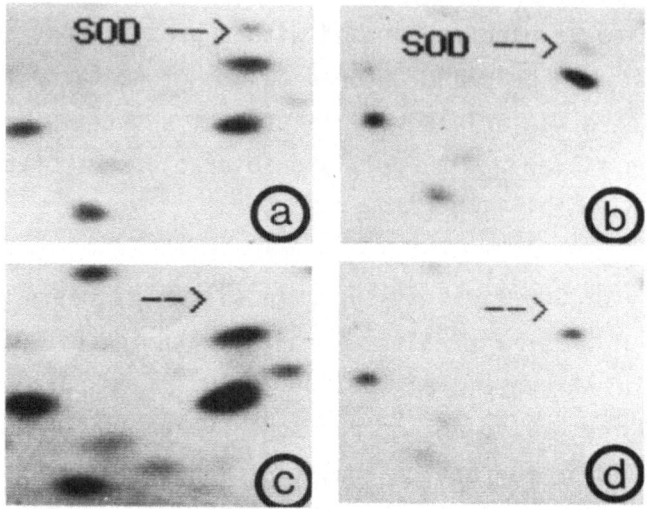

Fig. 2. *Comparison of the location of SOD-1 in the human/mouse hybrid and mouse parental cell lines.* Proteins extracted from the hybrid line (a,b) and the mouse line (c,d). Silver stained gels: (a) and (c); autoradiograms of cells labelled with a [^{14}C] amino acid mixture; (b) and (d). The arrow points to SOD-1. (Note the protein which appears below SOD-1 in silver stained gels but is absent in autoradiograms; other such proteins have been described (37)).

negative images of fluorograms and autoradiograms were superimposed at high magnification. The negative image of the fluorogram, which shows both [^{3}H] and [^{14}C] labelled proteins, and superimposed image of the autoradiogram, which shows only [^{14}C] labelled proteins, are displayed in Figure 3. Because proteins unique to the hybrid cell are labelled with [^{3}H] but not [^{14}C], they appear as white spots unobscured by black spots. Of 400 proteins visualized, only SOD-1 appeared as a unique protein in the hybrid cell.

Hybrid and mouse cell line proteins were examined further on separate gels, each containing [^{14}C] labelled proteins from

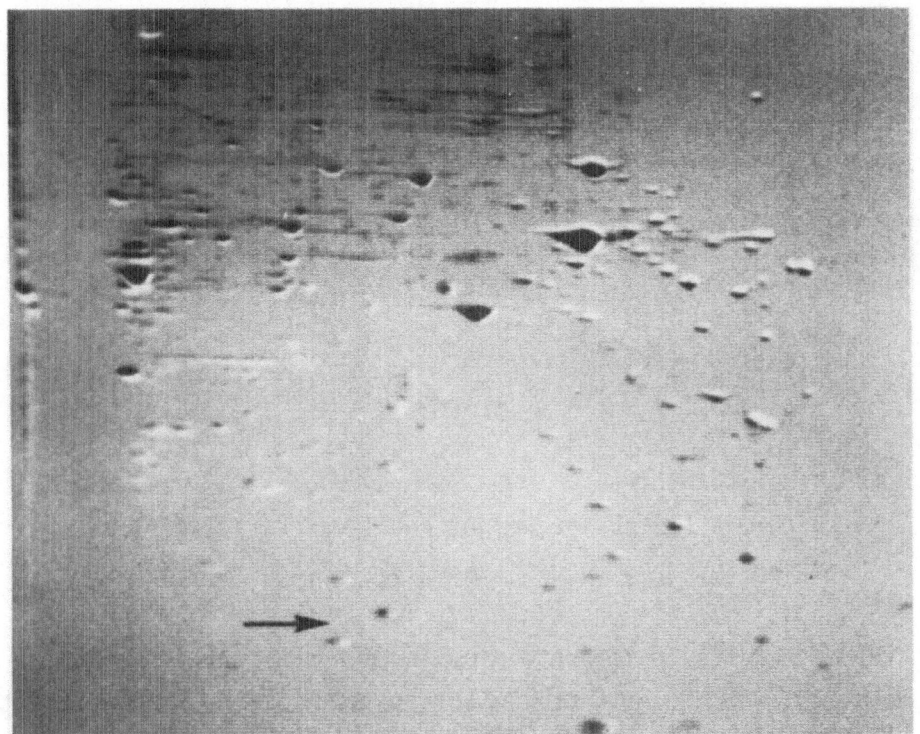

Fig. 3. *Double-label autoradiography of hybrid cell (labelled with [³H]) and mouse cells (labelled with [¹⁴C]).* The negative image of the fluorogram (white spots), which shows [¹⁴C] and [³H] labelled proteins, is superimposed by the autoradiogram (black spots), which shows [¹⁴C] labelled proteins only. SOD-1 (as indicated by the arrow) was seen in the fluorogram as a white spot which is too faint to be seen in this photograph. X-ray films were scanned on a 1000HS scanning microdensitometer (Optronics). A PDP 11/60 computer (Digital Equipment) with an IP5000 image processor (DeAnza Systems) was used for gel analysis.

one of the two cell lines. Five hundred proteins were measured in silver stained gel images and autoradiograms as described (35).

This analysis revealed three protein differences, in addition to SOD-1, between the hybrid and mouse cell lines: one density difference (Figure 4) and two apparent charge shifts (Figures 5 and

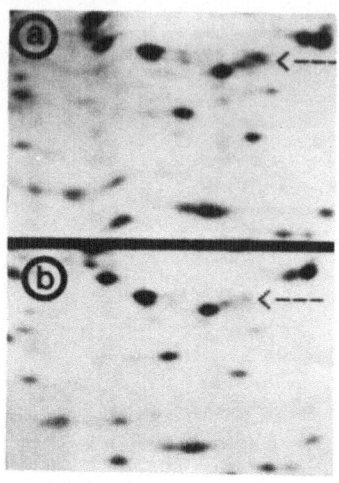

Fig. 4. *Autoradiograms showing protein density difference between hybrid and mouse cell lines.* (a) Proteins from the hybrid cell line. (b) Proteins from the mouse cell line. The molecular weight of this protein (arrow) is approximately 47 Kd, and the apparent pI is 4.5.

6). The approximate molecular weight and isoelectric point of the protein showing the density difference are 47 Kd and 4.5 (Figure 4). The density of this protein was twice as great in hybrid cells than in mouse cells. This protein could be a gene product of human chromosome 21 which co-migrates with a mouse protein. Fifty percent of human and rodent proteins comigrate on two dimensional gels (38) Alternatively, this could be a mouse protein whose quantity is affected by the presence of human chromosome 21 in the hybrid cell.

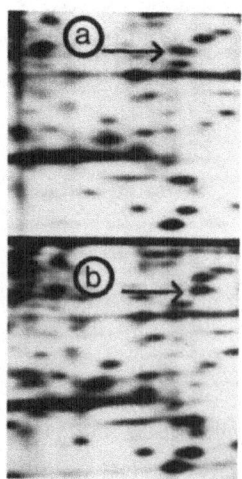

Fig. 5. *Silver stained gels of proteins from (a) hybrid cells and (b) mouse cells.* Arrows point to a protein with an apparent charge shift. The approximate molecular weight is 44 Kd, and the pI is 5.7 in the hybrid (a) and 5.6 in the mouse cells (b).

The apparent charge shifts are illustrated in Figures 5 and
6. A basic charge shift in the hybrid cell was seen for a protein of
approximate molecular weight 44 Kd and pI 5.6 and 5.7 (Figure 5).
An identity between these proteins is consistent with their having
equal normalized densities, their appearance only on silver stained
gels and not autoradiograms and their identical molecular weight.
As these proteins appeared only on silver stained gels, not on
autoradiograms, this variant would not have been detected by
double-label autoradiography.

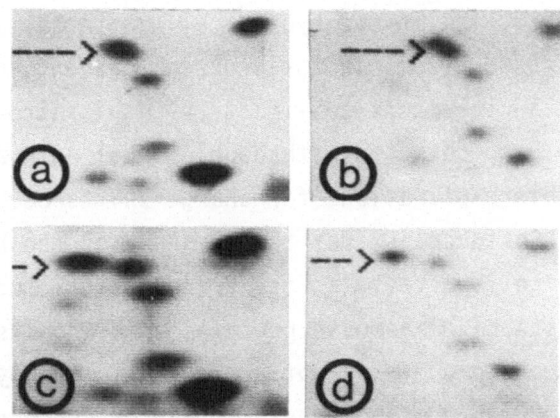

Fig. 6. *Proteins from the hybrid and mouse cell lines.* (a) Silver
stained gel of hybrid cell proteins; (b) autoradiogram of hybrid
cell proteins; (c) silver stained gel of mouse proteins; (d)
autoradiogram of mouse proteins. The illustrated proteins (arrow)
are of molecular weight approximately 20 Kd and apparent pI 5.1
(a and b) and 5.1 and 5.2 (c and d).

A second apparent protein charge shift between hybrid and
mouse cell lines is shown in Figure 6. Two protein species in the
hybrid cell electrophorese closely together but one is shifted
(basically) in mouse cell protein patterns. This interpretation
of the protein patterns is supported by the observation that the
sum of the densities of the two proteins in the mouse pattern
equals the density of the two closely positioned proteins in hybrid
cell autoradiograms (Figures 6 b and d). However, in silver stained

gels, the summed density of the proteins in the mouse cell was
twice the density of the protein in the hybrid cell (Figure 6 a
and b). The approximate molecular weight and isoelectric points
are 20 Kd and 5.1 (hybrid) and 5.1 and 5.2 (mouse). One possible
interpretation, given the fact that the labelling period was short
(four hours), is that the net synthetic rates of the proteins are
equal, but the degradation rate is greater in the hybrid cells.
The protein(s) in hybrid cells may be abnormal, and therefore
degrated more rapidly. The most acidic of the two proteins in
the mouse cell pattern appears in the same position as the two
closely positioned proteins in the hybrid cell patterns. Since
double labelling was used to search for proteins present in the
hybrid cell and not in the mouse cell patterns, this alteration
also could not have been detected with double-label autoradiography. Three possible explanations for charge shifts occurring
in the human/mouse hybrid cells are: the presence of the human
chromosome in mouse cells affects a post-translational modification
(such as phosphorylation) of a mouse protein, altering its isoelectric point; a mouse protein is not expressed in the hybrid
cell, and instead a human protein is expressed; or these proteins
have been altered due to mutation in the mouse genetic component
during the life of the hybrid cell line.

4. PROTEINS IN HUMAN FIBROBLASTS (GM2504, GM10 AND GM230)

Two dimensional protein patterns of human fibroblast cell
strains with one, two and three copies of chromosome 21 were
examined to identify proteins affected by chromosome 21. Over
500 protein spots were visible on fibroblast electrophoretograms.
For the survey, 400 proteins on silver stained gels·were analyzed
for positional and quantitative variation as described (26).
Triplicate gels were used for each cell strain. On the autoradiograms, 300 proteins were analyzed, as about 100 of the proteins
visible on silver stained gels were too faint on X-ray films for

reliable analysis (films were exposed for 29 days at -70°C). Early
and late passage GM2504 cells (trisomy 21), early and late passage
GM10 (normal) and early passage GM230 (monosomy 21) were analyzed.
In addition, silver stained gels and autoradiograms of proteins
from early passage fibroblasts of the strains GM2067 (trisomy 21),
GM41 (normal) and GM137 (monosomy 21) were also analyzed in this
study but their protein densities were not compared statistically.

Five qualitatively variable proteins were observed in these cell
strains with no correlation with chromosome 21 dosage. These variants,
illustrated in Figure 7 were: protein 84 (M$_r$ 64 Kd, pI 6.0), absent
in both monosomy 21 strains and the normal strain GM41; proteins
447 and 478 (M$_r$ 55 Kd for both, pI's 5.9 and 5.8, respectively),
present only in autoradiograms of the normal strain GM10; protein
284 (M$_r$ 28 Kd, pI 6.0), shifted slightly in the basic direction in
the normal strain GM10; and protein 486 (M$_r$ 31 Kd, pI 6.4), present
only in autoradiograms in the strains GM41 (normal) and GM230
(monosomy 21). These qualitative variants were not related to
chromosomal constitution and may be polymorphic proteins (40). Six-
teen proteins with mean densities significantly altered (2p<0.01)
in the trisomy 21 and/or monosomy 21 cell strains as compared to
normal cells were identified and listed in Table 3. The magnitude
of these changes is also given in Table 3 and the positions of these
proteins are shown in Figure 7. Proteins 19, 225, 244, 283 and 377
were significantly increased in trisomy 21 compared to normal cell
proteins. Increases ranged from 1.8 to 4.0 fold. Proteins 49, 81,
243, 245, 335, 405 and 455 were significantly decreased in trisomy
21 cells; the densities of trisomy 21 cell proteins ranged from
5 to 70% of the density of the corresponding normal cell protein
density. Of these seven variants, all but three (243, 405 and 455)
were similarly decreased for monosomy 21 cell proteins compared to normal
cell proteins. Three additional proteins (40,228 and 425) were significantly decreased for monosomy 21 cell proteins compared to those of

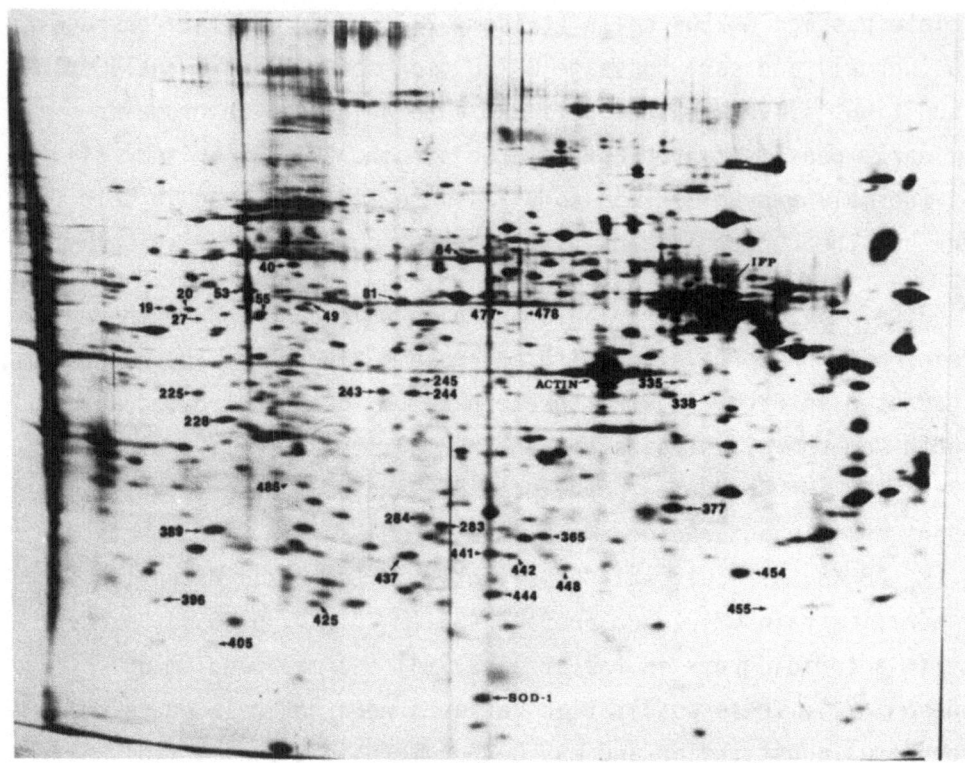

Fig. 7. *Silver stained gel of human skin fibroblast proteins*
(GM2504) separated by two dimensional electrophoresis. Cellular
proteins were extracted, electrophoresed and analyzed as described
(39). The acidic end is at the right and higher molecular weight
is at the top. Proteins 84, 284, 477, 478 and 486 are qualitatively
variable proteins. All other proteins labelled (except actin and
intermediate filament protein (IFP), normal proteins illustrated
for orientation) are those which displayed a significant density
variation in trisomy 21, monosomy 21 or aged cells. Approximate
molecular weights and isoelectric points are listed in Table 3
except for the following proteins: Actin: pI 5.6, M_r 43 Kd; IFP
pI 5.2, M_r 55 Kd; SOD-1: pI 5.9, M_r 16 Kd.

(Table 3). Arrows indicate whether the density was increased (↑)
or decreased (↓). "Auto" indicates the difference was found on
autoradiograms, and "ss" indicates it was found on silver stained
gel images. Values are given as ratios of mean normalized density
of the two cell types or conditions compared.

Table 3. *Proteins of variable density in trisomy 21 (GM2504), normal (GM10) and monosomy 21 (GM 230) fibroblasts*

Protein	Approximate Molecular Weight	Apparent Isoelectric Point	Trisomy 21 vs. Normal ("young")	Monosomy 21 vs. Normal ("young")	Normal "Aged" vs. "young"	Trisomy 21 "Aged" vs. "young"
19	54,000	6.7	↑ 1.8 (auto)		↑ 2.1 (auto)	
20	54,000	6.6				↓ 0.4 (ss)
27	53,000	6.6			↑ 8.6 (auto)	
40	64,000	6.4		↓ 0.5 (ss) ↓ 0.2 (auto)		
49	55,000	6.3	↓ 0.6 (ss) ↓ 0.4 (auto)	↓ 0.6 (ss)		
53	58,000	6.5			↓ 0.7 (auto)	↓ 0.4 (auto)
55	54,000	6.5				↓ 0.6 (auto)
81	55,000	6.1	↓ 0.7 (ss) ↓ 0.7 (auto)	↓ 0.8 (auto)		
225	41,000	6.6	↑ 1.6 (auto)			↓ 0.3 (auto)
228	39,000	6.5		↓ 0.1 (ss)		
243	43,000	6.1	↓ 0.7 (ss)			
244	43,000	6.0	↑ 4.0 (auto)			
245	43,000	6.0	↓ 0.4 (ss)	↓ 0.2 (ss)		
283	28,000	6.0	↑ 3.7 (ss)			
335	43,000	5.4	↓ 0.4 (ss)	↓ 0.2 (ss)	↓ 0.06(auto)	↓ 0.07(auto)
338	42,000	5.2			↑ 2.6 (ss)	
365	27,000	5.8				↓ 0.5 (auto)
377	31,000	5.4	↑ 1.8 (auto)	↓ 0.3 (ss)		
389	28,000	6.6			↓ 0.4 (ss)	↓ 0.5 (auto)
396	21,000	6.8		↑ 8.5 (ss) ↑11.2 (auto)		
405	19,000	6.6	↓ 0.2 (ss) ↓ 0.3 (auto)		↓ 0.3 (ss) ↓ 0.4 (auto)	
425	21,000	6.3		↓ 0.4 (ss)		
437	26,000	6.0			↓ 0.4 (ss)	↓ 0.5 (ss)
441	26,000	5.9				↓ 0.5 (ss) ↑ 1.8 (auto)
442	26,000	5.9				↓ 0.5 (auto)
444	25,000	5.9				↓ 0.5 (auto)
448	26,000	5.8				↓ 0.4 (ss)
454	26,000	5.1				↓ 0.5 (ss)
455	24,000	5.1	↓ 0.05 (ss)			

normal cells (values were from 10 to 50% those of normal cell
protein density) . Protein 396 was increased 10-fold in monosomy
21 cells; it was not altered in trisomy 21 cell patterns.

Of particular interest is protein 377, an uncharacterized
protein that varies in a fashion roughly proportional to a gene dosage
effect in both silver stained gels and autoradiograms; the density in
trisomic cells is greater than in normal cells, and still greater
than the density in monosomy 21 cells (Table 4).

Table 4. *Protein 377: Mean total density*

	Silver stain	Autoradiography
Trisomy 21 (GM2504)	59.0 ± 5.7	36.1 ± 1.2
Normal (GM10)	43.0 ± 3.1	19.8 ± 1.1
Monosomy 21 (GM230)	10.4 ± 4.8	13.4 ± 9.3

± standard error of the mean

Other than protein 377 and SOD-1, no other proteins displaying
gene dosage effects were detected. Proportionate gene dosage effects
in Down syndrome, in terms of enzyme acvities, have also been shown
for the activities of two enzymes mapped to chromosome 21, superoxide dismutase (17, 18), and the purine synthetic enzyme phosphoribosylglycineamide synthetase (21, 22). A non-proportionate gene
dosage effect has been found for response to interferon (19, 20). The
interferon receptor also maps to chromosome 21 (8). Since phosphoribosylglycineamide synthetase and the interferon receptor have
not been purified or characterized (their molecular weights and
isoelectric points are unknown), protein 377 may be one of them;
alternatively, it could represent a gene product specified
by chromosome 21 but not yet mapped there. However, it could
also be a protein which increases in quantity as a secondary

effect of the extra chromosome 21. A number of enzymes mapped to
other chromosomes have elevated activities in Down syndrome (23).

The chromosome 21 pair represents about 1.6% of the total
human chromosomal DNA (12). However, the short arms of chromosome
21 contain the rRNA genes (7). Because this region would not be
expected to code for cellular proteins, the fraction of protein-
coding chromosomal DNA represented by the 21st pair is 1.2%. If
the genes for fibroblast proteins are dispersed on the chromosomes
in proportion to their size, four proteins out of 400 examined
would be coded for by chromosome 21 and display a gene dosage
effect. Therefore, the two dimensional electrophoresis and protein
quantitation systems used in this study appear to have an efficiency of approximately 50%. However, given the small number of proteins
predicted, a more extensitve study involving analysis of 1,000
proteins per gel (12 proteins expected to map to chromosome 21) is
required to test the efficiency of chromosomal gene product
identification by two dimensional electrophoresis.

We have used two dimensional electrophoresis to examine the
extent of primary and secondary effects of trisomy 21 on cellular
proteins. There are two other reports of similar studies. Weil
and Epstein (41) studied 850 [^{35}S]-methionine labelled proteins
from five trisomy 21 and five normal fibroblast strains. From
visual inspection of autoradiograms, only four proteins were
found to be affected in trisomy 21 protein patterns compared to
normal. It was concluded that trisomy 21 does not cause extensive
secondary effects. Our data on 11 secondary protein alterations
per 400 proteins analyzed is consistent with their observations,
although we did detect more than twice as many secondary
alterations. We attempted to identify the four proteins reported
by Weil and Epstein to be altered. Because molecular weights and
pI's were not provided, we constructed "error boxes" in the

regions of our patterns in which these altered proteins might be.
However, because of overall pattern differences the location of
these proteins could not be determined on our gels.

Recently Klose *et al.* (42) visually examined 800 Coomassie
blue stained two dimensional gels of proteins from seven trisomy
21 and seven normal skin fibroblast strains. They reported five
consistent quantitative variations in each trisomy 21 as compared
to normal strain. The variations involved four decreases and one
increase in stain intensity. The affected proteins reported by
Klose *et al.* did not appear to coincide with ours; however, the
overall patterns of their gels were quite different, and molecular
weights and pI's were not provided.

Both of the other reported studies involving trisomy 21 and
normal cell strains (41, 42) were performed in a qualitative manner
in contrast to the quantitative analysis performed in our study.
It is important to make quantitative measurements when reporting a
study of quantitative protein alterations because many quantitative
changes are not apparent by eye, and apparent quantitative variants
may prove to be insignificant.

5. PROTEINS IN ADDITIONAL TRISOMY 21 AND NORMAL CELL STRAINS

To investigate the possibility that some of the quantitative
protein alterations detected in GM2504 (trisomy 21) and GM10
(normal) cells might be peculiar to those cell strains and not
characteristic of all trisomy 21 cells, additional cell strains
were examined. Three additional trisomy 21 fibroblast strains
(GM2767, CCL54 and CCL84) and three additional normal diploid
strains (GM969, GM1603 and GM1381) were evaluated. Low passage
cells were used for all samples in this study. Results are listed
in Table 5. A normal cell strain (GM10) used in the previous
study was included in this analysis. Mean densities that were

significantly different ($2p < 0.01$) are listed in Table 5. None of
the proteins found to be altered in the first study (GM2504 versus
GM10) were altered in every possible pairing of trisomy 21 versus
normal cells (there are twelve possible pairings for each protein).
Protein 19 was significantly altered only when GM10 was used as
the normal cell strain for the pairing. It may represent a protein
variation peculiar to GM10 and not affected by trisomy 21. The
other proteins showed quantitative changes in three to nine of the
12 possible combinations Protein 377, which had been shown to
express a gene dosage effect in the previous study, did the same
for all comparisons except those involving one normal cell strain
(GM1603). In conventional enzyme assays comparing two groups,
individual values frequently overlap, in spite of significantly
different means. Because of variability in gene expression and
quantitation, it may be difficult to be certain which protein
alterations are cell strain specific, which are related to trisomy
21 and which are due to experimental variation. In general, it must
be concluded that an extra chromosome 21 does not cause extensive
secondary effects on the production or degradation of protein
products in fibroblasts.

6.   PROTEINS ASSOCIATED WITH AGEING *IN VITRO*

Down syndrome has been cited as a model for ageing because
pathologic events which occur in this syndrome are associated with
advanced age in normal humans (2, 3). Because of these observations,
one trisomy 21 (GM2504) and one normal (GM10) fibroblast strain
were cultured for an additional thirty population doublings.
Protein patterns from these late passage cells were each compared
to those of their early passage counterparts. This study allowed us
to place a limit on the fraction of erroneously synthesized
proteins, predicted by "error catastrophe", which could have
accumulated in these aged cells and to see if there are qualitative
or quantitative protein alterations associated with *in vitro*

Table 5. *Ratios of mean densities of proteins in additional cell strains*

Protein	Original Observation GM2504/GM10	Additional trisomy 21 cells vs. original normal			New trisomy 21 strain GM2767 vs. each new normal strain			New trisomy 21 strain CCL54 vs. each new normal strain			New trisomy 21 strain CCL84 vs. each new normal strain		
		GM2767/GM10	CCL54/GM10	CCL84/GM10	GM2767/GM969	GM2767/GM1381	GM2767/GM1603	CCL54/GM969	CCL54/GM1381	CCL54/GM1603	CCL84/GM969	CCL84/GM1381	CCL84/GM1603
19	↑1.8(s)			↑2.7(a)									
49	↓0.6(s) ↓0.4(a)	↓0.4(s)	↓0.6(s)	↓0.3(s) ↓0.7(s)		↓0.7(a) ↓0.6(s)	↓0.8(a)	↑3.7(s)				↓0.4(s)	↓0.6(s)
81	↓0.7(s) ↓0.7(a)	↓0.7(a)		↓0.5(a)		↓0.4(a)	↓0.7(a)	↑1.5(a)	↓0.6(a)	↓0.7(a)	↓0.8(a)	↓0.4(a)	↓0.5(a)
225	↑1.6(a)					↓0.6(a)	↓0.5(a)	↑1.8(a)			↑1.8(a) ↓0.2(s)		↑0.8(a) ↓0.3(s)
243	↓0.7(s)	↑1.8(a)	↓0.2(s)				↓0.6(a)	↑2.4(a) ↓0.3(s)			↓0.8(a)	↓0.3(a)	↓0.8(a)
244	↑4.0(a)	↑1.5(a)	↓0.4(s)	↑1.7(s)	↑0.6(s)		↑2.0(a)	↓0.5(a)				↑6.3(s)	↑3.2(s)
245	↓0.4(s)	↑2.0(a)		↑0.2(s)					↑1.9(a)		↓0.4(s)		
283	↑3.7(s)	↑2.3(s) ↑6.2(a)	↑2.4(s) ↑5.9(a)		↑0.4(a)	↑2.7(s)			↑3.7(a)	↑1.8(s)		↑3.8(s)	
335	↓0.3(s)	↑0.2(s) ↓0.1(a)	↓0.1(s) ↓0.1(a)	↑0.1(s) ↑0.01(a)		↑1.5(s)	↑2.0(a)	↑0.4(s)					
377	↑1.8(a) ↓0.2(s) ↓0.3(a)	↑1.7(a)	↑1.3(s) ↑1.8(a)	↑1.3(s) ↑2.5(a)	↑1.3(a)			↑1.2(s) ↑1.4(a)	↑2.1(s) ↑2.0(a)		↑2.0(a)	↑2.2(s)	↑2.9(a)
405							↓0.2(a)			↓0.3(a)			↑0.2(a)
455	↓0.05(s)	↓0.1(s)	↓0.1(s)	↓0.04(s)									

Trisomy 21 fibroblast cell strains were GM2767, CCL54 and CCL84. Normal diploid fibroblast strains GM969, GM1381, and GM1603. Ratios given are for those comparisons that were significantly different ($2p < 0.01$). Data for proteins in silver stained gels are indicated as "s" and those in autoradiograms as "a".

ageing. According to the "error catastrophe" theory of cellular
ageing, errors in protein synthesis create protein synthetic
machinery with progressively lower accuracy, accelerating the
accumulation of aberrant proteins (43). A prediction of this
theory is that a mixture of normal and abnormal proteins are
present in cells near the end of their replicative lifespan. The
abnormal protein fraction should constitute a heterogeneous mixture
resulting from a random collection of amino acid substitutions.
Approximately one-third of random single base change mutations
causing amino acid substitutions should result in changes of net
protein charge (44). Interchange of the neutral amino acids and histidine, lysine, arginine, aspartic acid and glutamic acid result
in single charge changes, while interchange between aspartic acid
and histidine or glutamic acid and lysine results in a double charge
alteration. Protein charge variants will tend to appear in
discrete locations resulting from whole charge unit shifts rather
than fractional unit shifts. Such altered proteins should appear
as satellite spots, acidic and basic to the native protein (44).
No such satellite spots were observed in protein patterns of late
passage fibroblasts. To check for the presence of charge variants
of native proteins, twenty well-defined dense proteins were
examined for the presence of acidic or basic satellite spots in
both early and late passage normal cells (GM10). High resolution
density plots were made, and the image itself was examined after
constrast enhancement. No satellite spots appearing with ageing
were observed. An example of such a plot is shown in Figure 8.
In Figure 8, part A, optical density is plotted versus horizontal
distance for a protein in an autoradiogram of late passage fibroblasts (GM10). Figure 8, part B is a simulated representation of
the appearance of charge variant satellites.

This type of analysis allows the calculation of an upper bound
for the fraction of improperly synthesized proteins in aged fibro-

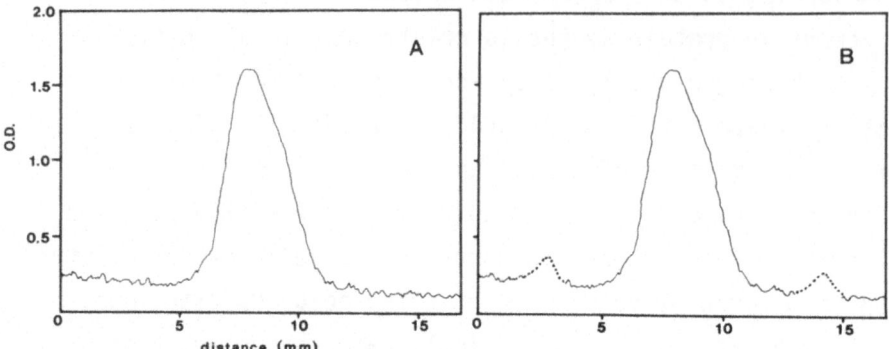

Fig. 8. *Plots of average optical density versus distance in an area including a dense, well-defined protein, and space to the left and right where single charge-substituted variants would exist.* (A) Plot for a protein from an autoradiogram of normal, late passage fibroblasts (GM10) proteins separated by two dimensional electrophoresis; (B) A simulation, based on plot A, with theoretical values expected if charge-substituted variants of the native protein were present.

blasts. The assumptions for this calculation are that approximately one-third of amino acid substitutions give rise to a change in isoelectric point and that charge-altered species will primarily focus in two discrete positions, one basic and one acidic to the native protein. Multiple amino acid substitutions and substitutions of one charged amino acid for another would lead to much smaller quantities of satellites located at greater distances from the native protein. Because the mean of the 10 faintest detectable protein spots was 1.45 density units (optical density x mm^2), and the mean of the 10 most intense protein spots was 350.5 units and because we detected no new satellite proteins with ageing, the maximum fraction of abnormal protein which could be present can be calculated as:

$$\text{maximum fraction abnormal} = \frac{a \times b}{c \times d} = \frac{2 \times 1.45 \text{ units}}{0.33 \times 350.5 \text{ units}} = 0.025$$

a = number of satellite positions per protein spot; b = density of faintest detectable protein spot; c = fraction of amino acid substitutions resulting in charge alteration; d = density of most intense protein spot.

Therefore, the mean level of amino acid substitutions for the 10 most intense proteins examined must be less than 2.5%.

Our finding of no new detectable protein spots with ageing *in vitro* corroborates the results of other studies. Wilson *et al.* (45) examined proteins among two dimensional patterns of [^{14}C]-leucine labelled proteins from young and old rat superior cervical ganglia. No new spots were detected in patterns from the old material.

Engelhardt *et al.* (46) examined proteins in two dimensional patterns of [^{35}S]-methionine labelled early and late passage human fibroblasts (two strains). No new proteins appeared in patterns of the late passage cells; however, Engelhardt *et al.* (46) found two proteins to be increased in density at least ten-fold with *in vitro* ageing. We could not identify these two proteins on our patterns.

Although there were no qualitative protein changes associated with *in vitro* ageing, there were several quantitative alterations. These proteins are listed in Table 3, and illustrated in Figure 7. Three proteins of GM10 (normal) fibroblasts were increased from 2 to 8-fold in density (2p<0.01) in silver stained gels and/or autoradiograms with *in vitro* ageing. Five other quantitative changes involved decreases with ageing in GM10 (see Table 3 and Figure 7. Late passage normal cell protein densities ranged from 6 to 70% of the density of proteins in early passage cells. In GM2504 (trisomy 21) fibroblasts, thirteen proteins were quantitatively different ((2p<0.01) for early passage versus late passage samples. All of these quantitative protein changes except one involved decreases with ageing. These late passage cell proteins were from 7 to 60% the density in early passage cells. Protein 441 showed a density which was two-fold less in silver stained gels of late passage cells, but was two-fold more in autoradiograms. Such a finding would be consistent with the production of an abnormal protein which is degraded more rapidly (the total amount silver stained is decreased), while the cell is attempting to compensate by increasing synthesis (increased tracer incorporation). Surveys of ratios of silver stained protein density to autoradiographic density may be important for ageing studies.

Three proteins that were quantitatively altered in late passage normal cells were also altered, in the same direction, for the trisomy 21 compared to normal cells. These were: protein 19 (2-fold increase); protein 335 (2.5-fold decrease for trisomy 21 versus normal, and 16-fold decrease for late versus early passage normal); and protein 405 (decreased by one-third fold). These proteins may be relevant to some of the features of premature ageing associated with Down syndrome.

Table 6. *Mean densities of proteins which change quantitatively with ageing in two human fibroblast strains*

| Protein | GM10 | | | GM2504 | | |
	Early Passage	Late Passage	Ratio[c]	Early Passage	Late Passage	Ratio
53(s)[a]	36.3 ± 3.0	26.3 ± 9.0	0.72	19.3 ± 3.4	14.4 ± 2.0	0.75
53(a)[b]	30.9 ± 0.5	20.4 ± 2.1	0.66**	55.1 ± 1.8	20.6 ± 1.6	0.36**
335(s)	49.6 ± 3.4	10.0 ± 9.0	0.20*	14.2 ± 4.5	6.1 ± 1.6	0.43
335(a)	38.9 ± 6.5	2.5 ± 0.7	0.06**	16.3 ± 3.1	1.2 ± 0.4	0.07**
389(s)	52.9 ± 7.1	21.8 ± 5.9	0.41**	54.1 ± 15.0	17.8 ± 3.7	0.33
389(a)	42.7 ± 6.2	23.7 ± 1.5	0.55*	107.5 ± 6.4	51.1 ± 8.5	0.57**
437(s)	50.4 ± 4.7	20.8 ± 3.9	0.41**	37.7 ± 11.2	22.7 ± 4.9	0.60
437(a)	54.0 ± 10.0	22.3 ± 12.9	0.41	69.2 ± 1.2	35.9 ± 0.4	0.52**

[a] s: silver stained gel
[b] a: autoradiogram
[c] (late passage)
    (early passage)
 * $2p < 0.05$
** $2p < 0.01$

Four quantitative protein alterations were observed in both the normal and trisomy 21 cell strains with ageing (Table 6). All four alterations were decreases in density, down to about one tenth. Their approximate isoelectric points and molecular weights are as follows: protein 53-pI 6.5, M_r 58 Kd; protein 335- pI 5.4, M_r 43 Kd; protein 389- pI 6.6, M_r 28 Kd; protein 437- pI 6.0, M_r 26 Kd.

In a recent review (47), alterations with cellular ageing are tabulated. More than twenty enzymes have been reported to increase and ten to decrease in activity as normal diploid cells approach the end of their *in vitro* lifespan. It is possible that the qualitative alterations described here are among those enzymes known to decrease in quantity with ageing.

7. SUMMARY

The effects of chromosome 21 and ageing *in vitro* on cellular proteins were studied by two dimensional electrophoresis, followed by silver staining and autoradiography. Protein densities were quantitated by computerized microdensitometry. Proteins from a human/mouse hybrid cell line containing human chromosome 21 were compared to those of the mouse parental cell line. One protein displayed a density difference, being twice as dense in the hybrid cell pattern. This may be a protein of human origin which co-migrates with a mouse protein. Two proteins displayed an apparent charge shift, raising the possibility that the presence of the human chromosome in the hybrid cell affects the charge of these two proteins; alternatively, they may represent a situation in which expression of a mouse protein is repressed in the hybrid cell, and a human protein is expressed instead. The protein spot corresponding to human superoxide dismutase, an enzyme mapped to chromosome 21, was identified in the human/mouse hybrid cell line with the use of purified enzyme.

Proteins patterns of human fibroblasts monosomic, disomic (normal), and trisomic for chrosome 21 were analyzed for both qualitative and quantitative variations. Five qualitative variants were found, but were not related to the numbers of chromosome 21 in the cell strains. These may be polymorphisms. A gene dosage effect, in terms of protein spot density on both silver stained gels and autoradiograms, was found for superoxide dismutase and one other protein (protein 377). Protein 377 may be one of the enzymes or the interferon receptor, which are mapped to chromosome 21 but as yet uncharacterized in terms of molecular weight and pI; it may also be some other gene product of chromosome 21. Based on the fact that chromosome 21 represents approximately 1.2% of the human genome, four of the 400 proteins analyzed would have been expected to display a gene dosage effect. Since two proteins displaying a gene dosage effect were observed, an efficiency of about 50% was achieved. Comparison of proteins from monosomic and trisomic cells to those in normal diploid cells revealed a small number of other significant quantitative differences. The magnitude of these differences varied from increases as great as 11-fold to decreases of as much as 1/20 the protein density of that found in normal cells. Some of these alterations were not apparent when protein patterns from other trisomy 21 and normal cells were examined. In general, trisomy for chromosome 21 does not cause widespread changes in cellular proteins.

Individuals with Down syndrome (trisomy 21) display many features of premature ageing. Because of this association, trisomy 21 and normal fibroblasts were aged *in vitro*. Proteins from the late passage cells were compared to those from early passage cells. Again, a limited number of significant quantitative changes were found; however, four of the proteins that displayed a decrease in density with *in vitro* ageing did so in both normal and trisomy 21 cell strains. The altered proteins were from 6 to 70% the density in early passage cells. These proteins may play a role in the phenomena of cellular ageing. No new proteins or satellite spots due to random amino acid substitutions were detected in the late passage cells. The error catastrophe hypothesis of cellular ageing predicts that random amino acid substitutions accumulate with ageing. The absence of satellite spots was confirmed by construction of high resolution density plots. These density studies demonstrated that error catastrophe, if it occurs, must operate at a level of less than 2.5%.

8. ACKNOWLEDGEMENTS

This work was supported by the National Institute of Mental Health and the National Institute on Alcohol Abuse and Alcoholism, and is in partial fulfillment of the Ph.D. degree requirements for M.L.V. at George Washingtom University, Washington, D.C.

9. REFERENCES

1) ZELLWEGER, H. and SIMPSON, J. (1977). Down's Syndrome. In: Chromosome of Man, p. 41, Lippincott, Philadelphia, Penn.

2) MARTIN, G.M. (1978). Genetic syndromes in man with potential relevance to the pathology of ageing. Birth Defects: Orig. Art. Ser. XIV, 1, 5.

3) GOLDSTEIN, S. (1978). Human genetic disorder that feature premature onset and accelerated progression of biological ageing. In: The genetics of ageing (ed. E.L. Schneider), p. 171, Plenum Press, New York.

4) JERVIS, G.A. (1970). Premature senility in Down's syndrome. Ann. N.Y. Acad. Sci., 171, 558.

5) ELLIS, W.G., McCULLOCH, J.R. and CORLEY, C.L. (1974). Presenile dementia in Down's syndrome: ultrastructural identity with Alzheimer's disease. Neurology, 24, 101.

6) OLSON, M.I. and SHAW, C. (1969). Presenile dementia and Alzheimer's disease in mongolism. Brain, 92, 147.

7) SCHMICKEL, R.D. and KNOLLER, M. (1977). Characterization and location of the human genes for ribosomal ribonucleic acid Pediat. Res., 11, 929.

8) TAN, Y.H., TISCHFIELD, J. and RUDDLE, F. (1973). The linkage of the genes for the human interferon-induced antiviral protein and indophenol oxidase-B traits to human chromosome G-21. J. Exp. Med., 137, 317.

9) MOORE, E.E., JONES, C., KAO, F.T. and OATES, D. (1977). Synteny between glycinamide ribonucleotide synthetase and superoxide dismutase (soluble). Am.J. Hum. Genet., 29, 389.

10) PATTERSON, D., GRAW, S. and JONES, C. (1981). Demonstration, by somatic cell genetics, of coordinate regulation of genes for two enzymes of purine synthesis assigned to human chromosome 21. Proc. Natl. Acad. Sci. USA, 78, 405.

11) VORA, S. and FRANKE, U. (1981). Assignment of the human gene for liver-type 6-phosphofructokinase isozyme (PFK_L) to chromosome 21 by using somatic cell hybrids and monoclonal antibodies. Proc. Natl. Acad. Sci. USA, 78, 3738.

12) MENDELSON, M.L., MAYALL, B.H., BOGART, E.B., MOORE, D.H. and PERRY, B.H. (1973). DNA content and DNA-based centrometric index of the 24 human chromosomes. Science, 179, 1126.

13) HOLLORAN, K.H., BERG, W.R. and MAHONEY, M.J. (1974). 21 monosomy in a retarded female infant. J. Med. Genet., 11, 386.

14) DAVIS, J.G., JENKINS, E.C., KLINGER, H.P. and WEED, R.G. (1976). A child with presumptive monosomy 21 (45, XY, -21) in a family in which some members are Gq-. Cytogenet. Cell Genet., 17, 65.

15) GRIPENBERG, U., ELFVING, J. and GRIPENBERG, L. (1972). A 45,XX,21- child: attempt at a cytological and clinical interpretation of the karyogram. J. Med. Genet., 9, 110.

16) KURNIT, D.M. (1979). Down syndrome: gene dosage at the transcriptional level in skin fibroblasts. Proc. Natl. Acad. Sci. USA, 76, 2372.

17) FEASTER, W.W., KWOK, L.W. and EPSTEIN, C.J. (1977). Dosage effects for superoxide dismutase-1 in nucleated cells aneuploid for chromosome 21. Am. J. Hum. Genet., 29, 563.

18) SCHITIU, S., SINET, P.M., LEJEUNE, J. and FREZAL, J. (1974). Dosage effects for superoxide dismutase-1 in nucleated cells aneuploid for chromosome 21. Humangenetik, 23, 65.

19) TAN, Y.H., SCHNEIDER, E.L., TISCHFELD, J., EPSTEIN, C.J. and RUDDLE, F.H. (1974). Human chromosome 21 dosage: effect on the expression of the interferon-induced antiviral state. Science, 186, 61.

20) EPSTEIN, C.J., EPSTEIN, L.B., WEIL, J. and COX, D. (1982). Trisomy 21: mechanisms and models. Ann. N.Y. Acad. Sci, in press.

21) BARTLEY, J.A. and EPSTEIN, C.J. (1980). Gene dosage effect for glycinamide ribonucleotide synthetase in human fibroblasts trisomic for chromosome 21. Biochem. Biophys. Res. Comm., 93, 1286.

22) SCOGGIN, C.H., BLESKAN, J., DAVIDSON, J.N. and PATTERSON, D. (1980). Gene expression of glycinamide ribonucleotide synthetase in Down syndrome. Clin. Res., 28, 31A (abstract).

23) FRANKE, U. (1981). Gene dosage studies in Down syndrome. In: Trisomy 21 (Down syndrome) Research Perspectives (eds, F.F. de la Cruz and P. Gerald), p. 237, University Park Press, Baltimore, Maryland.

24) MERRIL, C.R., GOLDMAN, D. and EBERT, M. (1981). Protein variations associated with Lesch-Nyhan syndrome. Proc. Natl. Acad. Sci. USA, 78, 6471.

25) MERRIL, C.R. and GOLDMAN, D. (1982). Quantitative two dimensional protein electrophoresis for studies of inborn errors of metabolism. Clin. Chem., 28, 1015.

26) GOLDMAN, D., MERRIL, C.R., POLINSKY, R.J. and EBERT, M. (1982). Lymphocyte proteins in Huntington's disease: quantitative analysis by use of two dimensional electrophoresis and computerized densitometry. Clin. Chem., 28, 1021.

27) LEAVITT, J., GOLDMAN, D., MERRIL, C.R. and KAKUNGA, T. (1982).
Changes in gene expression accompanying malignant transformation of human fibroblasts. Carcinogenesis, 3, 61.

28) O'FARRELL, P.H. (1975). High resolution two dimensional electrophoresis of proteins. J. Biol. Chem., 250, 4007.

29) SWITZER, R.C., MERRIL, C.R. and SHIFRIN, S. (1979). A highly
sensitive silver stain for detecting proteins and peptides
in polyacrylamide gels. Anal. Biochem., 98, 231.

30) MERRIL, C.R., SWITZER, R.C. and VAN KEUREN, M.L. (1979).
Trace polypeptides in cellular extracts and human body
fluids detected by two dimensional electrophoresis and a
highly sensitive silver stain. Proc. Natl. Acad. Sci. USA,
76, 4335.

31) MERRIL, C.R., GOLDMAN, D., SEDMAN, S. and EBERT, M.H. (1981).
Ultrasensitive stain for proteins in polyacrylamide gels
shows regional variation in cerebrospinal fluid proteins.
Science, 211, 1437.

32) MERRIL, C.R., GOLDMAN, D. and VAN KEUREN, M.L. (1982). Sim-
plified silver protein detection and image enhancement methods
in polyacrylamide gels. Electrophoresis, 3, 17.

33) VAN KEUREN, M.L., GOLDMAN, D. and MERRIL, C.R. (1981). Detection of radioactively labelled proteins is quenched by
silver staining methods: quenching is minimal for [^{14}C] and
partially reversible for [^{3}H] with a photochemical stain.
Anal. Biochem., 116, 248.

34) FRIDOVITCH, I. (1975). Superoxide dismutases. Ann. Rev. Biochem., 44, 147.

35) VAN KEUREN, M.L., GOLDMAN, D. and MERRIL, C.R. (1981). A
quantitative two dimensional electrophoretic survey of proteins affected by chromosome 21. In: Electrophoresis '81
(eds, R.C. Allen and P. Arnaud), p. 355, Walter de Gruyter &
Co., Berlin.

36) BROWN, W.T., DUTKOWSKI, R. and DARLINGTON, G.T. (1981). Loc-
    alization and quantitation of human superoxide dismutase
    using computerized 2-D gel analysis. Biochem. Biophys. Res.
    Comm., 102, 675.

37) BRAVO, R. and CELIS, J.E. (1982). Updated catalogue of HeLa
    cell proteins: percentages and characteristics of the major
    cell polypeptides labelled with a mixture of 16 ^{14}C -
    labelled amino acids. Clin. Chem., 28, 776.

38) McCONKEY, E.H. (1979). Double-label autoradiography for
    complex protein mixtures after gel electrophoresis. Anal.
    Biochem., 96, 39.

39) VAN KEUREN, M.L., GOLDMAN, D. and MERRIL, C.R. (1982). Pro-
    tein variations associated with Down's syndrome, chromosome
    21, and Alzheimer's disease. Ann. N.Y. Acad. Sci., in press.

40) GOLDMAN, D. and MERRIL, C.R., manuscript submitted.

41) WEIL, J. and EPSTEIN, C.J. (1979). The effect of trisomy 21
    on the patterns of polypeptide synthesis in human fibroblasts.
    Am. J. Hum. Genet., 7, 478.

42) KLOSE, J., ZEINDL, E. and SPERLING, K. (1982). Analysis of
    protein patterns in two dimensional gels of cultured human
    cells with trisomy 21. Clin. Chem. 28, 987.

43) ORGEL, L.E. (1963). The maintenance of the accuracy of protein
    synthesis and its relevance to ageing. Proc. Natl. Acad. Sci.
    USA, 49, 517.

44) STEINBERG, R.A., O'FARRELL, P.H., FRIEDRICH, U. and COFFINO,
    P. (1977). Mutations carrying charge alterations in regula-
    tory subunits of the cAMP-dependent protein kinase of
    cultured S49 lymphoma cells. Cell, 10, 381.

45) WILSON, D.L., HALL, M.E. and STONE, G.C. (1978). Test of
    some ageing hypotheses using two dimensional protein
    mapping. Gerontol., 24, 426.

46) ENGELHARDT, D.L., LEE, G.T.Y. and°MOLEY, S.F. (1978). Patterns
    of peptide synthesis in senescent and presenescent human fi-
    broblasts. J. Cell Phys., 98, 193.

47) HAYFLICK, L. (1981). Recent advanced in the cell biology of
    ageing. Mech. Ageing. Devel., 14, 59.

VARIATION IN EXPRESSION OF HUMAN MAJOR HISTOCOMPATIBILITY GENES IN MOUSE L CELLS AFTER DNA-MEDIATED GENE TRANSFER

J.A. Barbosa, M.E. Kamarck and F.H. Ruddle

Departments of Human Genetics and Biology, Yale University, New Haven Connecticut 06511 USA

1. INTRODUCTION

Somatic cell genetics and gene mapping have contributed significantly to our understanding of the organization and expression of the large and complex genomes of higher eukaryotes (1). The random segregation of donor chromosomes in somatic cell hybrids has allowed genes to be mapped to particular chromosomes by correlating their expression with the presence or absence of particular chromosomes. Recently, the availability of cloned DNA sequences and the techniques of Southern blot hybridization has circumvented the need for expression for successful gene mapping (2). From this work, a number of multigene families, i.e., groups of distinct but homologous gene sequences having similar function and structure, have been identified and chromosomally located. These gene families can either be closely linked or dispersed throughout the genome (3).

The human HLA-A,B,C loci represent a large multigene family which is localized within the major histocompatibility complex (MHC) on the short arm of chromosome 6 (4). These Class I antigens are of further interest because of the extensive polymorphism

379

maintained at each locus, and the high degrees of homology seen
among the different loci (5, 6). HLA-A,B,C antigens are integral
membrane glycoproteins of 44 Kd expressed on the surface of all
nucleated cells in non-covalent association with the 12 Kd water
soluble polypeptide β_2-microglobulin (β_2m)(7). These antigens are
responsible for the MHC restriction of T cell mediated cytolysis
of allogenic or virus-infected target cells (8, 9).

Although HLA-A,B,C antigens are present on most cells of the
adult, control of the level of expression on different cell types (10)
and of expression through development is not well understood (11).
Various gene transfer methods, such as somatic cell hybridization (12),
microcell-mediated gene transfer (13), chromosome mediated gene
transfer (14), DNA-mediated gene transfer (15), and microinjection (16,
17) have all been useful in the transfer of various portions of the
donor genome into a recipient cell background for the investigation
of expression of particular gene products. We have been examining the
quantitative expression of HLA and human β_2m in mouse-human somatic cell
hybrids (18). It has been previously shown that the proper expression
of HLA heavy chains in the membrane is dependent on its association with
heterologous mouse β_2m in somatic cell hybrids (19). Using specific
antibodies and the fluorescence-activated cell sorter (FACS), we
were able to fractionate and characterize separate hybrid subpopulations homogenously expressing HLA-A,B,C and human β_2M in
various combinations (i.e., HLA$^+$,β_2m$^+$; HLA$^+$,β_2m$^-$; HLA$^-$,β_2m$^+$;
HLA$^-$,β_2m$^-$). Genetic analysis showed that both gene products were
expressed constitutively in the fibroblast hybrids, i.e. selection
for or against surface antigen expression resulted in selection
for or against the chromosome carrying the gene coding for HLA
(chromosome 6) or β_2m (chromosome 15) (18). Our results demonstrated
that human and mouse histocompatibility heavy chains can associate
with either human or mouse β_2m for surface expression and further,

the presence of additional β_2m in the hybrid cell dramatically increases the surface expression of both human HLA-A,B,C and mouse H-2 antigens. These studies suggested that the transfer of human HLA genomic DNA into mouse recipient cells would result in the appropriate expression of the gene product.

A number of laboratories have isolated cDNA clones representing both the HLA and H-2 gene products (20-23). Southern blot hybridization of human or mouse genomic DNA with the cloned probe revealed 15-20 or more HLA or H-2 related gene sequences (24-28). Genomic sequences which hybridize to these Class I cDNA clones have recently been isolated from human and mouse libraries (29-32). These genomic sequences may represent functional histocompatibility loci, pseudogenes, closely related differentiation antigens, or other as yet serologically undefined gene products.

In work described here and elsewhere (33), we have utilized techniques of DNA-mediated gene transfer (DMGT), indirect immunofluorescence (IIF) and the FACS to screen a large number of human genomic clones for their ability to direct the synthesis of HLA-A,B,C antigens. Such analysis can be performed on transfected cell populations immediately following DMGT or on stable HAT^R populations generated by cotransfection with the Herpes simplex virus thymidine kinase (HSV-TK) gene. Using specific antibodies, we have identified HLA-A2 and HLA-B7 gene sequences.

In this paper, we will present data which suggest that the level of expression of human HLA seen in mouse L cells is roughly correlated to the number of intact HLA sequences present in the transfectants. Further, we will show that both the frequency of cotransfer of HLA and the level of expression in recipient cells was highly dependent on the HSV-TK plasmid used in cotransfections.

2.   GENERAL STRATEGY

The isolation of human genomic clones which hybridize with
the HLA-B7 cDNA probe PDP001 is described elsewhere (34). The
majority of clones chosen for analysis were isolated from a genomic
library produced by a partial EcoRI digest of DNA from the human
lymphoblastoid cell line JY (homozygous for HLA-A2 and HLA-B7).
These JY clones together form a representative selection of the
EcoRI bands seen in JY cellular DNA when it is hybridized with
the HLA-B7 cDNA (29).

A schematic representation of the general approach used to
identify and examine the expression of these human genomic
sequences is shown in Figure 1. Individual genomic clones (including
vector sequences) were cotransferred into the tk$^-$ mouse recipient
L cells using CaPO$_4$ DNA-mediated gene transfer (DMGT) with a
plasmid containing the HSV-TK gene in the presence or absence of
additional Ltk$^-$ carrier DNA. The cells remained in nonselective
medium until 60 hours after transfer at which time a portion was
taken for IIF and FACS analysis of HLA-A.B.C surface expression
using the mouse monoclonal antibody W6/32. This monoclonal
recognizes a public determinant shared by all known HLA-A,B,C
antigens (35). The remainder of cells were replated under HAT
selection and a HAT resistant mass population generated from
usually 100-1000 individual HAT^R clones could be harvested for
IIF and FACS analysis two weeks after gene transfer. Independent
HAT^R colonies were picked using cloning cylinders from sister
flasks which had not been harvested for IIF at previous time
points and were analyzed three or four weeks after transfer.

3.   ANALYSIS OF TRANSIENT HLA EXPRESSION

Previous work has shown that transferred genes are frequently
expressed shortly after transfection showing maximal frequencies of

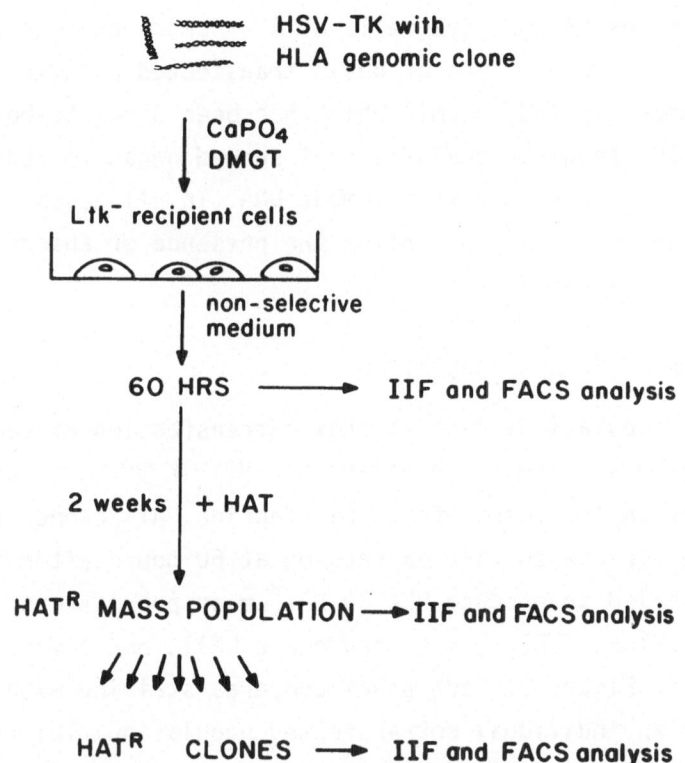

Fig. 1. *Approach used to screen human genomic clones for their ability to direct the synthesis of HLA-A,B,C surface antigens on mouse recipient Ltk⁻ cells following DNA-mediated gene transfer (DMGT).*

expression between 48-72 hours post-transfection (36-38). The 60
hour post-transfection expression of HLA-A,B,C antigens was
sensitively detected on Ltk⁻ recipient cells using IIF and the
FACS (33). Population of cells transfected with independent genomic
clones JY158 and JYB3.2 showed specific staining of a portion of
the cells. In a number of transfections with these two genomic
clones, frequencies of positive cells at 60 hours post-transfection
ranged from 1-25%. Populations of cells transfected with the remaining 21 clones, including LN11 which has been shown to be a
pseudogene by DNA sequence analysis (39) showed negative results
similar to control transfers with lambda DNA. In all cases, fluorescence microscopy was used to confirm the presence or absence of
any positively stained cells.

4. ANALYSIS OF HAT^R MASS POPULATIONS

HAT^R mass populations derived from cotransfection of each
genomic clone with a plasmid containing the HSV-TK gene were analyzed
for HLA expression two weeks after transfection. All clones shown
to be negative for HLA surface expression at 60 hours after transfection also failed to produce HLA on HAT^R mass populations. Results
obtained from JY158, JYB3.2, the pseudogene LN11, and lambda DNA
are presented in Figure 2. Each panel compares staining with W6/32
to control for an individual cotransfected population. Thirty-
three percent of JY158 and 38% of the JYB3.2 cotransfected HAT^R
mass population (Figure 2A, 2C) specifically stain for HLA-A,B,C
antigen expression. This contrasts with the total absence of

Fig. 2. *IIF and FACS analysis of HLA expression on HAT^R mass
populations derived from cotransfection of Ltk⁻ cells with cloned
human genomic DNA and HSV-TK.* Each panel presents a population of
transfected cells stained with monoclonal antibodies W6/32, W6/34,
and control myeloma supernatants (C). Photomultiplier voltages
were converted by logarithmic amplifier for display on the multi-
channel analyzer and cell number is presented on a linear scale.
Transfection with: a. JY 158; b. LN11; c. JY B3.2; d. λ DNA.

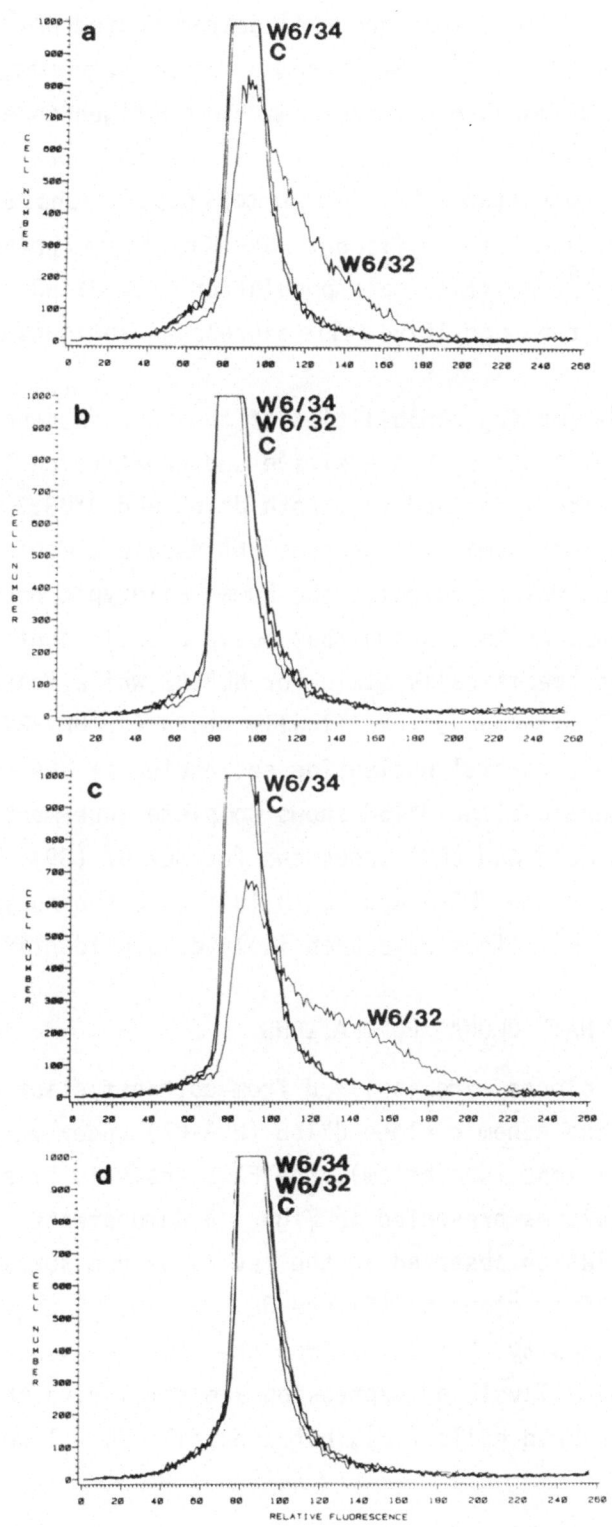

expression seen in the lambda and LN11 cotransfected HAT^R mass
population (Figure 2B and 2D). Staining with an inappropriate monoclonal
antibody W6/34 (35) which recognizes a surface antigen coded by a
gene on human chromosome 11 showed similar immunofluorescence as
control myeloma supernatants for all four mass populations. By varying
cotransfer conditions using different HSV-TK containing plasmids, we
have obtained HAT^R resistant mass populations for JY158 and JYB3.2
in which 85-95% of the tk positive cells express HLA antigens (see below).

We used the sorting capabilities of the FACS to generate
homogeneous populations of HLA positive and negative cells from
HAT^R mass populations derived from both JY158 and JYB3.2 cotransfections. Using both human multiparous antisera and mouse monoclonal antibodies which recognize the human allotypic determinants
HLA-B7 and HLA-A2, we have shown that mouse L cells containing
JYB3.2 sequences specifically stain for HLA-A2 while those
containing JYB158 sequences specifically stain for HLA-B7
determinants (33). Partial nucleotide sequencing of HLA sequences
contained in genomic clone JY158 shows complete agreement with
the known amino acid and cDNA sequences for HLA-B7 (39). Therefore,
the genes for both the HLA-A and HLA-B loci in the homozygous-A2,
-B7 JY lymphoid cell lines have been isolated and identified.

5. ANALYSIS OF HAT^R CLONAL POPULATIONS

Sixty HAT^R clones were isolated from cotransfection experiments
performed with the genomic clone JY158 (HLA-B7) under varying
cotransfer conditions (see below). The FACS analysis of six
representative clones presented in Figure 3 demonstrates the
significant variation observed in the levels of HLA surface expression. Cells from clone 6b (Figure 3e) had a 10 x higher
fluorescence intensity than cells from the low expressor 2b
(Figure 3b). These levels of expression are similar to those seen
in human-mouse hybrid cells (18), but are still 20 x lower than

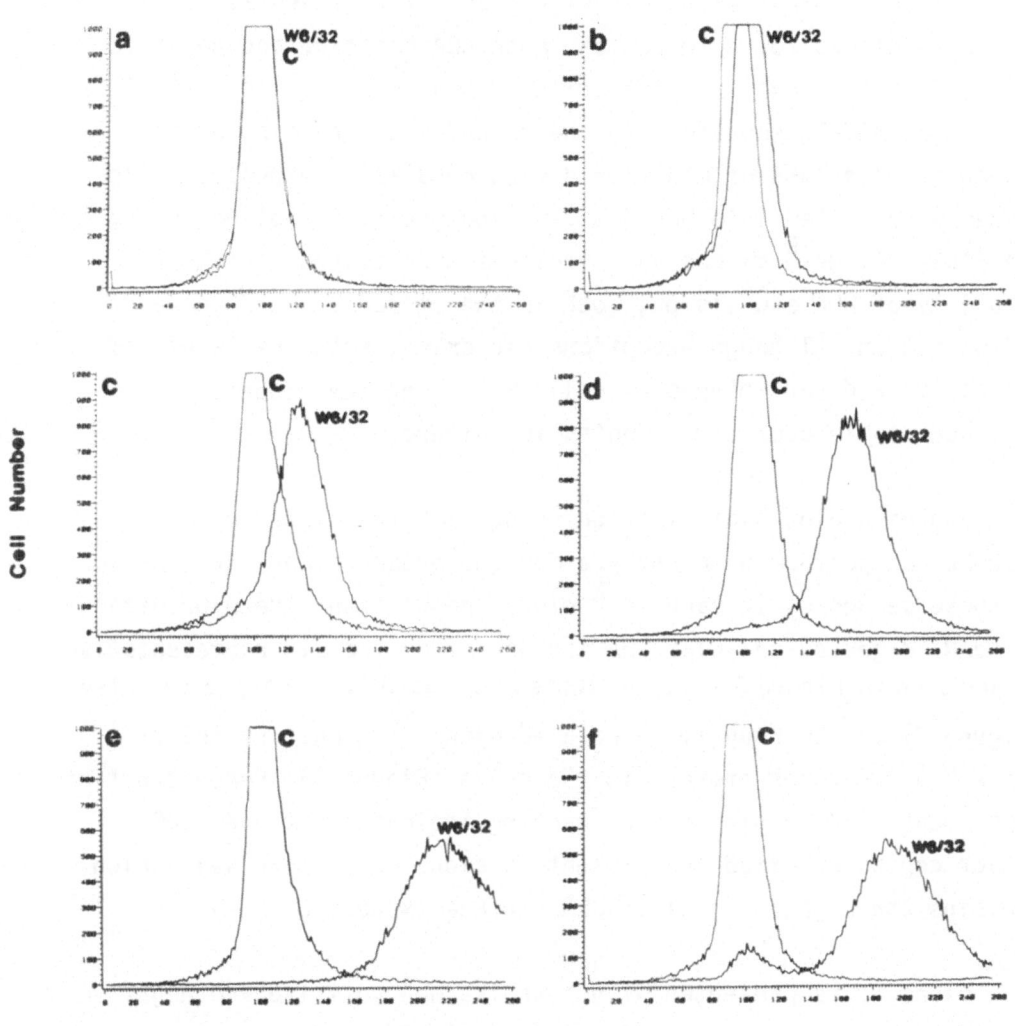

Relative Fluorescence

Fig. 3. *IIF and FACS analysis of HLA expression on HAT^R clones derived from cotransfection of Ltk$^-$ cells with HSV-TK and genomic clone JY158.* Each panel presents a different HAT^R clonal population stained with monoclonal antibody W6/32 compared to control staining with myeloma supernatant (C). FACS data are presented as in Figure 2.   a. HAT^R clone 1a;   b. HAT^R clone 2b;   c. HAT^R clone 19c; d.   HAT^R clone 10c; e.   HAT^R clone 6b; f. HAT^R clone 8c.

that seen on human HeLa or the parental lymphoblastoid cell line
JY (data not shown). Clone 8c (Figure 3f) is representative of a
number of clones found to contain both HLA positive and negative
cells. Such a situation might arise if HLA gene sequences stabilized
later than HSV-TK sequences in the transfected process, or if
sequences were lost or modified during population expansion. These
clones were sorted into HLA positive and negative sublines using
the FACS. All HAT^R clones and subclones were tested for their
stability of HLA expression, both in the presence and absence of
HAT selection. Although exceptions did exist, both the levels of
expression and percentages of cell expressing HLA sequences
remained stable over three months in culture.

Southern blot hybridization using nick-translated JY158 as
a probe was performed to analyze for the presence and organization
of these sequences in each HAT^R clonal population. The hybridization
of EcoRI digested cellular DNA isolated from the six representative
clones seen in Figure 3 are represented in Figure 4. HLA expression level
(Figure 3) can be roughly correlated with the number of intact
donor HLA sequences present in the cells (Figure 4). Reconstruction
experiments and densitometric scanning indicate that over 100
intact copies of JY158 are present in clone 6b (Figure 4e), which
contains the highest level of HLA surface expression.

Clone 8c is presented with both its HLA positive and negative
sorted populations (Figures 4f, 4f$^+$, 4f$^-$). The majority of
sequences which hybridize with the JY158 probe are absent from
the negative sorted population. However, Southern blot hybridization using nick-translated pORI1-TK revealed an identical EcoRI$_1$
digestion pattern, consistent with the integration of a single
pORI1-tk plasmid in all three populations (data not shown). Thus,
the majority of HLA sequences present in clone 8c may have existed
as a single unit which was lost en masse in the HLA negative subpopulation of cells during clonal expansion.

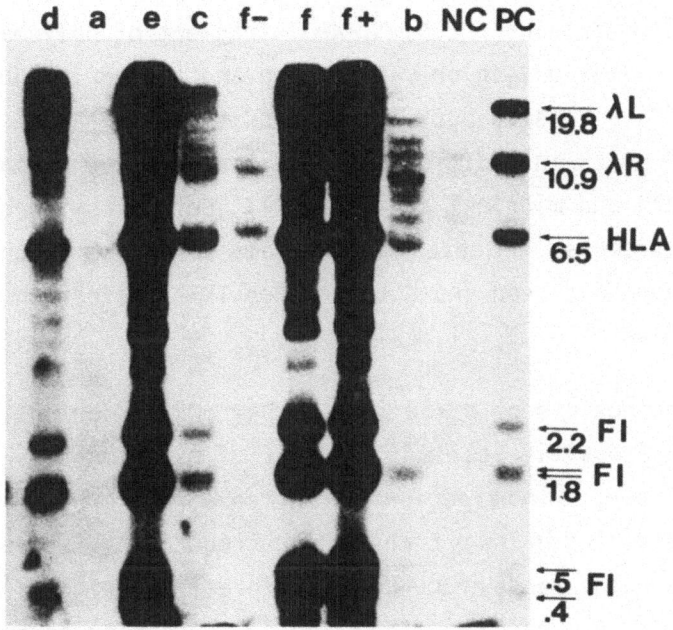

Fig. 4. *Southern blot hybridization analysis of the* HAT^R *clonal*
populations presented in Figure 3. Ten micrograms of DNA from each
clone was digested with *Eco*RI and the fragments were resolved on
a 0.8% agarose gel, transferred onto nitrocellulose filter paper
and hybridized with [^{32}P]-labelled JY158 (1-3 x 10^8 dpm/μg at 65°C)
(33). Lanes a through f correspond to panels a through f in
Figure 3. Lanes f$^-$ and f$^+$ represent the homogeneous HLA negative
and HLA positive subpopulations sorted from the original heterogeneous HAT^R clone 8c. Lane NC contains 10 μg of Ltk$^-$ DNA. Lane
PC contains 60 pg of Charon clone JY158 added to 10 μg Ltk$^-$ DNA.
The entire HLA coding sequence is present on the 6.5 kb *Eco*RI
fragment. F1 represents flanking material while λL and λR represent
the Charon 4a left and right arm, respectively.

6. VARIATION IN COTRANSFER FREQUENCY AND HLA EXPRESSION

Genomic clones JY158 and JYB3.2 were used in cotransfections of recipient Ltk⁻ cells with different plasmids containing the HSV-TK gene. Plasmid pTKX-1 contains the 3.5 kb *Bam*HI fragment of HSV-TK type 1 DNA in the unique *Bam*HI site of pBR322 (40). Plasmid pORT1-TK also contains the *Bam*HI-*Ava*I fragment of polyoma DNA which carries the viral origin or replication and the early and late gene promoters (41). Plasmid pgt1-TK contains the *Bam*HI-*Eco*RI fragment of polyoma DNA which in addition contains the complete coding sequence for the small and middle T antigens (41). Both plasmids have the viral late promoter oriented in the same direction as the HSV-TK promoter, and have been shown to enhance TK transfection frequencies of Ltk⁻ cells 10-100 x (41, 42).

Cotransfections of JY158 with either pORI1TK or pgt1TK resulted in 25-40% HLA positive cells in a number of experiments (Figure 5a). On the contrary, cotransfections performed with the plasmid pTKX-1, although resulting in lower transfection frequencies, showed cotransfer frequencies in the range of 85-95% in different experiments (Figure 5b). This was the case even though the ratio of cotransferred sequences (HLA; tk) in the donor DNA preparation was 5 x lower with pTKX-1 (50:1) than with either pORI1TK or pgt1TK (250:1). DMGT experiments performed in the absence of additional carrier Ltk⁻ DNA resulted in lower tk⁺ transfer frequencies (43), but showed similar cotransferred HLA expression

---

Fig. 5. *Cotransfer variation in HLA expression on HAT^R mass populations derived from cotransfections of Ltk⁻ cells with genomic clone JY158 and different plasmids containing the HSV-TK gene.* IIF and FACS data are presented as in Figure 2. Each panel presents a population of transfected cells stained with monoclonal W6/32 compared to control staining with W6/34 and/or myeloma supernatant (C). a. Cotransfection of 5 μg JY158 with 20 μg of pORI1TK in the presence of 20 μg Ltk-carrier DNA. b. Cotransfection of 5 μg JY158 with 100 μg pTKX1 in the presence of 20 μg of Ltk⁻ carrier DNA. Cotransfections of 5 μg JY158 with 20 μg of pgt1TK in the presence of 20 μg Ltk carrier DNA gave similar results as pORI1TK (data not shown).

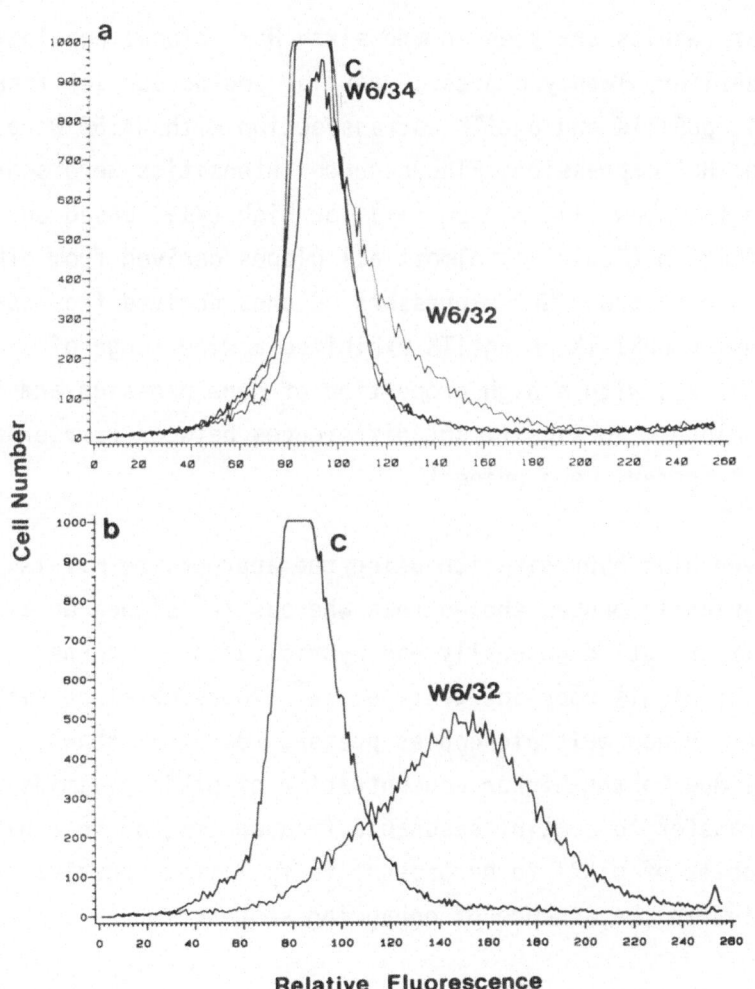

Thus, the presence of polyoma sequences which are known to enhance transcription and transfection frequencies do not give higher cotransfer frequencies of HLA expression (41, 42, 44). Plasmids containing both the HLA genomic sequences and the polyoma enhancing sequences are currently being constructed.

Similar results are seen in the sixty HAT^R clonal populations discussed earlier. Twenty clones of carrier and no carrier transfers from pTKX-1, pORI1TK and pgt1TK cotransfection with JY158 were analyzed for HLA expression. Fluorescence intensities were scored as negative (-), low (+), medium (++), or high (+++) based on FACS quantitation (Table 1). Almost all clones derived from pTKX-1 cotransfers were strong HLA expressors. Clones derived from cotransfers using pORI1TK or pgt1TK exhibited a wide range of expression levels with a high proportion of nonexpressing and low-expressing clones. No significant differences between carrier and no carrier transfers were evident.

Southern blot hybridization using the appropriate HSV-TK containing plasmid probe, showed that whereas tk$^+$ clones obtained with pORI1TK or pgt1TK generally had hybridization patterns indicative of single copy integration, tk$^+$ clones obtained with pTKX1 generally had multiple copies present (data not shown). This may be due to the higher concentration of pTKX1 plasmids used for transfer to obtain reasonable frequencies, or to a bias for more copies of pTKX1 to be present to facilitate survival in HAT medium in the absence of enhancing sequences.

7. DISCUSSION

The introduction and stable transformation of DNA sequences into mouse L cells by cotransfer with a viral TK gene is now a widely used procedure (45). We have used this technique and an analysis of the transient gene expression 60 hr after transfection

Table 1. *Comparison of HLA expression on HAT^R clones obtained with different HSV-TK plasmids*

	PLASMID		
	pORITK	pgt1TK	pTKX1
	1/10 HLA +++	1/10 HLA +++	9/10 HLA +++
			1/10 HLA +++; ++
Carrier	1/10 HLA +++; -	1/10 HLA +++; -	
	1/10 HLA ++; -		
	3/10 HLA +	4/10 HLA +	
	4/10 HLA -	3/10 HLA -	
		1/10 HLA +++	6/10 HLA +++
	1/10 HLA +++; -		1/10 HLA +++; -
		4/10 HLA ++	2/10 HLA ++
No carrier	3/10 HLA ++; -		
	2/10 HLA +	1/10 HLA +	
	4/10 HLA -	4/10 HLA -	1/10 HLA -

Sixty HAT^R clones were isolated from cotransfected experiments performed with the genomic clone JY158 and three different HSV-TK containing plasmids. Cotransfections in the presence of carrier Ltk$^-$ DNA (5 µg) were performed with 5 µg JY158 DNA and 1.5 ng pORITK, 1.5 ng pgt1TK, and 10 ng pTKX1, respectively. Cotransfections in the absence of carrier Ltk$^-$ DNA were performed with 1.5 µg JY158 DNA and 10 ng pORITK, 10 ng pgt1TK and 50 ng pTKX1, respectively. All plasmids have the TK containing 3.5 kb *Bam*HI fragment of HSV DNA in the plasmid pBR322. pORITK contains the *Bam*HI-*AVA*I fragment of polyoma virus DNA containing the origin of replication. pgt1TK contains the *Bam*HI-*Eco*RI fragment of polyoma virus DNA which codes for the origin of replication, the small T and middle T antigens. Indirect immunofluorescence was performed with the monoclonal antibody, W6/32 and fluorescence intensity, quantitated by the signal preamplifier of the FACS, was scored from - to +++ (see Figure 3). Both subpopulations within a heterogeneous clone have been scored.

to screen a large number of human genomic clones for their ability
to direct the expression of HLA surface antigens on mouse L cells
recipients. Transient expression analysis provides a simple and
rapid method to identify functional members of this large multigene
family. In addition, using the FACS sorting capabilities, we have
been able to apply direct selection on cells expressing HLA at
this time point, to enrich for and finally develop stable homogeneous populations of HLA expressing mouse cells, independent
of cotransfer with HSV-TK and HAT selection (data not shown). This
direct selection will allow for the introduction of HLA into cell
lines which do not have appropriate biochemical deficiencies.

     The production of stable mouse cell lines containing individual
HLA genes may facilitate investigation of the role of these genes
in complex immune responses. By cotransferring the human genomic
clones, JYB3.2 (HLA-A2) and JY158 (HLA-B7) with the HSV-TK gene,
it was possible to obtain stable HAT^R populations, a portion of
which also expressed HLA surface antigens. By utilizing FACS
selection, stable homogeneous populations of mouse cells expressing
these gene products were produced. These lines could serve as
standards for testing human tissue typing reagents and may be
useful as immunogens in the production of anti-allotype antisera in mice.

     HSV-TK containing plasmids with particular segments of DNA
from SV-40 (46, 47), polyoma (41, 42, 44) and certain retroviruses
(48) have been shown to increase tk transfection frequencies 10
to 100 fold. We have observed striking differences in both cotransfer frequencies and the levels of HLA expression in HAT^R
populations derived from cotransfer of the JY158 HLA genomic
sequence with three different HSV-TK containing plasmids (Figure 5,
Table 1). Although transfection with plasmid pTKX1 gave low tk transfer frequencies, 85-95% of HAT^R cells expressed high levels of HLA.

Transfection with plasmids pORI1TK and pgt1TK which contained the polyoma origin region, showed a 10-50 x higher frequency of HAT^R colony formation, but resulted in lower cotransfer frequencies and lower HLA expression levels. For efficient survival in HAT, cells cotransfected with pTKX1 may be selected which contain a larger amount of DNA in transgenomes containing multiple copies of HSV-TK (15). When enhancing sequences are present, higher transcription of the TK gene may result in the frequent survival of cells containing only a single copy of HSV-TK, as well as fewer cotransferred sequences. Southern blot analyses of cotransfected clonal lines with both HLA and TK specific probes support this hypothesis. Plasmids containing both the HLA genomic sequences and the polyoma enhancing sequences are currently being constructed.

In sixty individual HAT^R clonal lines, the level of cotransferred HLA expression was seen to correlate very well with the number of intact HLA gene copies. The levels of HLA expression in transfectants was similar to that seen in mouse x human somatic cell hybrids (18). But the amount of HLA surface antigen expression seen in transfected clone 6B, which contained greater than 100 copies of JY158, was still 20 x lower than that seen on the parental lymphoblastoid cell line JY, or transformed epithelial line HeLa, and approximately 10 x lower than the total H-2 antigen expression in the transformants. In contrast to our data, mouse H-2 antigens transfected into Ltk$^-$ cells have been shown to be expressed at levels comparable to the endogeneous H-2 (31, 49). The nature of this specific control of histocompatibility antigen expression in the mouse background is not yet understood. To this end, we are introducing these HLA genomic sequences into human tk$^-$ fibroblasts which are negative for HLA-A2 or HLA-B7 antigens.

In summary, the transfer of human genomic sequences directing the expression of HLA-A2 and HLA-B7 antigens into mouse L cell recipient cells results in human surface antigen expression on the mouse cell surface. The levels of this expression correlates well with the number of intact HLA sequences present in the trans- formants, but are still lower than those seen in the parental donor cells. This work further suggests that the microinjection of genomic HLA clones in the production of transgenic mice expressing human histocompatibility antigens holds promise for the analysis of complex immunological and developmental events mediated at the cell surface (50).

8. ACKNOWLEDGEMENTS

We thank Gloria Schoolfield for her technical assistance, Suzy Pafka for her photographic expertise, and Marie Siniscalchi for manuscript preparation. Dr. P.A. Biro and Dr. S.M. Weissmann provided the genomic clones and helpful discussion. This investigation was supported by the National Institute of Health grant GM09966 to F.H.R. J.A.B. was supported by NIH Training Grant 5T32-HD07149-03. M.E.K. was supported by Postdoctoral Fellowships from the Leukemia Society of America and by NRSA Fellowship A106499-01.

9. REFERENCES

1) RUDDLE, F.H. (1981). A new era in mammalian gene mapping: somatic cell genetics and recombinant DNA methodologies. Nature, 294, 115-120.

2) D'EUSTACHIO, P., PRAVTCHEVA, D., MARCU, K. and RUDDLE, F.H. (1980). Chromosomal localization of the structural gene cluster encoding murine immunoglobin heavy chains. J. Exp. Med., 151, 1545-1550.

3) D'EUSTACHIO, P. and RUDDLE, F.H. (1982). Mammalian gene families. Science, in press.

4) FRANCKE, U. and PELLEGRINO, M.A. (1977). Assigment of the
    major histocompatability complex to a region of the short arm
    of human chromosome 6. Proc. Natl. Acad. Sci. USA, 74, 1147.

5) DAUSSET, J. (1981). The major histocompatibility complex in
    man. Science, 213, 1469.

6) BACH, F.H. and VAN ROOD, P. (1976). The major histo-
    compatibility complex - genetics and biology. N. Eng. J. Med.,
    295, 806.

7) PLOEGH, H.L., ORR, H.T. and STROMINGER, J.L. (1981). Major
    histocompatibility antigens: The human (HLA-A,B,C) and murine
    (H-2K, H-2D) class I molecules. Review. Cell, 24, 287.

8) ZINKERNAGEL, R.M. and DOHERTY, P.C. (1980). MHC-restricted
    cytotoxic cells; studies on the biological role of polymorphic
    major transplantation antigens determining T cell restriction
    specificity, function and responsiveness. Adv. Immunol., 27,
    51.

9) McMICHAEL, A.J., TING, A., ZWEERINK, H.J. and ASKONAS, B.A.
    (1977). HLA-restriction of cell mediated lysis of influenza
    virus-infected human cells. Nature, London, 270, 524.

10) WILLIAMS, K.A., HART, D.N.J., FABRE, J.W. and MORRIS, P.J.
    (1980). Distribution and quantitation of HLA-A,B,C and DR
    (Ia) antigens on human kidney and other tissues. Transplan-
    tation, 29, 274.

11) GOODFELLOW, P.N., BARNSTABLE, C.J., BODMER, W.F., SNARY, D.
    and CRUMPTON, M.J. (1976). Expression of HLA system anti-
    gens on placenta. Transplantation, 22, 595.

12) RUDDLE, F.H. (1980). Somatic cell genetics; past, present,
    and future. In: Genes, Chromosomes and Neoplasia. (eds.
    F.E. Arrighi, P.N. Rao, and E. Stubblefield), p. 7, Raven
    Press, New York.

13) FOURNIER, R.E.K. and RUDDLE, F.H. (1977). Microcell mediated
    transfer of murine chromosomes into mouse, Chinese hamster and
    human somatic cells. Proc. Natl. Acad. Sci. USA, 74, 319.

14) KLOBUTCHER, L.A. and RUDDLE, F.H. (1981). Chromosome mediated gene transfer. Ann. Rev. Biochem., 50, 533.

15) SCANGOS, G. and RUDDLE, F.H. (1981). Mechanisms and applications of DNA mediated gene transfer in mammalian cells - a review. Gene, 14, 1.

16) GRAESSMANN, A., GRAESSMANN, M. and MUELLER, C. (1980). Micro-injection of early SV-40 DNA fragments and T antigens. Methods in Enzymology (Acad. Press N.Y.), 65, 816.

17) GORDON, J.W., SCANGOS, G.A., PLOTKIN, D.J., BARBOSA, J.A. and RUDDLE, F.H. (1980). Genetic transformation of mouse embryos by microinjection of purified DNA. Proc. Natl. Acad. Sci. USA, 77, 7380.

18) KAMARCK, M.E., BARBOSA, J.A. and RUDDLE, F.H. (1982). Somatic cell genetic analysis of HLA-A,B,C and human β_2 microglobulin expression. Somat. Cell Genet., 8, 305.

19) ARCE-GOMEZ, B., JONES, E.A., BARNSTABLE, C.J., SOLOMON, E. and BODMER, W.F. (1978). The genetic control of HLA,B,C antigens in somatic cell hybrids: Requirement for β_2 micro-globulin. Tissue Ant., 11, 96.

20) SOOD, A.K., PEREIRA; D. and WEISSMAN, S.M. (1981). Isolation and partial nucleotide sequence of a cDNA clone for human histocompatibility antigen HLA-B by use of an oligodeoxy-nucleotide primer. Proc. Natl. Acad. Sci. USA, 78, 616.

21) PLOEGH, H.L., ORR, H.T.and STROMINGER, J.L. (1980). Molecular cloning of a human histocompatibility antigen cDNA fragment. Proc. Natl. Acad. Sci. USA, 77, 6081.

22) KVIST, S., BREGEGERE, F., RASK, L., CAMI, B., GAROFF, H., DANIEL, F., WIMAN, K., LARHMNAR, D., ABASTADO, J.P., GACHELIN, G., PETERSON, P.A., DOBBERSTEIN, B. and KOURISKLY, P. (1981). cDNA clone coding for part of a mouse H-2^d major histocompatibility antigen. Proc. Natl. Acad. Sci. USA, 78, 2772.

23) STEINMETZ, M., FRELINGER, J.G., FISHER, D., HUNKPILLER, T., PEREIRA, D., WEISSMAN, S.M., UEHARA, H., NATHENSON, S. and HOOD, L. (1981). cDNA clones encoding mouse transplantation antigens: Homology to immunoglobin genes. Cell, 24, 125.

24) BIRO, P.A., PEREIRA, S.. SOOD, A.K., DeMARTINVILLE, B., FRANCKE, U. and WEISSMAN, S.M. (1981). The structure of the human major histocompatibility locus. In: ICNUCLA Symposium on Molecular and Cell Biology. (eds. E. Janeway, C.A. Seteare, and H. Wigzell), vol XX. Immunoglobulin Idiotype, Acad. Press, New York, p. 315.

25) ORR, H.T., BACH, F.H., PLOEGH, H.L., STROMINGER, J.L., KAVATHAS, P. and DEMARS, R. (1982). Use of HLA loss mutants to analyse the structure of the human major histocompatibility complex. Nature, London, 296, 454.

26) STEINMETZ, M., MOORE, K.W., FRELINGER, J.G., SHER, I.T., SHEA, J.F., BOYSE, E.A. and HOOD, L. (1981). A pseudogene homologous to mouse transplantation antigens is encoded by eight exons that correlated with protein. Cell, 25, 683.

27) CAMI, B.F., BREGEGERE, F., ABASTADO, T.P. and KOURILSKY, P. (1981). Multiple sequences related to classical histocompatibility antigens in the mouse genome.. Nature, 291, 673.

28) MARGULIES, D. H., EVANS, G.A., FLAHERTY, L. and SEIDMAN, J.G. (1982). H-2-like genes in the TLa region of mouse chromosome 17. Nature, London, 295, 168.

29) BIRO, P.A., PAN, J., DAS, H., SOOD, A.K., WEISSMANN, S.M., BARBOSA, J.A., KAMARCK, M.E. and RUDDLE, F.H. (1982). Part II. Analysis of genes of the human major histocompatibility complex. In: Annals of the New York Acad. Sci., in press.

30) JORDON, B.R., BREGERERE, F. and KOURILSKY, P. (1981). Human HLA gene segment isolated by hybridization with mouse H-2 cDNA probe. Nature, 296, 521.

31) EVANS, L.A., MARGULIES, D.H., CAMERINI-OTERO, D., OZATA, K. and SEIDMAN, J.G. (1982). Structure and expression of a mouse

major histocompatibility antigen H-2L^d. Proc. Natl. Acad. Sci.
USA, <u>79</u>, 1994.

32) STEINMETZ, M., WINOTO, A., MINARD, K. and HOOD, L. (1982).
Clusters of genes encoding mouse transplantation antigens.
Cell, 28, 489.

33) BARBOSA, J.A., KAMARCK, M.E., BIRO, P.A., WEISSMAN, S.M.,
and RUDDLE, F.H. (1982). Identification of human genomic
clones coding the major histocompatibility antigens HLA-A2
and HLA-B7 by DNA-mediated gene transfer. <u>Proc. Natl. Acad.</u>
<u>Sci. USA</u>, <u>79</u>, 6327.

34) BIRO, P.A., REDDY, V.B., SOOD, A., PEREIRA, D. and WEISSMAN,
S.M. (1981). Isolation and analysis of human major histocompatibility complex genes in recombinant DNA. Proceedings
of the Third Cleveland Symposium on Macromolecules (ed. A.G.
Walton), Elsevier Scientific Publishing Co., Amsterdam, p.
41.

35) BARNSTABLE, C.J., BODMER, W.F., BROWN, G., GALFRE, G.,
MILSTEIN, C., WILLIAMS, A.F. and ZIEGLER, A. (1978).
Production of monoclonal antibodies to group A erythrocytes,
HLA and their human cell surface antigens - new tools for
genetic analysis. <u>Cell</u>, <u>14</u>, 9.

36) MILMAN, G. and HERZBERG, M. (1981). Efficient DNA transfection
and rapid assay for thymidine kinase activity and viral antigenic determinants. <u>Somatic Cell Genet.</u>, <u>7</u>, 161.

37) CHANG, L.-J.A., GAMBLE, C.L., IZAQUIRRE, C.A.,MIDEN, M.D.,
MAK, T.W. and McCULLOCK, E.A. (1982). Detection of genes
coding for human differentiation markers by their transient
expression after DNA transfer. <u>Proc. Natl. Acad. Sci. USA</u>,
<u>79</u>, 146.

38) BARBOSA, J.A. and RUDDLE, F.H. (unpublished).

39) BIRO, P.A. and WEISSMAN, S.M. (unpublished).

40) ENQUIST, L.W., VAN DE WOUDE, G.F., WAGNER, M., SMILEY, J.R.,
and SUMMERS, W.C. (1979). Construction and characterization

of a recombinant plasmid encoding the gene for the thymidine kinase of Herpes simplex type I virus. Gene, 7, 335.

41) YAMAIZUMI, M., HORWICH, A.L. and RUDDLE; F.H. (1982). Expression and stabilization of microinjected plasmids containing HSV-TK and polyoma viral DNA in mouse cells, submitted for publication.

42) BARBOSA, J.A.and RUDDLE, F.H. (unpublished).

43) HUTTNER, K.M., BARBOSA, J.A., SCANGOS, G,A., PRAVTCHEVA, D. and RUDDLE, F.H. (1981). DNA-mediated gene transfer without carrier DNA. J. Cell Biol., 91, 153.

44) DE VILLIERS, J.and SCHAFFNER, W. (1981). A small segment of polyoma virus DNA enhances the expression of a cloned β-globin gene over a distance of 1400 base pairs. Nucl. Acid Res., 9, 6251.

45) WIGLER, M., SWEET, R., SIM, G.K., WOLD, B., PELLICER, A., LACY, E., MANIATES, T., SILVERSTEIN, S.and AXEL, R. (1979). Transformation of mammalian cells with genes from procaryotes and eucaryotes. Cell, 16, 777.

46) CAPPECCHI, M.R. (1980). High efficiency transformation by direct microinjection of DNA into cultured mammalian cells. Cell, 22, 479.

47) BANERJI, J., RUSCONI, S. and SCHAFFNER, W. (1981). Expression of β globin gene is enhanced by remote SV-40 DNA sequences. Cell, 27, 299.

48) LEVINSON, B., KHOURY, G., VAN DE WOUDE, G. and GRUSS, P. (1982). Activation of SV-40 genome by 72 base pair tanden repeat of Maloney sarcoma virus. Nature, 295, 568.

49) GOODENOW, R.S., McMILLAN, M., ORN, A. and NICOLSON, M. (1982). Identification of a Balb/c H-2L^d gene by DNA-mediated gene transfer. Science, 215, 677.

50) GORDON, J.W.and RUDDLE, F.H. (1981). Integration and stable germ line transmission of genes injected into mouse pronuclei. Science, 214, 1244.

EXPRESSION OF DEVELOPMENT-PHASE SPECIFIC ALKALINE PHOSPHATASE ISOENZYMES IN CULTURED CANCER CELLS

W.H. Fishman

Cancer Research Center
La Jolla Cancer Research Foundation
La Jolla, California

1. INTRODUCTION

The history of the demonstration of development-phase specific alkaline phosphatase (AP) isoenzymes goes back to 1976 when Lillian Fishman and her colleagues (1) reported their observations on placental AP. They found that the microvilli of placenta earlier than 10 weeks produced a different alkaline phosphatase than the one which characterized the second and third trimester placentas. The early placental alkaline phosphatase is heat-sensitive, inhibited by L-homoarginine, not L-phenylalanine, and migrates on electrophoresis faster than the term isoenzyme. The latter is very heat stable, inhibited by L-phenylalanine but not L-homoarginine, and is represented by a myriad of electrophoretic variants and distinctive immunologic determinants. Interestingly, the two isoenzymes overlapped in their expression between 10 and 12 weeks of pregnancy and both have the very same syncitiotrophoblast location.

The discovery of the early placental isoenzyme provided a plausible interpretation for the presence in choriocarcinoma tissue of a heat-sensitive alkaline phosphatase - presumably the first trimester isoenzyme. Choriocarcinoma cell lines may express either one of these isoenzymes or a combination of both.

In the meanwhile, the Regan isoenzyme, a term placental type of alkaline phosphatase was being investigated extensively in ovarian cancer (2) and other malignancies. Frequently, tumours were found to exhibit a heat-sensitive and L-homoarginine-sensitive isoenzyme with immunologic determinants of liver and bone alkaline phosphatase. This moiety was referred to as non-Regan alkaline phosphatase.

A great advantage to experimental work on the expression of these genes was achieved with the cloning of two sublines of HeLa, one (TCRC-1) expressing Regan isoenzyme and the other (TCRC-2) which produces non-Regan isoenzyme (3). L. Fishman and her colleagues demonstrated that HeLa TCRC-2 was indistinguishable from early placental alkaline phosphatase in a number of immunologic tests (1). These facts made plausible the view that cancer cells are capable of expressing development-phase specific genes to varying degrees.

Another oncodevelopmental isoenzyme prepared from rodent or human intestine is the product of a third gene and is reexpressed in a number of human tumours (4).

From the evolutionary point of view (Figure 1), the ancestral gene is the tissue unspecific gene being the most widespread in the animal kingdom, and being expressed in all placentas except for those of a few of the higher primates and man. The intestinal gene which appears in all vertebrate mammalian species is believed to arise from the ancestral gene by gene duplication. The human type placental gene produces a family of alleles of a number and variety not hitherto encountered with other isoenzyme families.

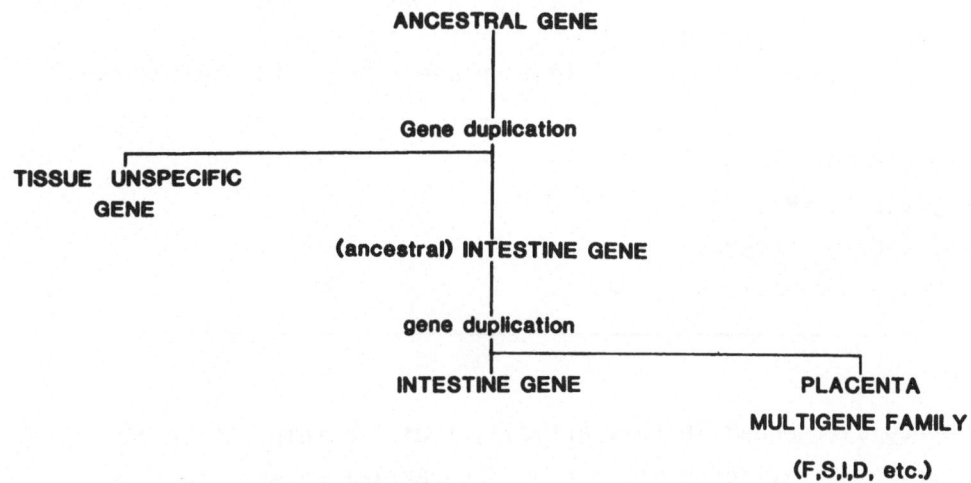

Fig. 1. *Hypothetical genealogy of alkaline phosphatase isoenzymes genes.*

Accordingly, if "ontogeny repeats phylogeny", it is understandable that the ancestral tissue unspecific gene is expressed in the early placenta and is replaced by the human term placental gene (see Stigbrand *et al.* (5)) after the first trimester.

Modulation of placental gene expression has been particularly useful in attempting to understand the control of enzyme biosynthesis. Thus, prednisolone, butyrate and NaCl added to the media of different lines of cancer cells induce the formation, to varying degrees, of placental isoenzymes (Table 1).

Currently we have investigated four questions: 1) Where does the accummulation of induced enzyme occur in the cell? 2) What are the dynamics of its biosynthesis, transport and membrane incorporation? 3) What can the use of monoclonal antibodies tell us about the domains of the molecule coded by the relevant gene in placental and in neoplastic cells? 4) Under what circumstances can one expect to observe concordant expression of first trimester onco-

Table 1. *Degree of induction of placental type alkaline phospha-*
tase in HeLa TCRC-1 cells produced by agents added to the medium

Modulating agent	Fold increase - (alkaline phosphatase)
Prednisolone (2 μg/ml)	2-4
Butyrate (2 mM)	2-4
NaCl (50 mM)/butyrate	15-20
Prednisolone/NaCl/butyrate	30

trophoblast genes? In this article, I will attempt to present
experimental evidence which provides partial answers to these
questions.

2. WHERE DOES THE INDUCED ENZYME ACCUMULATE IN THE CELL?

Until 1981, the accepted location of alkaline phosphatase in
cancer cells was the cell-surface membrane. Earlier, additional
sites in the endoplasmic reticulum, Golgi apparatus and mitochondria were described in ovarian cancer cells (2) and in cultured
bladder cancer cell lines.

Tokumitsu and his co-workers (6,7) have published evidence
that intracytoplasmic alkaline phosphatase is cryptic and can be
visualized after the fixed cells have been exposed to dilute
saponin. In preparations of a variety of cultured cancer cells
(HeLa, BeWo, KMK-2, etc.), using both immunohistochemical and
cytochemical methods, the intracytoplasmic sites of endoplasmic
reticulum, mitochondria and Golgi apparatus were regularly observed
It was possible to ascertain the sites of accumulation of enzyme
induced by the addition of modulating substances to the medium.
These were chiefly the cell surface membrane and Golgi apparatus.

3. WHAT ARE THE DYNAMICS OF THE BIOSYNTHESIS, TRANSPORT AND
   MEMBRANE INSERTION OF INDUCED TERM PLACENTAL ALKALINE
   PHOSPHATASE?

In approaching this question, the proteolytic enzyme, bromelain, has proved most useful. When cancer cells which exhibit term
placental or intestinal type isoenzymes on their cell surfaces are
treated with bromelain, the enzyme is cleaved from the membrane
completely (8). This is not the case with tissue unspecific alkaline phosphatase.

If one introduces [^{35}S]-methionine into the medium of cells
whose alkaline phosphatase is being induced and harvests them at
regular intervals, one can measure the time required for the newly
synthesized enzyme to travel from the site of its origin, the
ribosomes of endoplasmic reticulum, to its destination in the cell
surface membrane. According to Hanford and Fishman, this transit
time is approximately 55 minutes (9).

As expected, the synthesis can be prevented by adding cyclohexamide or puromycin to the medium. Under these circumstances,
the enzyme cannot be seen in the endoplasmic reticulum 10 minutes
later. Inhibitors of protein transport such as monensin and of
glycosylation, such as tunicamycin, produce the expected effects.
Accordingly, this enzyme transport system recommends itself in
studies of the dynamics of membrane components with regard to their
biosynthesis, transport via the Golgi and insertion in the plasma
membrane.

4. MONOCLONAL ANTIBODY STUDIES

A comparison of the products of the term placental alkaline
phosphatase gene and the gene in neoplastic cells which produces
a similar isoenzyme is complicated. This is because the placental

alkaline phosphatase is multiallelic, producing six common phenotypes and a myriad of rarer phenotypic forms. Similarly, the neoplastic placental-type alkaline phosphatases exist in electrophoretic forms which rarely match exactly their normal counterparts.

The monoclonal antibody technology offers a route to a
comparison of the major domains of the gene products of interest
The following is a progress report of our studies in this field.

Table 2. *Monoclonal antibodies to alkaline phosphatase isoenzymes
and their specificities*

Designation	Immunogen	Placenta F I S D	Intestine	Liver	References
F11	SS-PLAP	- + + +	-	-	(10)
A9	pooled PLAP	+ ? + ?	+	-	(11)
H5	HeLa TCRC-1 cells	+ ? + ?	-	-	(12)

Table 3. *Binding of anti-alkaline phosphatase monoclonal antibodies to cancer cells*

Cell[a]	AP phenotype	Monoclonal antibody binding		
		F11	A9	H5
HeLa TCRC-1	placental	+	+	+
SKG IIIa	placental/liver	+	+	+
SKG IIIb	liver	-	-	-
SNG	intestinal	-	-	-

[a]SNG is an endometrial cancer: all others listed are cervical
cancers.

An enzyme antigen immunoassay (EAIA) has been devised (11)
which takes advantage of the catalytic site activity of the antigen as the signal molecule. The sequence illustrated in Figure 2
permits a convenient identification of cell clones which produce
the desired antibodies. With the exception of the F11 antibody
which was identified with the ELISA technology, the other mono-

clonal antibodies (A9, H5 and others) are being recognized by
EAIA and ELISA (Figure 2).

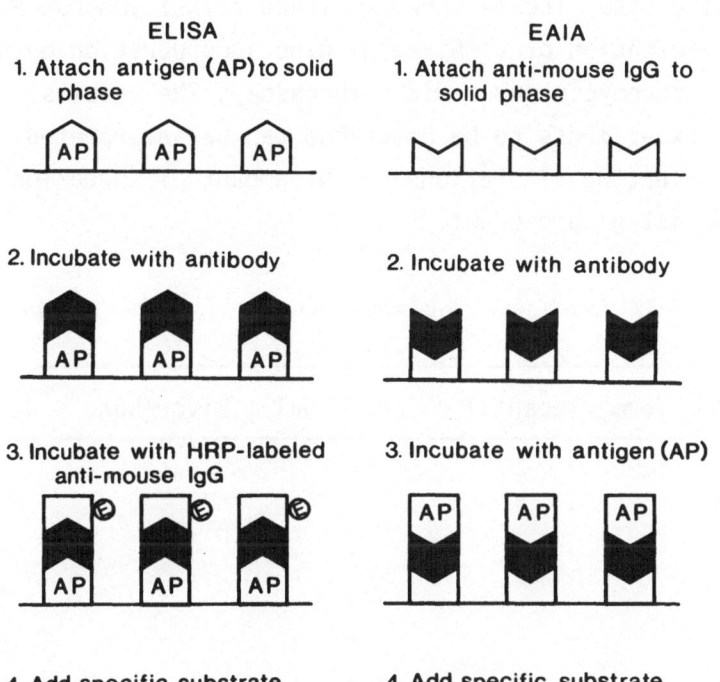

ELISA

1. Attach antigen (AP) to solid phase

2. Incubate with antibody

3. Incubate with HRP-labeled anti-mouse IgG

4. Add specific substrate

EAIA

1. Attach anti-mouse IgG to solid phase

2. Incubate with antibody

3. Incubate with antigen (AP)

4. Add specific substrate

Fig. 2. *Comparison of ELISA and EAIA.*

These antibodies show interesting specificities (Tables 2
and 3). The F11 antibody recognizes the phenotypes I, S and D but
not F, while A9, and H5 recognize the F phenotype. The intestinal
domain is shared with term placental alkaline phosphatase according
to the A9 antibody. All three antibodies appear to react equally
with HeLa TCRC-1 and pooled placental alkaline phosphatase.

5.   MODULATION OF TROPHOBLAST GENE EXPRESSION

Strong evidence has now accumulated to support the statement that the placental-type alkaline phosphatase which is produced by cancer cells reflects specific enzyme protein synthesis rather than a conformational change favouring greater catalytic efficiency at the active site. The evidence included rocket immunoelectrophoresis, prevention of [^{35}S]-methionine incorporation by cyclohexamide or puromycin and radio-immunoassay. The results of modulation experiments to be described can be interpreted, therefore, as reflecting alterations in the amount of induction of an oncotrophoblast gene product.

Table 4. *Cell lines synthesizing single alkaline phosphatase isoenzymes*

Cell line[a]	Term placental	Intestinal	Liver-bone	Reference
HeLa TCRC-1	+	−	−	(3)
HeLa S3-10	+	−	−	(13)
DoT	+	−	−	(14)
HeLa D98AH2	−	+	−	(15)
HeLa S3-5	−	+	−	(16)
HeLa TCRC-2	−	−	+	(3)
KMK-2	−	−	+	(17)

[a]KMK-2 is a gastric cancer cell line. All others listed are cervical cancers.

The induction conditions in use in our laboratory are listed in Table 1 and the cell lines which synthesize single alkaline phosphatase isoenzymes in Table 4. Attention should also be given, (15), to the HeP-2 and FL-amnion cell lines whose uninduced phenotype is intestinal alkaline phosphatase but which is essentially replaced by term placental alkaline phosphatase on induction.

6.   COORDINATE EXPRESSION OF FIRST TRIMESTER GENES?

The production of early and term placental alkaline phosphatase, human chorionic gonadotropin β-subunit (β-hCG) and pregnancy-specific $β_1$ glycoprotein (SP_1) was demonstrated in a newly established cell line SKG-IIIa (18). While treatment of these cells with sodium butyrate caused an increase of all these oncodevelopmental proteins, prednisolone diminished their production but induced term placental alkaline phosphatase.

Although there may be other possible interpretations of these observations, it would appear that the first trimester genes were induced coordinately. If so, it may be explained by the fact that SKG IIIa is a freshly established cell line in contrast to HeLa cell lines, none of which in our experience exhibit this degree of coordinate expression. It is reasonable to expect also that the SKG IIIa data may reflect more closely the phenotypic characteristics of its epidermoid carcinoma origin.

7.   CONCLUSIONS

Three genes code for the alkaline phosphatase isoenzymes in mammalian tissues: the ancestral tissue unspecific gene, the later intestinal gene and the most recent in evolution, the term placental gene. Gene products closely resembling these are expressed in human cancer and in human cancer cell lines. Isoenzyme monophenotypic cell lines have been identified. These have provided valuable model systems for studying modulation of oncotrophoblast gene expression. Their examination has led to the estimation of the transit time of term placental alkaline phosphatase from its ribosomal site of synthesis to its insertion in its terminal plasma membrane location. Also, these model cell systems have aided in the appreciation of the "cryptic" nature of cytoplasmic alkaline phosphatase. The use of monoclonal antibodies should permit an exploration of the structural and antigenic domains of placental and neoplastic alkaline phosphatases. Such information is basic to an understanding of genetic regulation in the case of oncotrophoblast genes discussed in this article.

8. ACKNOWLEDGEMENTS

This work was supported by grants (CA-21967, PO1-CA 28896, CA-31378 and P30-CA 30199) of the National Cancer Institute, National Institutes of Health. The cooperation is acknowledged of my colleagues L. Fishman, W. Hanford, S.-I Tokumitsu, R. Jemmerson and S. Nozawa.

9. REFERENCES

1) FISHMAN, L., MIYAYAMA, H., DRISCOLL, S.G. and FISHMAN, W.H. (1976). Developmental phase-specific alkaline phosphatase isoenzymes of human placental and their occurrence in human cancer. Cancer Res., 36, 2268.

2) SASAKI, M. and FISHMAN, W.H. (1973). Ultrastructural studies on Regan and non-Regan isoenzymes of alkaline phosphatase in human ovarian cancer cells. Cancer Res., 33, 3008.

3) SINGER, R.M. and FISHMAN, W.H. (1974). Characterization of two HeLa sublines: TCRC-1 produces Regan isoenzyme and TCRC-2, non Regan-isoenzyme. J. Cell Biol., 60, 777.

4) HIGASHINO, K., KUDO, S., OTANI, R., YAMAMURA, Y., HONDA, T. and SAKURAI, J. (1975). A hepatoma-associated alkaline phosphatase, the Kasahara isoenzyme, compared with one of the isoenzymes of FL amnion cells. An. N. Y. Acad. Sci., 259, 337.

5) STIGBRAND, T., MILLAN, J. and FISHMAN, W.H. (1982). The genetic basis of alkaline phosphatase isoenzyme expression. In: Isozymes CTMBR (ed. M.C. Rattazzi), Vol. 6, Alan R. Liss, Inc., New York.

6) TOKUMITSU, S., TOKUMITSU, K. and FISHMAN, W.H. (1981). Immuno-
    cytochemical demonstration of intracytoplasmic alkaline phos-
    phatase in HeLa TCRC-1 cells. J. Histochem. and Cytochem., 29,
    1080.

7) TOKUMITSU, S., TOKUMITSU, K. and FISHMAN, W.H. (1981). Intra-
    cellular alkaline phosphatase activity in cultured human
    cancer cells. Histochemistry, 73, 1.

8) KOTTEL, R.H. and HANFORD, W.C. (1980). Differential release
    of membrane-bound alkaline phosphatase isoenzymes from tumour
    cells by bromelain. J. Biochem. Biophys, Methods, 2, 325.

9) HANFORD, W.C. and FISHMAN, W.H. (1982). unpublished obser-
    vations.

10) MILLAN, J.L., STIGBRAND, T., RUOSLAHTI, E. and FISHMAN, W.H.
    (1982). An allotype-specific monoclonal antibody to placental
    alkaline phosphatase; its characterization and use in the
    study of cancer related phosphatase polymorphism. Cancer Res.,
    42, 2444.

11) JEMMERSON, R. and FISHMAN, W.H. (1982). Convenient selection
    of monoclonal antibodies to isoenzymes of placental alkaline
    phosphatase using the antigens catalytic activity in enzyme
    antigen immunoassay (EAIA). Analytical Biochemistry, in press.

12) SINGER, R.M. and FISHMAN, W.H. (1975). Specific isozyme pro-
    files of alkaline phosphatase in prednisolone-treated human
    cell populations. In: Isoenzymes III: Developmental Biology,
    Academic Press, Inc., New York, p. 753.

13) TAKEYA, M., TOKUMITSU, S. and KOHNOE, K. (1980). Application
    of the enzyme histochemical method for the isolation of a
    HeLa variant inducibly producing the Kasahara isoenzyme of
    alkaline phosphatase. Gann., 71, 710.

14) KOTTEL, R.H. and FISHMAN, W.H. (1981). Placental alkaline
    phosphatase isoenzyme expression by the non-HeLa cervical
    carcinoma cell line. Biochem. J., 200, 461.

15) KOTTEL, R.H. and FISHMAN, W.H. (1978). Modulation of HeLa
     cell alkaline phosphatase isoenzyme expression by prednisolone
     In: Carcinoembryonic Proteins (eds, B. Norgaard-Pedersen and
     N.H. Axelson), p. 571, Blackwell Scientific Publ., Oxford.

16) KOHNOE, K., TOKUMITSU, S. and TAKEYA, M. (1980). Alkaline
     phosphatase from two HeLa S3 subclones: one producing Regan
     and the other Kasahara Isoenzyme. Oncodevelop. Biol. and
     Med., 1, 199.

17) TOKUMITSU, S-I., TOKUMITSU, K., KOHNOE, K. and TAKEUCHI, T.
     (1979). Characterization of liver-type alkaline phosphatase
     from human gastric carcinoma cells (KMK-2) *in vitro*. Cancer
     Res., 39, 4732.

18) NOZAWA, S., UDAGAWA, Y., OOTA, H., KURIHARA, S. and FISHMAN,
     W.H. (1982). A newly established uterine cervical cancer
     cell line (SKG-III) with Regan isoenzyme, human chorionic
     gonadotrophin β-subunit and pregnancy-specific $β_1$ glycoprotein
     phenotypes. Cancer Res., in press.

NONHISTONE PROTEIN ANTIGENS IN RAT HEPATOMAS

L.S. Hnilica, W.N. Schmidt and R.C. Briggs

Department of Biochemistry and the A.B. Hancock, Jr. Memorial Laboratory, Vanderbilt University School of Medicine, Nashville, Tennessee 37232, USA

1. INTRODUCTION

Application of two dimensional polyacrylamide gel electrophoresis to the studies on chromosomal nonhistone proteins revealed their exceptional heterogeneity (1, 2). Assuming that at least some of these proteins participate in the process of differentiation and regulation of genetic transcription, they should be cell specific. Because of their unique selectivity and sensitivity, immunological procedures are well suited for probing the cell and tissue specificity of chromosomal proteins and their structural associations. Antibodies to histones (3) and to chromosomal nonhistone proteins (4-7) were used to study chromatin structure or cellular distribution and specificity of the respective antigens. Specific antisera localized nuclear protein antigens to selected areas of polytene chromosomes (5) or to the nucleoli of normal and malignant cells (8). Isolated intact (9, 10) or dehistonized (11, 12) chromatin, nucleoli (8, 13) or nuclear membrane proteins (14, 15) were all found immunogenic when injected into rabbits or other animals. Some of these antisera recognized cell or tissue-specific antigens and their changes during cell differentiation (11, 16) or carcinogenesis (17-19).

2. ANTISERA TO DEHISTONIZED CHROMATIN

Extraction of chromatin with concentrated salt-urea solutions pH 5.0 or 6.0 results in almost complete removal of histones (20). Such dehistonized chromatin was first used as immunogen by Chytil and Spelsberg (11). The antisera, elicited in rabbits, were specific for chick oviducts chromatin and did not react appreciably with chromatins (intact or dehistonized) from other chicken organs. We have shown in our laboratory that antisera to dehistonized chromatins from rat hepatomas, rat liver, chicken reticulocytes, HeLa cells, human granulocytes, etc. recognize antigens specific for the tissue or cell type from which the immunogen was isolated (4, 7). We have also found that at least some of this immunological specificity can be attributed to recognition by the antisera of complexes between DNA and a select group of chromosomal nonhistone proteins (4, 7).

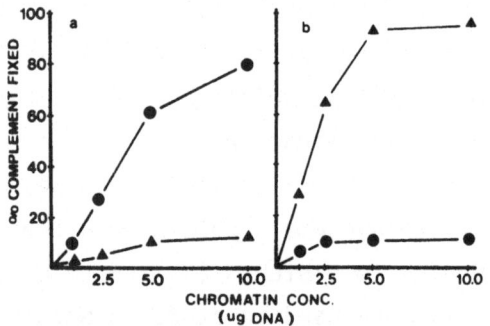

Fig. 1. *The immunological specificity of rat liver and Novikoff hepatoma chromatins.* The complement fixation assay was performed in the presence of antisera to dehistonized chromatin from normal rat liver (a) or Novikoff hepatoma (b). Normal rat liver (●) or Novikoff hepatoma (▲) chromatins.

Immunological specificity of the normal rat liver-hepatoma chromatins is illustrated in Figure 1, where the complement fixation of antisera to dehistonized rat liver or Novikoff hepatoma chromatins is compared. Either antiserum reacted only in the presence of chromatin (dehistonized or intact) isolated from tissue source of the immunogen. The antiserum to Novikoff hepatoma did not react with chromatins of various other cell types but it was crossreactive with chromatins of several other hepatomas (AS 30D, Morris hepatomas) and other transplantable rat tumours (21, 22).

The complement fixing activity could be abolished by separating the DNA from chromosomal proteins by dissociation in concentrated salt-urea solution, pH 8.0 and extensive ultracentrifugation. Neither the DNA pellet nor the protein supernatant were reactive with the antiserum. Reconstitution of chromosomal proteins to the DNA (either in the pellet or purified rat spleen DNA) restored the immunological reactivity and specificity of the reconstituted complex. Similarly, digestion of the DNA with nuclease abolished the complement fixing activity of the treated chromatin (Figure 2). Substitution of DNA in the reconstituted

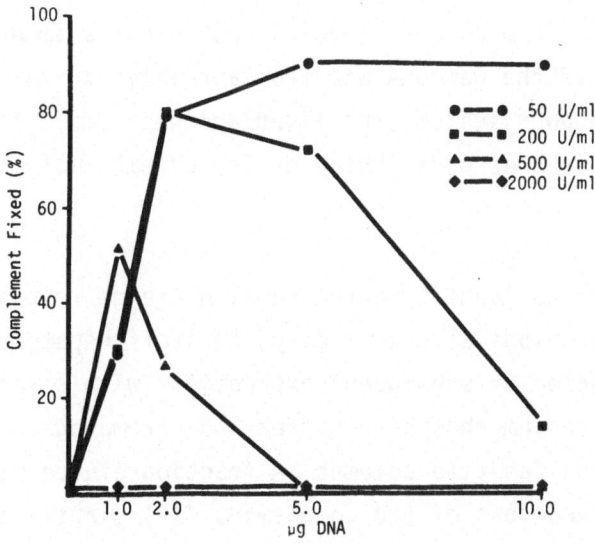

Fig. 2. *Micrococcal nuclease digestion of HeLa chromatin showing loss in activity after extensive digestion.* Each tube contained 100 μg chromatin DNA and the indicated quantity of enzyme. Incubation was at 37°C for 30 min. The complement fixation was performed in the presence of antiserum to dehistonized HeLa chromatin.

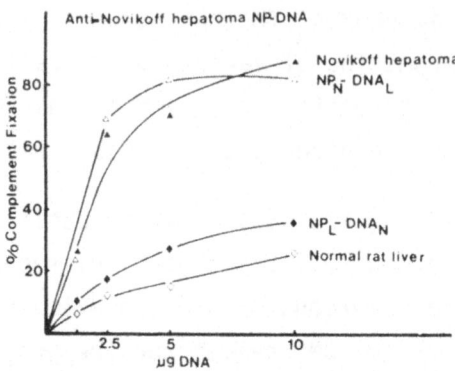

Fig. 3. *Complement fixation of normal and reconstituted nonhistone protein (NP)-DNA complexes from rat liver and Novikoff hepatoma in the presence of antiserum to dehistonized Novikoff hepatoma chromatin.* The chromosomal nonhistone proteins NP were isolated by the method described in Ref. 17. Novikoff hepatoma chromatin (▲), reconstituted complex of Novikoff hepatoma NP and normal rat liver DNA (△), normal rat liver chromatin (◇), reconstituted complex of rat liver NP and Novikoff hepatoma DNA (◆). The tissue specificity is determined by the source of NP protein.

chromatin for DNA from another species restored its immunological reactivity only if the new DNA was from a related animal species (24). In a homologous system (rat liver and hepatoma), the immunological specificity was contributed by the chromosomal proteins (Figure 3).

In an effort to identify chromosomal proteins responsible for the observed immunological specificity, Novikoff hepatoma chromatin was fractionated by subsequent extractions with concentrated urea containing sodium phosphate buffer and chromatography of the extracts on hydroxylapatite columns. A fraction eluted with 50 mM phosphate contained most of the complement fixing activity (with antiserum to dehistonized Novikoff hepatoma chromatin) and was further fractionated by filtration on Biogel P200 column. Immunological analysis of the fractions and their polyacrylamide gel patterns revealed that three Novikoff hepatoma proteins (M_r 39, 49 and 56 Kd as determined by polyacrylamide gel electrophoresis) were responsible, at least partially, for the observed immunological specificity of the Novikoff hepatoma chromatin (25, 25).

3.   ANALYSIS OF THE p39, p49 and p56 ANTIGENS

After the identification of these three antigens to be among those recognized by the antisera to dehistonized Novikoff hepatoma chromatin, we then elicited new antiserum directed to the mixture of p39, p49 and p56 obtained by gel filtration on Biogel P200. This second generation antiserum was specific by complement fixation assays (Figure 4). Immunoabsorbtion of this antiserum with

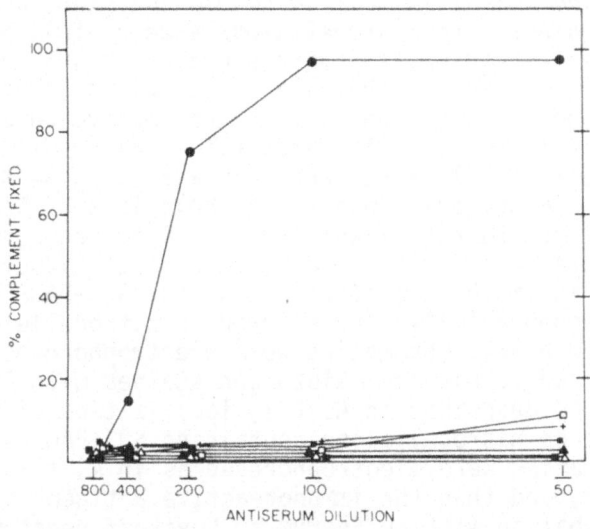

Fig. 4. *Complement fixation of antiserum to Novikoff hepatoma antigens p39, p49 and p56 in the presence of 10 μg/ml (as DNA) of Novikoff hepatoma (●), mature rat liver (▲), fetal rat liver (*), 24-hr regenerating rat liver (+), immature (22 days old) rat liver (□), fetal rat brain (■) and fetal rat kidney (△) chromatins.* Reproduced with permission from Ref. 31.

normal rat liver or other tissues chromatins did not abolish its complement fixing activity. However, this antiserum no longer recognized complexes between chromosomal nonhistone proteins and DNA (26). Because of this change, immunoassays other than complement fixation could be utilized for the detection of the Novikoff

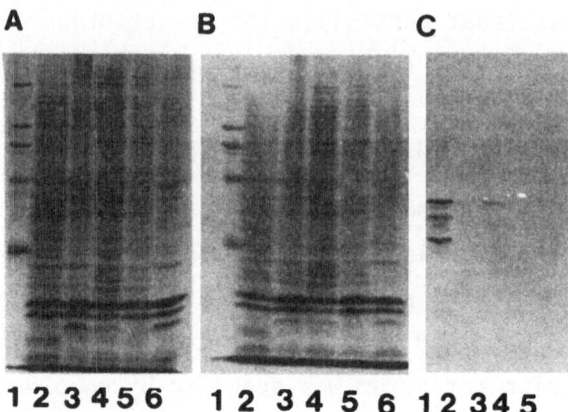

Fig. 5. *Identification of Novikoff hepatoma antigens in chromoso-*
mal proteins transferred to nitrocellulose sheets. (A), SDS-PAGE
of various chromatins. Chromatins were prepared in SDS gel sample
buffer, sonicated, and electrophoresed as described in Ref. 31.
Each lane contained 25 µg (as DNA) of the respective chromatin.
Lane 1, high molecular weight standards (Bio-Rad) (myosin, M_r 20
Kd; β-galactosidase, M_r 116.5 Kd, phosporylase β, M_r 94 Kd; bovine
serum albumin, M_r 68 Kd; ovalbumin, M_r 43 Kd); lane 2, Novikoff
hepatoma; lane 3, mature rat liver; lane 4, 24-hr regenerating rat
liver; lane 5, fetal rat liver; lane 6, mature rat kidney chromatins. The origin is at the top of the gels. (B),SDS-PAGE of various
chromatins electrophoretically transferred to nitrocellulose and
stained with Amido black. Chromatins were electrophoresed as in A
and then transferred to nitrocellulose and stained with Amido
black. Lanes are as described in A. (C), localization of immunoreactive antigens on nitrocellulose containing SDS-PAGE-separated
chromatins. Chromatins were electrophoresed as in A, transferred
to nitrocellulose, and then the immunoreactive antigens were
localized by incubation with antiserum to Novikoff hepatoma proteins
p39, p49, and p56 (1:500 dilution) followed by the PAP procedure
as described in Ref. 31. Lane 1, Novikoff hepatoma, lane 2, mature
rat liver; lane 3, 24-hr regenerating rat liver; lane 4, fetal
rat liver; lane 5, mature rat kidney chromatins. Reproduced with
permission from Ref. 31.

hepatoma antigens p39, p49 and p56. The immunotransfer procedure
of Towbin *et al.* (27), together with the immunodetection of anti-
gen-antibody complexes with the peroxidase-antiperoxidase method
of Sternberger (28), combined in our hands the outstanding resolution power of polyacrylamide gel electrophoresis with the sensitivity and selectivity of immunochemical reactions (29). Using

this approach, we were able to demonstrate a number of antigens, detectable by antisera to dehistonized Novikoff hepatoma chromatin, which were absent from normal rat liver (29).

Immunochemical analysis of the distribution in various tissues of the Novikoff hepatoma antigens p39, p49 and p56 is shown in Figure 5. The p39 and p49 antigens are not present in any normal adult or fetal tissue analyzed. Absorption of the antiserum with

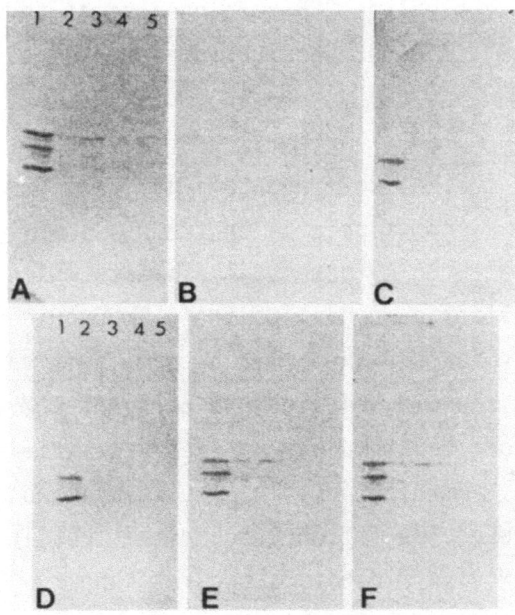

Fig. 6. *Immunoabsorbtion of antiserum to Novikoff hepatoma proteins p39, p49 and p56 with various chromatins.* Antiserum was absorbed twice. Absorbed antisera were then incubated with nitrocellulose sheets containing SDS-PAGE-separated chromatins (25 µg as DNA of each chromatin), and immunoreactive species were then visualized with the PAP reaction. Each nitrocellulose sheet contains (lanes left to right), Novikoff hepatoma, rat liver, 24-hr regenerating rat liver, fetal rat liver and rat kidney chromosomal proteins with the origin at the top of the sheet. The absorbed antisera used to stain each sheet were: (A) unabsorbed; (B) absorbed with Novikoff hepatoma chromatin; (C) absorbed with rat liver chromatin; (D) absorbed with 24 hr regenerating rat liver chromatin; (E) absorbed with rat kidney chromatin; (F) absorbed with fetal rat liver chromatin. Reproduced with permission from Ref. 31.

chromatins isolated from adult, regenerating or fetal rat liver or
adult kidney confirmed these results (Figure 6). The p56 antigen
was found in normal and regenerating rat liver, albeit in quantities
smaller than in the hepatoma. It was absent, however, from fetal
liver and adult kidney. Although concentrated in chromatin, the
Novikoff hepatoma antigens p39, p49, and p56 were also found in
the cytoplasm (30). A survey of several rat tumours and cell lines
revealed that all malignant cells of the epithelial origin (carcinomas) contained significant quantities of the p39 and p49 antigens. Chromatins of the Walker 256 carcinosarcomas and two transplantable rat sarcomas did not show the presence of p39 and p49
antigens (31).

The absence of the Novikoff hepatoma antigens p39 and p49
from transplantable sarcomas pointed to the possibility that these
proteins may be related to cytoskeletal elements which are known
to differ in composition according to their embryonic origin (32,
33). This notion was further supported by similarities in amino
acid composition of the p39, p49 and p56 antigens and cytokeratins
(26, 34). The Novikoff hepatoma p39 antigen was purified to homogeneity and used for immunization. The resulting monospecific
antiserum was then used for immunocytochemical localization of
the p39 antigen in Novikoff hepatoma and rat kangaroo PTK cells.
In all experiments the immunolocalization of this antiserum was
consistent with the reported cellular distribution of intermediate
filaments. Although the original antiserum decorated both the
Novikoff hepatoma and PTK2 cells, after further purification by
affinity chromatography. This antiserum became specific for the
Novikoff hepatoma cells only (34). Results of these experiments
are summarized in Figure 7. As can be seen in this Figure, the
affinity purified antiserum decorated not only the Novikoff hepatoma cytoplasm but the nucleus as well. This poses and interesting
question whether some of the cytoskeletal proteins may form a

continuous network extending through and perhaps beyond the nuclear membrane.

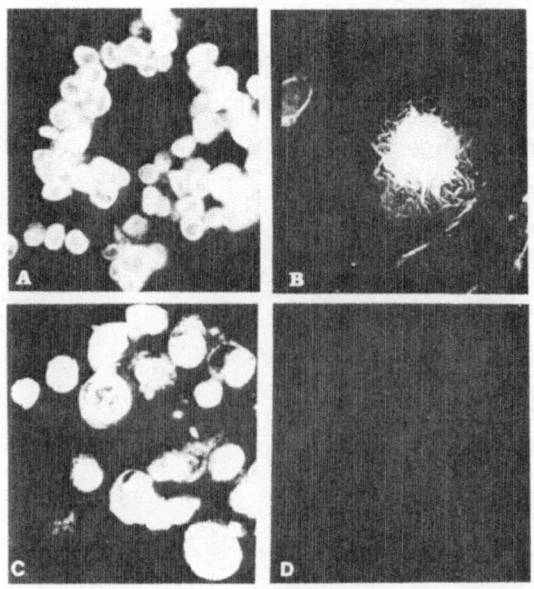

Fig. 7. *Immunofluorescence microscopy of Novikoff hepatoma cells (A and C) or rat kangaroo PTK2 cells (B and D).* In A and B cells were treated with antiserum prepared to purified antigen p39 but showing activity for all the Novikoff antigens on nitrocellulose transfers. In C and D cells were treated with antibodies to p39 purified by affinity chromatography. Exposure time for all pictures was identical. (A, B, and D, x 1, 250; C, x 1,560). Reproduced with permission from Ref. 34.

4. NONHISTONE PROTEIN ANTIGENS IN EXPERIMENTAL HEPATOCARCINO- GENESIS

Livers of rats on hepatocarcinogenic diet exhibit gradual morphological and biochemical changes, increasing with the severity of neoplastic transformation (35). When the complement fixation assay, using antiserum to Novikoff hepatoma dehistonized chromatin, was performed with chromatins isolated from livers of rats sacrificed at various times intervals on hepatocarcinogenic diet [N,N-dimethyl-p(m-tolylazo)aniline, 3'MDAB], a gradual increase of complement fixing activity could be observed with the progression of the experiments (Figure 8). It is noteworthy that a significant increase in the immunological reactivity was evident early in the feeding schedule, before any detectable neoplasia. Control animals and rats on a diet containing α-naphthyl isothiocyanate, a compound producing extensive proliferation of bile duct cells but no tumours did not exhibit any changes in the immunoreactivity of their liver chromatins (Figure 8).

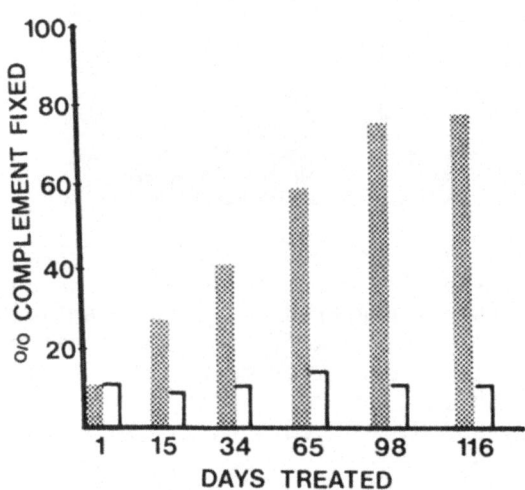

Fig. 8. *Time course of increasing fixation of chromatins isolated from liver of rats maintained on 3'-MDAB containing diet.* The complement fixation assay was performed in the presence of anti-sera against Novikoff hepatoma dehistonized chromatin.⬜, Rats maintained on 3'-MDAB; ▓▓▓, rats maintained on γNIT.

To investigate whether changes similar to those detectable by complement fixation assays could be also observed in electro-phoretically separated and immunotransferred nuclear proteins, Fisher rats were fed 3'-MDAB containing diet and sacrificed at regular intervals. Chromatin, isolated from their livers was

electrophoresed, the separated proteins were transferred to nitrocellulose sheets and stained with the peroxidase-antiperoxidase
method (28-30) using antiserum to dehistonized chromatin isolated
from 3'-MDAB produced hepatoma. As can be seen in Figure 9,
considerable changes in immunological reactivity of chromosomal
nonhistone proteins begin to appear as early as 3-6 weeks on the
azo dye diet. The number of antigens reactive with the tumour
antiserum reached an apparent maximun in 27 weeks and did not
change appreciably until the death of the animals caused by
massive hepatic neoplasia. The immunological reactivity could not
be abolished by absorbing the antiserum with normal, fetal or
regenerating rat liver chromatins (19).

Although direct comparison of the complement fixation assays
with the immunotransfer data is presently impossible, our results
indicate that considerable changes in antigenic character of
numerous chromosomal proteins take place during the process of
carcinogenesis.

5. CONCLUSIONS

Our investigations show that changes of normal cells to
malignancy are accompanied by alterations in immunological character of numerous chromosomal nonhistone proteins. While the
complement fixation assays detect preferentially conformational
differences, probably due to interactions between chromosomal
proteins and DNA, the immunotransfer technique facilitates the
detection of individual antigens as resolved by polyacrylamide gel
electrophoresis. Unfortunately, most chromosomal nonhistone
proteins cannot be successfully resolved in the absence of anionic
detergents, such as sodium dodecyl sulphate. This represents a
potential limitation of the immunotransfer method since any antigens which become immunologically denatured by the detergent will
go undetected. However, as was first reported by Stumph *et al.* (36)

Fig. 9. (Following page) *Identification of azo dye hepatoma antigens in various chromatins prepared from the livers or hepatomas
of rats fed azy dye-containing diet.* (A), SDS-PAGE of various
chromatins. Chromatins were prepared in sodium dedecyl sulphate
sample buffer and electrophoresed as described in Ref. 19. Each
lane contained 25 μg (as DNA) of the respective chromatin. Lane 1,
normal (control-fed) liver; lane 2, 3-week azo dye-treated liver;
lane 3, 6-week azo dye treated liver; lane 4, 9-week azo dye-
treated liver; lane 5, 13-week azo dye-treated liver; lane 6, 16-
week azo dye-treated liver; lane 7, 19-week azo dye-treated liver;
lane 8, 28-week azo dye treated liver; lane 9, 28-week azo dye-
treated liver; lane 10,30-week azo dye-treated liver; lane 11, 31-
week azo dye-treated liver; lane 12, 32-week azo dye-treated liver;
lanes 13 to 16, azo dye hepatoma chromatins isolated after 23, 26,
28 and 30 weeks of treatment, respectively; lane 17, high-molecular
weight standards (Bio-Rad) (myosin, M_r 200 Kd; β-galactosidase,
M_r 116.5 Kd; phosphorylase β, M_r 94 Kd; bovine serum albumin, M_r
68 Kd; ovalbumin, M_r 43 Kd. With the chromatins for lanes 9 and
15, animals were removed from azo dye-containing diet after 21
weeks of feeding and fed control chow for an additional seven
weeks until autopsy and chromatin preparation. (B), localization
of immunoreactive antigens on nitrocellulose containing SDS-PAGE-
separared chromatins. Chromatins were prepared and electrophoresed
as in A, transferred to nitrocellulose, and the immunoreactive
antigens were localized by incubation with antiserum to azo dye
hepatoma dehistonized chromatin (1:200 dilution), followed by the
PAP procedure. Lanes are exactly as indicated in A, with the
omission of molecular weight standards (lane 17). Reproduced with
permission from Ref. 19.

——▶

and confirmed by others, most nuclear proteins retain their immunological reactivity even after exposure to anionic detergents. With
this limitation in mind, the immunotransfer technology in combination
with one or two dimensional polyacrylamide gel electrophoresis
offers the most powerful approach for the detection and identification of immunoreactive chromosomal proteins.

Presently, we do not know whether all the specific antigens
observed in various hepatomas represent new protein species whose
biosynthesis was induced by the process of carcinogenesis or
whether at least some of them may reflect specific post-translational modifications.

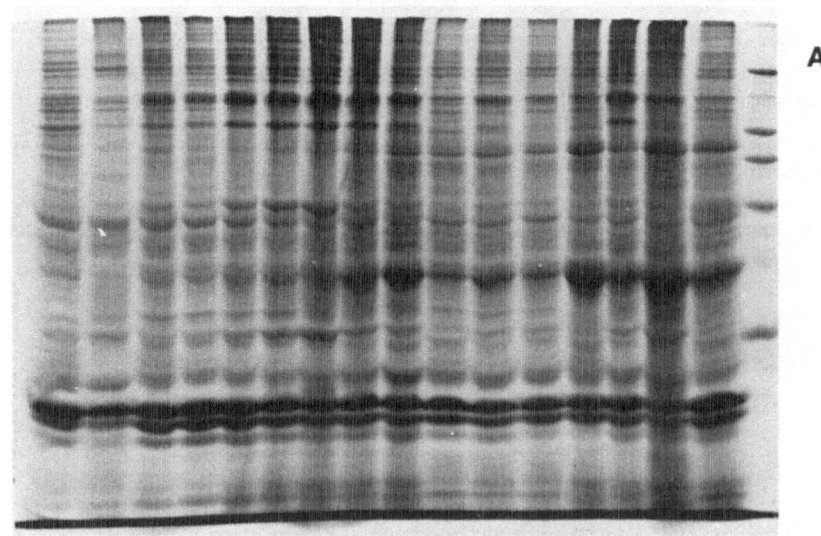

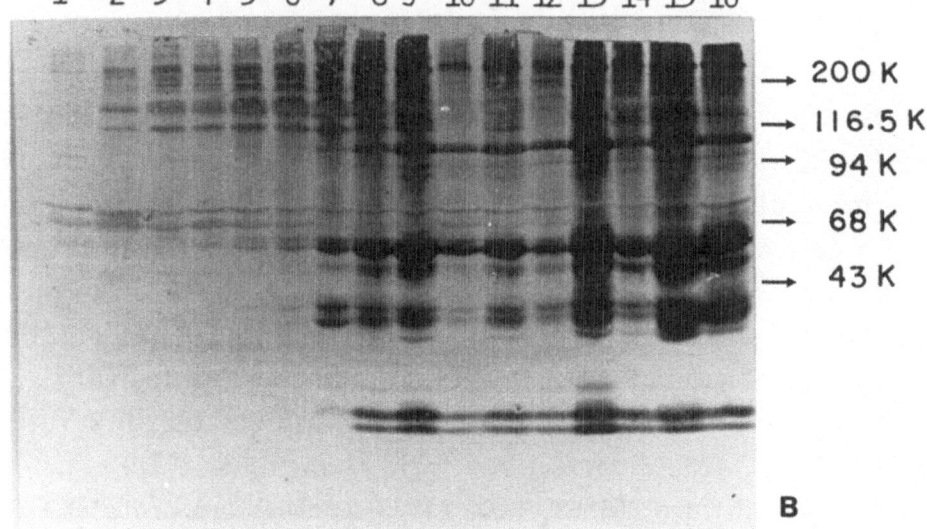

6. ACKNOWLEDGEMENTS

Original research described in this article was supported by the National Cancer Institute Grant CA 26412. The authors are indebted to Ms. Doris Harris for her help in preparing this manuscript.

7. REFERENCES

1) PETERSON, J.L. and McCONKEY, E.H. (1976). Non-histone chromosomal proteins from HeLa cells. A survey by high resolution, two dimensional electrophoresis. J. Biol. Chem., 251, 548.

2) BRAVO, R. and CELIS, J.E. (1982). Up-dated catalogue of HeLa cell proteins: percentages and characteristics of the major cell polypeptides labelled with a mixture of 16 ^{14}C -labelled amino acids. Clin. Chem., 28, 766.

3) BUSTIN, M. (1978). Histone antibodies - structural probes for chromatin and chromosomes. In: The Cell Nucleus (ed. H. Busch) Vol. IV, p. 195, Academic Press, New York.

4) HNILICA, L.S., CHIU, J.F., HARDY, K., FUJITANI, H., and BRIGGS, R.C. (1978). Antibodies to nuclear protein fractions. In: The Cell Nucleus (ed. H. Busch), Vol. V, p. 307, Academic Press, New York.

5) SILVER, L.M. and ELGIN, C.R. (1978). Immunological analysis of protein distributions in Drosophila polytene chromosomes. In: The Cell Nucleus (ed. H. Busch), Vol. V, p. 215, Academic Press, New York.

6) YEOMAN, L.C., WOOLF, L.M., TAYLOR, C.W. and BUSCH, H. (1978). Nuclear antigens of tumour cell chromatin. In: Biological Markers of Neoplasia: Basic and Applied Aspects (ed. R.W. Ruddon), p. 409, Elsevier, New York.

7) HNILICA, L.S. and BRIGGS, R.C. (1979). Nonhistone proteins antigens. In: Cancer Markers: Developmental and Diagnostic Significance (ed. S. Sell), p. 463, Humana Press, Clifton, New Jersey.

8) DAVIS, F.M., BUSCH, R.K., YEOMAN, L.C. and BUSCH, H. (1978). Differences in nucleolar antigens of rat liver and Novikoff hepatoma ascites cells. Cancer Res., 38, 1906.

9) ZARDI, L., LIN, J.C. and BASERGA, R. (1973). Immunospecificity to nonhistone chromosomal proteins of anti-chromatin antibodies. Nature, New Biol., 245, 211.

10) ZARDI, L. (1975). Chicken antichromatin antibodies: specificity to different chromatin fractions. Eur. J. Biochem., 55, 231.

11) CHYTIL, F. and SPELSBERG, T.C. (1971). Tissue differences in antigenic properties of nonhistone protein-DNA complexes. Nature, New Biol., 233, 215.

12) WAKABAYASHI, K. and HNILICA, L.S. (1973). The immunospecificity of nonhistone protein complexes with DNA. Nature, New Biol., 242, 153.

13) MARASHI, F., DAVIS, F.M., BUSCH, R.K., SAVAGE, H.E. and BUSCH, H. (1979). Purification and partial characterization of nucleolar antigen-1 of the Novikoff hepatoma. Cancer Res., 39, 59.

14) WILSON, E.M. and CHYTIL, F. (1976). Antigenic properties of rat liver nuclear membrane. Biochim. Biophys. Acta, 426, 88.

15) GERACE, L., BLUM, A. and BLOBEL, G. (1978). Immunocytochemical localization of the major polypeptides of the nuclear pore complex-lamina fraction. J. Cell Biol., 79, 546.

16) CHYTIL, F., GLASSER, S.R. and SPELSBERG, T.C. (1973). Alterations in liver chromatin during perinatal development of the rat. Develop. Biol., 37, 295.

17) CHIU, J.F., HUNT, M. and HNILICA, L.S. (1975). Tissue-specific DNA protein complexes during azo dye hepatocarcinogenesis. Cancer Res., 35, 913.

18) CHIU, J.F., DECHA-UMPHAI, W., MARKERT, C. and LITTLE, B.W. (1979). Immunospecificity of nuclear nonhistone protein-DNA complexes in colon adenocarcinoma. J. Nalt. Cancer Inst., 63, 313.

19) SCHMIDT, W.N., GRONERT, B.J., PAGE, D.L., BRIGGS, R.C. and
    HNILICA, L.S. (1982). Antigenic changes in nonhistone proteins
    during azo dye hepatocarcinogenesis. Cancer Res., 42, 3164.

20) SPELSBERG, T.C., HNILICA, L.S. and ANSEVIN, A.T. (1971).
    Proteins of chromatin in template restriction. III. The macro-
    molecules in specific restriction of the chromatin DNA.
    Biochim. Biophys. Acta, 228, 550.

21) WAKABAYASHI, K., WANG, S. and HNILICA, L.S. (1974). Immuno-
    specificity of nonhistone proteins in chromatin. Biochemistry,
    13, 1027.

22) CHIU, J.F., CRADDOCK, C., MORRIS, H.P. and HNILICA, L.S.
    (1974). Immunospecificity of chromatin nonhistone protein-
    DNA complexes in normal and neoplastic growth. FEBS Lett., 42,
    94.

23) CAMPELL, A.M., BRIGGS, R.C., ZIMMER, M.S., KRAJEWSKA, W.M.,
    CHIU, J.F. and HNILICA, L.S. (1978). Antigenic complexes of
    chromosomal nonhistone proteins with DNA. In: Biological
    Markers of Neoplasia: Basic and Applied Aspects (ed. R.W.
    Ruddon), p. 369, Elsevier, New York.

24) WANG, S., CHIU, J.F., KLYSZEJKO-STEFANOWICH, L., FUJITANI,
    H. and HNILICA, L.S. (1976). Tissue-specific chromosomal non-
    histone protein interactions with DNA. J. Biol. Chem., 251,
    1471.

25) DUNN, J.H., LYALL, R.M., BRIGGS, R.C., CAMBELL, A.M. and
    HNILICA, L.S. (1980). DNA-binding specificity of a chromatin
    nonhistone protein fraction of HeLa cells. Biochem. J., 185,
    277.

26) FUJITANI, H., CHIU, J.F. and HNILICA, L.S. (1978). Purifica-
    tion of nuclear antigens in Novikoff hepatoma. Proc. Natl.
    Acad. Sci. USA, 75, 1943.

27) STRYJECKA-ZIMMER, M., SCHMIDT, W.N., BRIGGS, R.C. and HNILICA,
    L.S. (1982). Immunological specificity of Novikoff hepatoma
    chromatin: isolation of three antigenic proteins. Int. J.

Biochem., 14, 591.

28) TOWBIN, H., SHAEHLIN, T. and GORDON, J. (1979). Electrophoretic transfer of proteins from polyacrylamide gels to nitrocellulose sheets: procedure and some applications. Proc. Natl. Acad. Sci. USA, 76, 4350.

29) STERNBERGER, L.A. (1979). The unlabelled antibody peroxidase-antiperoxidase (PAP) method. In: Immunochemistry (ed. L.S. Sternberger), p. 104, Prentice-Hall, Engelwood Cliffs, New Jersey.

30) GLASS, W.F., BRIGGS, R.C. and HNILICA, L.S. (1981). Identification of tissue-specific nuclear antigens transferred to nitrocellulose from polyacrylamide gels. Science 211, 70.

31) SCHMIDT, W.N., STRYJECKA-ZIMMER, M., BRIGGS, R.C., GLASS, W.F. and HNILICA, L.S. (1981). Tissue specificity and cellular distribution of Novikoff hepatoma antigenic proteins p39, p49 and p56. J. Biol. Chem., 256, 8117.

32) SCHMIDT, W.N. and HNILICA, L.S. (1982). Distribution of Novikoff ascites hepatoma antigens p39, and p49 in various tumorigenic cell lines. Cancer Res., 42, 1441.

33) LAZARIDES, E. (1982). Intermediate filaments: a chemically heterogeneous, developmentally regulated class of proteins. Ann. Rev. Biochem., 51, 219.

34) SCHMIDT, W.N., PARDUE, R.L., TUTT, M.C., BRIGGS, R.C., BRINKLEY, B.R. and HNILICA, L.S. (1982). Identification of cytokeratin antigens in Novikoff ascites hepatoma. Proc. Natl. Acad. Sci. USA, 79, 3138.

35) FARBER, E. (1981). The sequential analysis of cancer induction with chemicals. Acta. Pathol. Jpn., 31, 1.

36) STUMPH, W.E., ELGIN, S.C.R. and HOOD, L. (1974). Antibodies to proteins dissolved in sodium dodecyl sulphate. J. Immunol., 113, 1752.

CONTRIBUTORS

ADAMS, S.L., Departments of Anatomy, Human Genetics and Micro-
    Biology, School of Medicine, University of Pennsylvania,
    Philadelphia, USA. p. *315*
BARBOSA, J.A., Department of Biology, Yale University, New Haven,
    Connecticut, USA. p. *379*
BARKER, P.E., Department of Biology, Yale University, New Haven,
    Connecticut, USA. p. *71*
BELLATIN, J., Medical Research Council, Laboratory of Molecular
    Biology, Hills Road, Cambridge, UK. p. *263*
BERTHELOT, F., Laboratoire de Biochimie Cellulaire, Collège de
    France, Paris, France. p. *1*
BISHOP, Department of Microbiology and Immunology, University of
    California, San Francisco, California, USA. p. *193*
BOETTIGER, D., Departments of Anatomy, Human Genetics and Micro-
    Biology, School of Medicine, University of Pennsylvania,
    Philadelphia, USA. p. *315*
BRAVO, R., European Molecular Biology Laboratory, Meyerhofstrasse
    1, 6900 Heidelberg, FRG. p. *263*
BRIGGS, R.C., Department of Biochemistry and the A.B. Hancock, Jr.
    Memorial Laboratory, Vanderbilt University School of Medicine,
    Nashville, Tennessee, USA. p. *415*
BRZESKI, H., Department of Biochemistry, University of Strathclyde,
    Strathclyde, UK. p. *87*
BURNY, A., Department of Molecular Biology, University of Brussels
    (ULB), 1640 Rhode-St-Genèse, Belgium. p. *231*
CARMON, Y., Department of Cell Biology, The Weizmann Institute of
    Science, 76100 Rehovot, Israel. p. *71*
CELIS, J.E., Division of Biostructural Chemistry, Department of
    Chemistry, Aarhus University, DK-8000 Aarhus C, Denmark. p. *263*
COSSU, G., Istituto di Istologia ed Embriologia Generale,
    Universita' di Roma, Italy. p. *101*
COUEZ, D., Department of Molecular Biology, University of Brussels
    (ULB), 1640 Rhode-St-Genèse, Belgium. p. *231*
CROIZAT, B., Laboratoire de Biochimie Cellulaire, Collège de
    France, Paris, France. p. *1*

433

CZOSNEK, H., Department of Cell Biology, The Weizman Institute of
   Science, 76100 Rehovot, Israel. p. *71*
DESCHAMPS, J., Department of Molecular Biology, University of
   Brussels, (ULB), 1640 Rhode-St-Genèse, Belgium. p. *231*
EDDE, B., Laboratoire de Biochimie Cellulaire, Collège de France,
   Paris, France. p. *1*
EVANS, M.J., Department of Genetics, University of Cambridge,
   Downing street, Cambridge, UK. p. *87*
FEY, S.J., Division of Biostructural Chemistry, Department of
   Chemistry, Aarhus University, DK-8000 Aarhus C, Denmark. p. *263*
FISHMAN, W.H., Cancer Research Center, La Jolla Cancer Research
   Foundation, La Jolla,California, USA. p. *403*
FORCHHAMMER, J., The Fibiger Laboratory, Ndr. Frihavnsgade 70,
   DK-2100 Copenhagen Ø, Denmark. p. *291*
FRANKS, L.M., Imperial Cancer Research Fund, Lincoln's Inn Fields,
   London, WC2A 3PX, UK. p. *141*
GHYSDAEL, J., Laboratoire de Chimie Biologique, Département de
   Biologie Moléculaire, Université Libre de Bruxelles, 1640
   Rhode-St-Genèse, Belgium. p. *209*
GOLDMAN, D., Laboratory of Clinical Science, National Institute
   of Mental Health, Bethesda, Maryland, USA. p. *349*
GRAESSMANN, A., Institut f. Molekularbiologie und Biochemie der
   Freien Universität Berlin, Arnimalle 22, 1 Berlin 33, FRG.
   p. *247*
GRAESSMANN, M., Institut f. Molekularbiologie und Biochemie der
   Freien Universität Berlin, Arnimalle 22, 1 Berlin 33, FRG.
   p. *247*
GROS, F., Laboratoire de Biochimie Cellulaire, Collège de France,
   Paris, France. p. *1*
HNILICA, L.S., Department of Biochemistry and the A.B. Hancock, Jr.
   Memorial Laboratory, Vanderbilt University School of Medicine,
   Nashville, Tennessee, USA. p. *415*
HOLTZER, H., Departments of Anatomy, Human Genetics and Microbiology,
   School of Medicine, University of Pennsylvania, Philadelphia,
   USA. p. *315*
JEGGO, P., Genetics Division, National Institute for Medical Res.,
   Mill Hill, London NW7 IAA, UK. p. *177*
KAMARCK, M.E., Departments of Human Genetics and Biology, Yale
   University, New Haven, Connecticut, USA. p. *379*
KETTMANN, R., Department of Molecular Biology, University of
   Brussels (ULB), 1640 Rhode-St-Genèse, Belgium. p. *231*
LATCHMAN, D., Eukaryotic and Molecular Genetics Res. Group,
   Department of Biochemistry, Imperial College, London, UK. p. *87*
LOVELL-BADGE, R.H., MRC Mammalian Development Unit, University
   College, London, UK. p. *87*
MACDONALD-BRAVO, H., European Molecular Biology Laboratory,
   Meyerhofstrasse 1, 6900 Heidelberg, FRG. p. *291*
MAMMERICKX, M., National Veterinary Institute, 1180 Uccle,
   Belgium. p. *231*

MARBAIX, G., Department of Molecular Biology, University of
    Brussels, (ULB), 1640 Rhode-St-Genèse, Belgium. p. *231*
MERRIL, C.R., Laboratory of General and Comparative Biochemistry,
    National Institute of Mental Health, Bethesda, Maryland, USA.
    p. *349*
MOSE LARSEN, P. Division of Biostructural Chemistry, Department
    of Chemistry, Aarhus University, DK-8000 Aarhus C, Denmark. p.
    263
NUDEL, U., Department of Cell Biology, The Weizmann Institute of
    Science, 76100 Rehovot, Israel. p. *71*
PACIFICI, M., Department of Anatomy, School of Medicine, University
    of Pennsylvania, Philadelphia, USA. p. *315*
PAULIN, D., Département de Biologie Moléculaire, Institut Pasteur,
    Paris, France. p. *1*
PORTIER, M.M., Laboratoire de Biochimie Cellulaire, Collège de
    France, Paris, France. p. *1*
PRIVALSKY, M.L., Department of Microbiology and Immunology,
    University of California, San Francisco, California, USA. P.
    193
RADMAN, M., Institut Jacques Monod, C.N.R.S., Université Paris 7,
    Tour 43, Pl. Jussieu, Paris, France. p. *177*
RUDDLE, F.H., Department of Biology, Yale University, New Haven,
    Connecticut, USA. p. *71* and *379*
SCHMIDT, W.N., Department of Biochemistry and the A.B. Hancock, Jr.
    Memorial Laboratory, Vanderbilt University School of Medicine,
    Nashville, Tennessee, USA. p. *415*
SEALY, L., McArdle Laboratory for Cancer Research, University of
    Wisconsin, Madison, Wisconsin, USA. p. *193*
SHANI, M., Department of Cell Biology, The Weizmann Institute of
    Science, 76100 Rehovot, Israel. p. *71*
SIMMONNEAU, L. Unité de Recherches Gérontologiques, INSERM, U 118,
    29 Rue Wilhem, Paris, France. p. *117*
STACEY, A., Department of Genetics, University of Cambridge,
    Downing Street, Cambridge, UK. p. *87*
VAN KEUREN, M.L., E. Roosevelt Institute for Cancer Research,
    4200 East Ninth Avenue, Denver, Colorado, USA. p. *349*
VENNSTROM, B., Department of Microbiology, University of Uppsala,
    Uppsala, Sweden. p. *193*
WAGNER, R., Département de Biologie Moléculaire, Université Libre
    de Bruxelles, 1640 Rhode-St-Genèse, Belgium. p. *177*
WARREN, L., The Wistar Institute, 36th Street at Spruce, Philadel-
    phia, USA. p. *101*
WHALEN, R.G., Département de Biologie Moléculaire, Institut Pasteur,
    25, Rue du Dr. Roux, Paris, France. p. *45*
YAFFE, D., Department of Cell Biology, The Weizmann Institute of
    Science, 76100 Rehovot, Israel. p. *71*

INDEX